THE PLANTS OF JORDAN

An annotated checklist

THE PLANTS OF JORDAN

An annotated checklist

Hatem Taifour and Ahmed El-Oqlah

Edited by Shahina A. Ghazanfar

Kew Publishing

Royal Botanic Gardens, Kew

Royal Botanic Gardens Kew

First published in 2017 by
Royal Botanic Gardens, Kew,
Richmond, Surrey, TW9 3AB, UK
www.kew.org

Distributed on behalf of the Royal Botanic Gardens, Kew in North America by the University of Chicago Press, 1427 East 60th Street, Chicago, IL 60637, USA

ISBN 978-1-84246-641-4

British Library Cataloguing in Publication Data
A catalogue record for this book is available from the British Library

Design and page layout by Christine Beard
Kew Publishing
Royal Botanic Gardens, Kew

Print managed by Jellyfish Print Solutions

MIX
Paper from responsible sources
FSC www.fsc.org FSC® C013604

Contents

Foreword

In the Name of Allah. The Most Merciful. The Most Compassionate.

It gives me great pleasure and honour to write the foreword to this very important and eagerly awaited piece of work, *The Plants of Jordan: an annotated checklist*.

The Royal Botanic Garden at Tal Al-Rumman in Jordan was specifically founded to conserve Jordan's native plants, and showcase them to the public in such a way that they would be revered and respected. However, it quickly became apparent that a large proportion of the information needed to develop the Garden was quite fragmented and insufficient or missing entirely. Thus, as a first step in achieving the overarching goal of RBG, it was imperative that RBG gathers all available data in order to begin the documentation process that would facilitate the identification and prioritization of propagation of the plants needed to establish the Garden. This process was quite painstakingly slow, as the data was dispersed across the globe and some of which was held in old archival herbarium specimens. Thus, the information had to be repatriated, cross referenced and analysed before making it final. It proved to be precariously difficult to classify and categorize which plants occur in Jordan and whether or not they are native. Thus, quite a few workshops and many meetings were held by RBG with the experts in the field to accomplish this task.

Although, there have been several efforts in the past which have tackled this issue, none were as meticulously detailed nor rigorously reviewed as this valuable piece we have here today. This is the very first annotated checklist for the vascular plants of Jordan that brought together both national and international experts that worked arduously together over the years to produce this publication. It is an important reference for Jordan that will hopefully inspire many a young researcher to study Jordan's flora in more depth. It will also provide a solid framework for conservationists in the region to rely on. Nonetheless as with all checklists and the world of nomenclature it is constantly dynamic and will forever be reviewed and updated.

The Royal Botanic Garden at Tal Al-Rumman in Jordan, is privileged to gift Jordan this valuable reference which will become the basis on which future taxonomists and botanists alike will work on and continually update.

HRH Princess Basma Bint Ali
July 2016

Preface

The Plants of Jordan: an annotated checklist is a comprehensive list of vascular plants known from Jordan following the classification of the Angiosperm Phylogeny Group (APG III 2009; APG IV 2016). It is primarily based on plant collections present at various international and national herbaria (see list of herbaria cited), surveys carried out by the Royal Botanic Garden of Jordan, the Royal Society for the Conservation of Nature (RSCN), published and unpublished records. All families and species have been updated in their classification and nomenclature.

With the establishment of the Royal Botanic Garden of Jordan in 2005 under the patronage of H.H. Princess Basma Bint Ali, it was befitting that a checklist be produced which would form the basis of a descriptive flora of the country. The EU funded project (BOT ERA 2012-2014) awarded to the Royal Botanic Garden Jordan and Royal Botanic Gardens Kew as a consortium partner, helped in furthering the collections, training staff, preparation and publication of the Jordan Plant Red List (Vol. 1), and the final preparation of this Checklist.

The Plants of Jordan: an annotated checklist covers 112 families and 2531 species (including ferns and gymnosperms) recorded for Jordan. It was decided during the preparation of the Checklist that at least one specimen would be cited for each taxon, or if no specimen was cited then a reference to the species name would be given. This has resulted in some old citations especially those from HUJ dating back to the 1880's. Where specimens have not been seen, we have placed the entry in the Checklist based on good judgement and references in regional publications. The Checklist may not be comprehensive at this stage, but as a first working list we can only improve on it.

This checklist would have never taken off had it not been for the full support of Her Highness Princess Basma Bint Ali whose enthusiasm is pleasantly catching. I would like to thank Tariq Abu Talib, Executive Director of the Royal Botanic Garden Jordan for his support and help in more ways than one. Thank you to Hatem Taifour and Ahmad El-Oqla, who have excellent field experience and knowledge of plants, and who are the main leaders of this Checklist. I would also like to acknowledge the contributions made by Yaseen Ananbeh, another very experienced field botanist, and Jamil Lahham, Yarmouk University, an excellent botanist and a friend of long standing: thank you both. The support of the other team members of the Royal Botanic Garden of Jordan is gratefully acknowledged.

I am extremely grateful to colleagues at the Royal Botanic Gardens Kew for their help on the EU project which initiated this Checklist. In particular I would like to thank Renata Borosova, Tivvy Harvey, Joanna Osborne, Michael van Slageren and Emma Williams. It is my pleasure to acknowledge the support of many colleagues and friends both in Jordan and UK who have in some way or another played a part in the completion of this Checklist, and with whom I have spent pleasant times in the field in Jordan. In particular I would like to thank Giles Brandson, Habiba Barbara, Sabah Saifan, and Maysa Shishani for making the project interesting for all of us. I am very grateful to Kamal Neimat for his hospitality and for giving us his time so freely and introducing us to the beautiful country of Jordan.

Finally, I hope that the *The Plants of Jordan: an annotated checklist* would be used with as much interest and enthusiasm as has gone into its production. We hope that it will be useful in the field as well in the laboratories for the identification of plants, and will be used by students, foresters, conservationists and all enthusiasts of natural history. The Royal Botanic Garden of Jordan will welcome comments on new distribution records, new findings and corrections to this list.

I wish the Royal Botanic Garden of Jordan the best of wishes and fully support them in their endeavours to document, protect and conserve the native plants of Jordan.

Shahina A. Ghazanfar,
Royal Botanic Gardens, Kew, January 2017

Acknowledgements

The Royal Botanic Garden of Jordan sincerely appreciates the cooperation and support of many organizations and people who contributed to the *The Plants of Jordan: an annotated checklist*.

Our data was facilitated through access to documents, databases and herbaria at the Royal Society for the Conservation of Nature, National Center for Agricultural Research and Extension, Jordan Badia Research and Development Program, Forestry Department of the Ministry of Agriculture, Hashemite Fund for Development of Jordan Badia, Yarmouk University, Jordan University of Science and Technology and Al-Bayt University. Our data was strengthened through cooperation from the International Centre for Agricultural Research in the Dry Areas, Royal Botanic Gardens Kew, American University of Beirut, and Hebrew University of Jerusalem. Particular thanks go out to HRH Princess Basma Bint Ali and Tariq Abu Taleb, for their support at key stages.

Funds for producing an earlier version were received through the BOT-ERA project, "Reinforcing Cooperation between the Royal Botanic Garden of Jordan and the European Research Area (ERA)," under grant agreement No. 294416 of the European Union's Seventh Framework Programme for coordination and support action.

Finally, many thanks to Royal Botanic Gardens Kew, for collaborating in this landmark publication.

Introduction[1]

Jordan does not have a countrywide flora. The western part of Jordan is covered by Flora Palaestina (Zohary, 1966, 1972; Feinbrun-Dothan 1978, 1986), and by Post & Dinsmore (vols I & II, 1932, 1933). A couple of illustrated field guides illustrate the flora of Jordan (Al-Eisawi 1998; Maani, 2008), but neither is comprehensive. The most recent documentation of the native plants of Jordan is a checklist of vascular plants for Jordan by Al Eisawi (2013) which is countrywide but no specimens are cited for the taxa included.

Through the BOT-ERA project (Reinforcing cooperation between the Royal Botanic Garden of Jordan and European Research Area, 2012–2014, funded by the EC Seventh Framework Programme (FP7), INCO.2011-6.2 KBBE), an up-to-date checklist of the plants of Jordan was produced which incorporated recent name changes and new distribution records for the country. This Checklist is a further update, incorporating several published and unpublished lists, and survey reports to include all names that have been used, and are in use for the flora of Jordan. All published and unpublished sources are listed under References and Bibliography. *The Plants of Jordan: an annotated checklist* aims to include all taxa of vascular plants (pteridophytes, gymnosperms and angiosperms) found in Jordan, either as native or naturalized aliens. With a few exceptions, cultivated trees, shrubs and herbs are not included. At least one specimen is cited for each taxon; where no specimen is cited, reference to Al Eisawi (2013) or to the RSCN database (database of wild plants kept by the Royal Society for the Conservation of Nature) is made, and for some species references to other sources are also included.

Several national and international herbaria have been consulted to validate collections and plant names: these are listed at the end of Introduction. Names of plants that have no cited specimens remain to be authenticated.

The checklist follows the classification of the Angiosperm Phylogeny Group (APGIII, 2009) and incorporates subsequent changes (www.mobot.org/mobot/research/apweb) and updates (APG IV, 2016). Families are listed alphabetically, as are the genera and species within each family. For each accepted name its authority and original citation is given; basionyms and synonyms relevant to this region are provided for most taxa. Species names follow largely the accepted names and synonymy as treated under Kew's World Checklist series (http://apps.kew.org/wcsp/), Tropicos (www.tropicos.org/) and The Plant List (www.theplantlist.org/); for some taxa, the Euro+Med plant Database (ww2.bgbm.org/EuroPlusMed/), International Legume Database and Information Service ILDIS, (www.tropicos.org/Project/Palaestina) and Flora Palaestina (2nd Edition — incomplete) have been followed. For Salsoleae (Amaranthaceae) Akhani et al. (2007) is followed. There are several names that are recorded as "unresolved" in The Plant List as research on these taxa is underway; for such taxa, the most recent published taxonomic treatment or an older accepted published treatment is followed. For families of monocotyledonous plants, accepted names and synonymy are as given in eMonocot (http://e–monocot.org/). All names are databased in the BRAHMS database at the Royal Botanic Gardens Kew, and will be kept updated as part of the Middle East Plant Database.

---

[1] By Shahina A. Ghazanfar, Hatem Taifour & Ahmad El-Oqlah

Checklists are never final! There will be name changes with new research and, as more field work is carried out, new distribution records for Jordan will be found. As this checklist is also present in an electronic format, it is relatively easy to keep the plants of Jordan up to date, and we hope that it will form the basis of a national Flora of Jordan.

FLORISTICS, VEGETATION AND BIOGEOGRAPHY

A quantitative analysis of the flora shows 2498 taxa in 813 genera belonging to 112 families of flowering plants. The families Asteraceae (287 spp), Fabaceae (317 spp), Poaceae (266 spp), Brassicaceae (153 spp), Lamiaceae (123 spp) and Caryophyllaceae (122 spp) are the largest with more than 100 species in each. *Astragalus* (53 spp), *Trifolium* (50 spp), *Silene* (40 spp), *Euphorbia* (31 spp) and *Allium* (31 spp) are the largest genera with more than 30 species in each. An analysis of the distribution of the flora in the different ecogeographic regions shows the northern and southern mountains of western Jordan to be the most species rich.

Biodiversity surveys carried out by Jordan's Royal Society for the Conservation of Nature (RSCN) in designated nature reserves show Ajloun Forest Reserve and Dana Biosphere Reserve to be the most species rich. There are seven species endemic to Jordan and ± 106 near endemics, (endemic to Palestine and/or Syria). Threats to biodiversity and habitats in Jordan include urbanization, wood cutting, overgrazing, and extractor mining.

In addition to the present checklist, the Jordan Plant Red List, Volume 1, was completed by the Royal Botanic Garden of Jordan (RBG) in 2014, in collaboration with the International Union for the Conservation of Nature. This landmark achievement is the first ever IUCN Red List of Jordanian plants. Volume 2, now under preparation, aims to complete a comprehensive assessment of the entire checklist of the plants of Jordan. RBG is also preparing a National Biodiversity Strategy and Action Plan and a Wild Socioeconomic Plant Conservation Strategy for the protection and conservation of species and crop wild relatives.

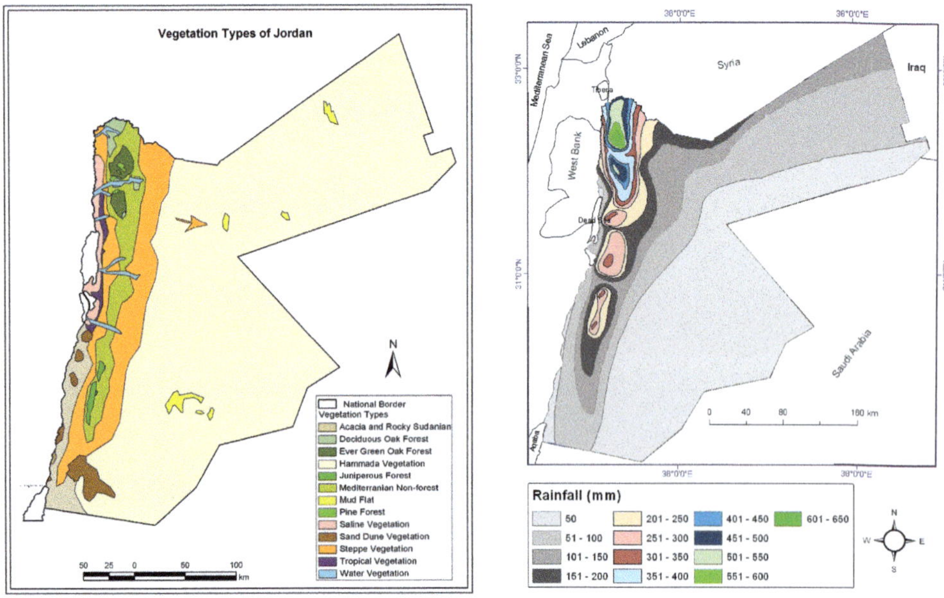

Fig. 1. Vegetation map of Jordan (Source: RSCN, after Eisawi 1996), and rainfall map of Jordan.

Volume 1 of the Jordan Plant Red List includes 1072 vascular plants, assessed on the basis of taxonomy, distribution, population size and trends, habitats and ecology, and threats, in accordance with IUCN Red List Categories and Criteria Version 3.1.

Five quantitative criteria were used to determine the status of each plant species: declining population (past, present and/or projected); geographic range size, and fragmentation, decline or fluctuations; small population size and fragmentation, decline, or fluctuations; very small population or very restricted distribution; and quantitative analysis of extinction risk (Population Viability Analysis). Data were compiled for each species, reviewed and the information presented to national specialists for ratification.

Of the 1072 plant species studied, 147 were found to be Not Applicable in the IUCN Red List assessment, as they were unresolved, misapplied, or did not exist in IUCN's Species Information System (SIS). Conservation assessments of the remaining 925 species were completed. One species, *Salvia fruticosa*, is recorded as Regionally Extinct (Region-Jordan) (RE), 19 species are Critically Endangered (CR), 54 are Endangered (EN), 33 are Vulnerable (VU), 11 are Near Threatened (NT), one is Data Deficient, and 806 are of Least Concern (LC).

Vegetation and habitats

The vegetation of the region of Palestine and Jordan has been described by several authorities, some with special reference to rangeland and forestry. Worth mentioning are those of Baierle et al. (1988), Danin (1983), Frey & Probst (1977), Frey & Kürschner (1989), Halevy & Orshan (1972), Hunting Technical Services (1956) (reports on forest and soils of Jordan), Kasapligil (1956), Zohary (1962), Zohary & Orshan (1956).

However, major contributions to the knowledge of the flora and vegetation of Jordan have been made during the last two decades with regional studies by Al-Eisawi (1982; 1985; 1986a; 1986b; 1988; 1996, 1998), Al-Eisawi & Hatough (1987), Boulos (1977), Boulos & Al-Eisawi (1977; 1997), Boulos & Lahham (1977), Boulos, Jallad & Lahham (1975; 1977), El-Oqla & Lahham (1985), Qishwani (1999).

A detailed description of the vegetation and ecology of Jordan has been described by Eisawi (1996) who classified the vegetation into 13 vegetation types (Fig. 1). These were: 1, Pine forest; 2, Evergreen oak forest; 3, Deciduous oak forest; 4, Juniperus forest; 5, Mediterranean non-forest vegetation; 6, Steppe vegetation; 7, Hammada vegetation; 8, Saline (halophytic) vegetation; 9, Tropical vegetation; 10, Sand dune vegetation; 11, Acacia and Sudanian rocky vegetation; 12, Water (hydrophytic) vegetation; 13, Mud flats (Qaa or Bajada).

The following outline of the main vegetation types in Jordan is adapted from Eisawi (1996).

1. Pine Forest

Land classification: Mountain ranges in northern parts of Jordan.
Altitude range: 550–1000 m. **Annual rainfall:** 400–600 mm.
Representative Governorates: Ajloun; Jerash; Balqa; Dibeen.
Vegetation: Open to dense forest with *Pinus halepensis* forming emergent trees; typical Mediterranean vegetation forms the best forests in Jordan, reaching climax forest in some places dominated by *Pinus halepensis* (Aleppo pine) found naturally over 700 m. In some places *P. halepensis* is replaced by *Quercus coccifera* as a result of degradation. Subemergent associated trees and shrubs species: *Arbutus andrachne, Qyercus coccifera, Pistacia palaestina.*

Low shrubs: *Calycotome villosa, Cistus villosus, C. salvifolius, Smilax aspera*. Herbaceous cover of *Fumana arabica, Thesium hergeri, Helianthemuim lavandulaefolium*; orchids: *Ophrys, Lemodorum, Cephalanthera*.
Biogeography: Mediterranean.

2. **Evergreen Oak Forest**

Land classification: Mountainous areas of northern, central and southern Jordan; red soil (Terra Rossa) of hard limestone parental rock and calcareous or yellow (Rendzina) soils of soft limestone parental rock.
Altitude range: 600–1200 m. **Annual rainfall:** 400–500 mm.
Representative Governorates: Amman; Irbid; best represented near Ajloun; Tafila.
Vegetation: This vegetation type is found at high altitudes (>700 m). Open to dense forest with *Quercus coccifera* as the dominant tree; associates *Pistacia palaestina, Pyrus syriaca, Arbutus andrachne, Crataegus azarolus, Phillyrea latifolia, Ceratonia siliqua. C. siliqua* is associated with *Juniperus phoenicea*, *Pistacia atlantica* and *Amygdalus korshinski* in the southern mountains.
Low shrubs: *Amygdalus communis, Asparagus aphyllus, Lonicera etrusca, Sarcopoterium spinosum, Rhamnus palaestinus, Rubia olivieri, Calycotome villosa, Cistus villosus, Clematic cirrhosa*.
Herbaceous: *Dactylis glomerata, Anemone coronaria, Echinops spp., Lecokia cretica, Orchis anatolica, Poa bulbosa, Cyclamen persicum, Linum pubescens, Adonis palaestina.*
Biogeography: Mediterranean.

3. **Deciduous oak forest**

Land classification: The name of this vegetation type is derived from the deciduous oak species. Three to four species of *Quercus* are used by different workers to designate the vegetation: *Q. aegilops*, *Q. boissieri*, *Q. ithaburensis* and *Q. infectoria*. The taxonomic status of these taxa in Jordan is not clear, although some authorities may consider two or even one valid taxon. Irrespective of the taxonornic treatment, here, *Q. ithaburensis* is used as the indicative species for the deciduous oak forests in Jordan.

The deciduous oak forests occur at a lower altitude than all other forests in Jordan and mostly grow on red or brown soil of hard limestone parental rock. Most of these forests are not well protected, hence they are subject to degradation. The deciduous oak forests are found at the lower borders of the evergreen oak forests, mixing together for a limited area until they become pure stands of each type alone. Scattered old deciduous trees occur within the Mediterranean region in Jordan, which show the degradation of these forests. The composition of deciduous oak forest varies in relation to the amount of humidity and temperature. For example, the forests at Um-Qeis and Malka are associated with *Pistacia palaestina*, *Ceratonia siliqua*, *Styrax officinalis*, *Cercis siliquastrum* and *Amygdalus communis*; while the forests in Al-Alouk ares associated with *Pistacia atlantica*, *Amygdalus communis*, *Rhamus palaestinus* and *Retama raetam* at the margins since the latter forest is at the Irano-Turanian border which receives less rainfall and has higher temperatures. In some places like As-Baihi in Salt area, Burma, the slopes at the Southern side of King Talal Dam and in Halawa between Ajloun and Irbid, the deciduous oak forest is mixed with remnants of wild olive (*Olea europaea*) trees and shrubs. At lower borders of the deciduous oak trees, especially at the Western slopes of Jordan Mountains descending to the Jordan Valley, shrubs of *Ziziphus nummularia* mix with oak trees, finally becoming pure stands of *Ziziphus nummularia* until

they reach the Jordan Valley. This is clearly seen at the slopes from Halawa to Al-Girim and at the Western slopes of Um-Qeis. Contrary to that, at the Eastern sides of the deciduous oak forest, shrubs of *Retama raetam* mix with oak trees until they become pure stands of *Retama raetam* when humidity rainfall and altitude become less, and the temperature is increased.
Representative Governorates and localities: Um-Qeis and Yarmuk river; Malka;Dair Abu Said to the borders of Ishtafaina forest in Ajloun; Al-Alouk near Jarash; As-Baihi near Salt; over the hills facing the Zarka river near King Talal Dam; West Mahes on the hills facing Wadi Shuaib and in Al-Bahi ith, West Amman.
Vegetation: Three strata, not always clear, can be found with the following dominant species: Trees: *Quercus ithaburensis*, *Styrax officinalis*, *Ceratonia siliqua*, *Pistacia atlantica*, *Olea europaea*. Shrubs: *Crataegus azarolus*, *Olea europaea*, *Amygdalus communis*, *Calycotome villosa*, *Rhamnus palaestinus*, *Retama raetam*. Low shrubs, climbers, grasses and geophytes: *Alcea* spp., *Carlina hispanica*, *Euphorbia hierosolymitana*, *Salvia* spp., *Sarcopoterium spinosum*, *Asparagus aphyllus*; grasses: *Dactylis glomerata*, *Poa bulbosa*; geophytes: *Drimia maritima*, *Colchicum* spp., *Tulipa* spp.
Biogeography: Mediterranean.

4. **Juniper woodland** (Juniperus forest vegetation, Eisawi 1996)

Land classification: Southern mountains.
Altitude range: 600–1700 m. **Annual rainfall:** 200–400 mm.
Representative Governorates: Tafila: Dana, Shobak, Wadi Musa (best represented at Dana Biosphere Reserve); western slopes extending to Petra and Wadi Musa where few trees are found in Wadi Rum.
Vegetation: Open woodland of *Juniperus phoenica* as the dominant tree, associated with veteran trees of *Cupressus sempervirens* (*C. sempervirens* is rare and occurs only in this locality). Above 1300 m juniper forest is taken over by evergreen oak forest dominated by *Quercus coccifera* and a mixed forest of *Quercus* and *Juniperus* where the two dominant species meet. Main associates: *Pistacia atlantica, Rhamnus palaestinus, Thymelaea hirsuta, Daphne linearifolia, Amygdalus korschinskyi, Colutea istria, Crataegus azarolus;* a few trees of *Ceratonia siliqua* are also present.
Shrubs: *Globularia arabica, Helianthemum vesicarium, Sarcopoterium spinosom, Osyris alba, Noaea mucronata, Achillea santolina, Artemesia sieberi, Zosima absinthifolia*; grasses: *Dactylis glomerata*.
Biogeography: Mediterranean.

5. **Mediterranean shrubland** (Mediterranean non forest vegetation, Eisawi 1996)

Land classification: Northern and southern mountains and foothills; Mediterranean region not covered by forests; a degraded forest.
Altitude range: 900–1500 m. **Annual rainfall:** ?400–600 mm.
Representative Governorates: Tafila, Karak, Jarash, Ajloun.
Vegetation: The secondary degraded forest and shrubland generally borders the forest and is result of overgrazing and anthropogenic influences. Dominant shrubs: *Rhamnus palaestinus*, *Sarcopoterium spinosum*, *Calicotome villosa* and *Cistus* spp. in the north, and *Artemisia sieberi* in the south with other associates.
Biogeography: Mediterranean.

6. **Mixed shrubland** (Steppe[2] vegetation Eisawi 1996)

Land classification: This vegetation forms a strip surrounding the Mediterranean shrubland region. excl. wooded areas and cultivations. It extends north to join the Syrian desert north of Mafraq; its southern limit is Nagb, the end of the railway, sometimes extending eastwards to join the Hammada; the eastern limit is the Hejaz railway.
Altitude range: 600–1500 m. **Annual rainfall:** 100–300 mm.
Representative Governorates: Tafila, Shobak, Ma'an, Karak.
Vegetation: Dominated by large shrubs with occasional tree species; composition varies in the north and south with respect to the Mediterranean Region. Dominant shrubs mainly *Artemesia sieberi* and scattered trees as *Pistacia atlantica* and *Crategus azarolus* in the north and south. Shrubs: *Retama raetam, Ziziphus lotus, Z. nummularia, Ferula communis* (dominant in the north), *Pistacia atlantica, Anabasis syriaca, Artemisia sieberi, Sarcopoterium spinosum* (dominant in NE and S Mediterranean borders); other species *Noaea mucronata, Gypsophila arabica, Astragalus spinosus;* geophytes: *Crocus moabiticus, Asphodelus aestivus, Drimia maritima, Moraea sisyrinchium.*
Biogeography: Irano-Turanian; Mediterranean or Saharo-Sindian in parts.

7. **Acacia woodland** (Acacia & rocky Sudanian vegetation, Eisawi 1996)

Land classification: Base of granite mountains, rocky areas, rocky gorges, stony and rocky beds.
Altitude range: 200–400 m. **Annual rainfall:** 100–200 mm.
Representative Governorates: Aqaba; Ma'an; Tafila; Karak. Wadi Araba, Wadi Yatum, Wadi Rum.
Vegetation: Characterised by scattered *Acacia* trees, getting denser at the base of mountains forming open *Acacia* woodlands; associated species include: *Acacia raddiana, Acacia tortilis, Anabasis articulata, Hammada scoparia, Cassia italica, Zygophyllum dumosum, Caralluma, Traganum nudatum, Fagonia* spp., *Reaumuria hirtella, Gymnocarpos decandrum, Helianthemum lippii, Asteriscus graveolens, Sclerocephalus arabicus, Anastatica hierochuntica.* At the southern end of the Dead Sea, increased water availability has resulted in forested areas especially near Gor Fifa and Ghora Safi dominated by *Ziziphus spina-christi, Maerua crassifolia, Salvadora persica, Balanities aegyptiaca, Calotropis procera, Acacia tortilis, A. raddiana* and *Ochradenus baccatus.*
Biogeography: Arabian regional subzone and Nubo-Sindian local centre of endemism.

8. **Hammada and sand dune vegetation** (Hammada vegetation & Sand dune vegetation types, Eisawi 1996)

Land classification: Eastern desert of Jordan; few volcanic hills, wadi fans; gravel plains; stony and pebble plains; sandy plains, sand dunes or sand and gravel plains (flint stones, basaltic volcanic tuffs, limestones).
Altitude range: 600–700 m. **Annual rainfall:** 50–100 mm.
Representative Governorates: Aqaba; Ma'an; Amman; Zarqa; Mafraq
Vegetation: Wadis dominated by *Acacia tortilis, A. raddiana, Tamarix* spp., *Artemisia judaica* and *A. monosperma, Retama raetam, Nitraria retusa, Prunus arabicus, Atriplex halimus, Lycium*

[2] The term 'steppe' for this vegetation is somewhat misapplied, as the term 'steppe' is normally used for a dry, cold ecoregion characterized by grassland plains without trees apart from those near rivers and lakes. The vegetation type in Jordan classified as 'steppe' is dominated by shrubs and perhaps better named as a mixed shrubland. — S.A. Ghazanfar (ed.).

europaeum, Artemisia sieberi, Achillea fragrantissima, Phlomis brachyodon, Tamarix spp., *Peganum harmala, Astragalus* spp., *Anabasis articulata, Atractylis mutica, Moraea sisyrinchium.*
Gravelly areas: *Seidlitzia rosmarinus, Spergularia diandra, Herniaria hirsuta, Aaronsohnia factorovskyi, Anthemis deserti, Asteriscus pygmaeus, Mesembryanthemum nodiflorum, Filago desertorum, Gymnarrhena micrantha, Trigonella stellata*. Grasses: *Stipa capensis, Bromus* spp.
Areas with small stones and pebbles: *Salsola vermiculata, Anabasis articulata, Linum album, Thymus bovei, Paracaryum rugulosum, Zilla spinosa, Halogeton alopecuroides, Diplotaxis harra, Euphorbia retusa, Alcea chrysantha, Atriplex leucoclada, Lepidium aucheri.*
Sandy places and sand dunes: *Haloxylon persicum, Seidlitzia rosmarinus, Atriplex spp., Artemisia sp., Anabasis articulata, Achillea fragrantissima, Halocnemum strohilaceum, Ephedra transitoria, Deverra triradiata, Calligonum tetrapterum, Zilla spinosa.*
Biogeography: Saharo-Sindian regional zone.

9. **Dry Tropical vegetation** (Tropical vegetation type, Eisawi 1996)

Land classification: Jordan valley slopes, hill slopes, wadi fans, edges of runnels and cultivations.
Altitude range: 200–400 m. **Annual rainfall:** 100–200 mm.
Representative Governorates: Balqa; Madaba: Dead Sea area, Lower Jordan valley; Aqaba: Gulf of Aqaba. Extending from Dair 'Alla in the North to the Gulf of Aqaba in the south, Southern Ghor, Ghor Safi and Ghor Faifa.
Vegetation: Vegetation confined to alluvial soils of the Rift valley greatly altered with many cultivated areas. Vegetation consists of: Trees: *Acacia raddiana, Ziziphus spina-christi, Balanites aegyptiaca, Maerua crassifolia.* Lianas: *Cocculus pendulus.* Shrubs: *Salvadora persica, Moringa peregrina, Calotropis procera, Ochradenus baccatus, Aerva* spp., *Forsskaolea tenacissima, Capparis decidua.*
Biogeography: Saharo-Sindian regional zone.

10. **Halophytic vegetation**

Land classification: Saline and sandy marshes, dry saline soils and wet saline soils.
Altitude range: minus 400–500 m. **Annual rainfall:** 100–200 mm.
Representative Governorates: Zarqa, Azraq, Ma'an Bayer and Jafer, Ajloun. Dry saline: Ghor Safi, Ghor Fifa, Wadi Araba, El-Jafr, Wadi Sirhan. Wet marsh: eastern slopes of the Dead Sea, El Jirm & Zor Abdullah, Azraq pools, coastal Aqaba.
Vegetation: Vegetation tolerant to various degrees of soil salinity. Characteristic species of dry saline areas: Trees & large shrubs: *Tamarix tetragyna, T. macrocarpa, Nitraria retusa.* Shrubs: *Anabasis setifera, Atriplex halimus, A. turcomanica, Suaeda fruticosa, S. palaestina.* Species of wet saline areas: Trees: *Tamarix nilotica.* Shrubs: *Seidlitzia rosmarinus, Inula viscosa.* Grasses & reeds: *Panicum turgidum, Phragmites communis.*
Biogeography: Saharo-Sindian regional zone.

AZONAL VEGETATION

Mud flats and hydrophytic vegetation (Water vegetation & mud flats, Eisawi 1996)

Mudflats are found where soil is silty and clayey, and which after drying becomes hard and generally lacks vegetation except at the edges. Examples are Qaa' Al-Jafr and Qaa' Al Oisi, Qaa' Al-Azraq, Qaa' Shbaikah and Qaa' Ad-Dahik. Common species are: Large shrubs: *Capparis ovata, Nitraria retusa, Tamarix* spp. Small herbaceous shrubs: *Frankenia hirsuta, Suaeda* spp. Grasses: *Aeluropus littoralis.*

Riverine and hydrophytic vegetation is found around Jordan R., Zarka R., around pools in Wadi Shuaib, Wadi Mujeb, Wadi Al-Hasa and Azraq Oasis. Main species are: Trees and large shrubs: *Nerium oleander*, *Salix* spp., *Tamarix* spp., *Vitex angus-castus*. Sedges and grasses: *Cyperus longus*, *Phragmites australis*, *Typha domingensis*.

Biogeography

Jordan is influenced by two major phytochoria, the Irano-Turanian and Saharo-Sindian, both subdivided into regional phytochoria specific to the Middle Eastern region (Leonard 1989). Species richness and endemism is high in the Mediterranean part of the Irano-Turanian region of Jordan (also treated separately as the Mediterranean region). Endemism is restricted mainly to the southern Mediterranean and Irano-Turanian regions as the southern mountains have been isolated by large wadis in the north and west.

Irano-Turanian regional centre of endemism (IT):

(Endemism: 19 genera, 59 species)

IT 1 — SE Turkey, most of Syria, parts of Lebanon, Israel, Palestine, N Sinai, N Jordan, N Iraq. Representatives of IT endemic genera in Jordan: *Cousinia*, *Gundelia* (Asteraceae); *Heterocaryum*, *Paracaryum* (Boraginacae), *Ducrosia* (Apiaceae).

Species endemic to IT 1 found in Jordan: *Alysum marginatum, Cardaria draba, Heterocaryum subsessile, Hyocyamus pusillus, Ixioliron tataricum, Ziziphora tenuior, Scorzonera pusilla, Anchusa strigosa, Artemisia sieberi, Ducrosia anethifolia, Gundelia tournefortii, Linum album, Minuartia picta, Pistacia khinjuk.*

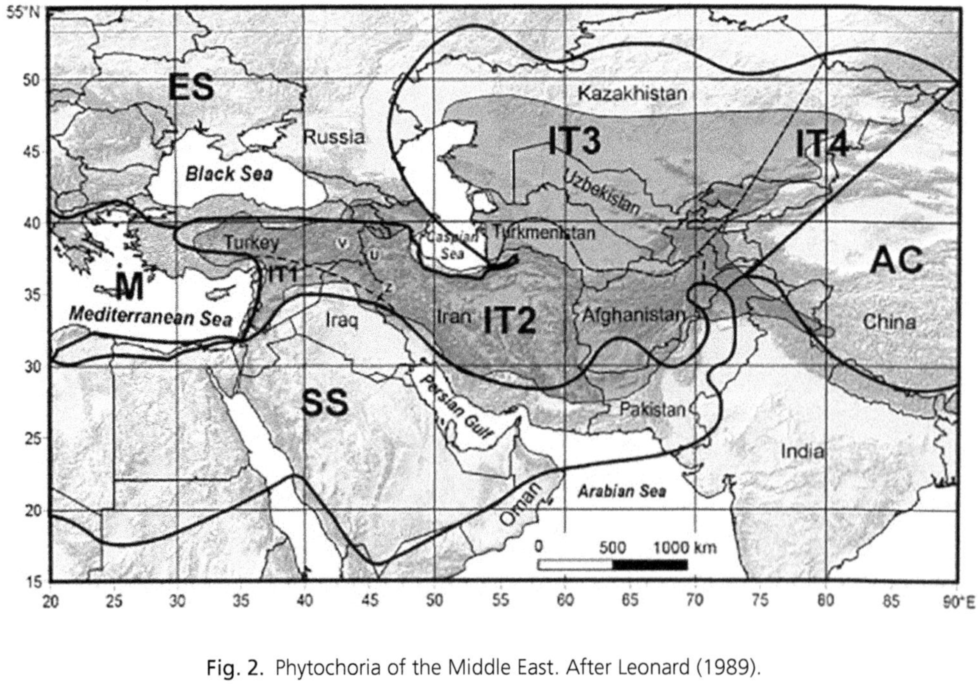

Fig. 2. Phytochoria of the Middle East. After Leonard (1989).

Saharo-Sindian regional zone (SS)

Saharo-Sindian (SS1), Arabian regional subzone (SS2), Nubo-Sindian local centre of endemism (SS3).
Endemism: 6 genera, 17 species.
Representatives of SS endemic genera in Jordan: *Anastatica, Eremobium, Gymnarrhena, Moltkiopsis, Neurada, Notoceras, Savigna.*
Species endemic to Saharo-Sindian (SS) found in Jordan: *Anastatica hierochuntica, Gymnarrhena micrantha, Neurada procumbens, Notoceras bicorne, Savigna parviflora.*

Arabian regional subzone + Nubo-Sindian local centre of endemism[3] — (SS 2 + SS 3) endemic genera in Jordan: *Anisosciadium.*

Species endemic to Arabo-Sindian (SS 2 + SS 3) found in Jordan: *Anisosciadium isosciadium, A. lanatum, Haloxylon salicornicum, Zygophyllum propinquum, Paronychia Arabica, Trigonella glabra.*
SS3 — Represented in Jordan by species such as *Ziziphus nummularia, Acacia pachyceras*, and *Cleome amblyocarpa*, which are considered as Sudanian relicts that migrated from western Arabia through flow cut channels during pluvial times during the late Pliocene (c. 2.5 MYBP).

Mediterranean regional centre of endemism (M)

Occupies most of western Jordan, mountains and Jordan valley.
Representative vegetation: coniferous forests of Aleppo pine (*Pinus halepensis*) and Callabrian pine (*Pinus brutia* — not in Jordan); and dry oak (*Quercus* spp.) woodlands and shrubland formations.
Olive (*Olea europea*), carob (*Ceratonia siliqua*), oak (*Quercus coccifera* — most widespread), pistachio (*Pistacia palaestina, P. lentiscus*), and *Arbutus andrachne* are the principal species. In addition, *Phillyrea latifolia, Pistacia terebinthus, Calicotome villosa, Genista acantha, Rhamnus oleoides, Myrtus communis, Laurus nobilis, Styrax officinalis*, and *Spartium junceum* also occur.
Vegetation is degraded, in particular *Ceratonia siliqua*, which is represented only by a small population.

---

[3] Nubo-Sindian province of the Sudanian region of Zohary (1973)

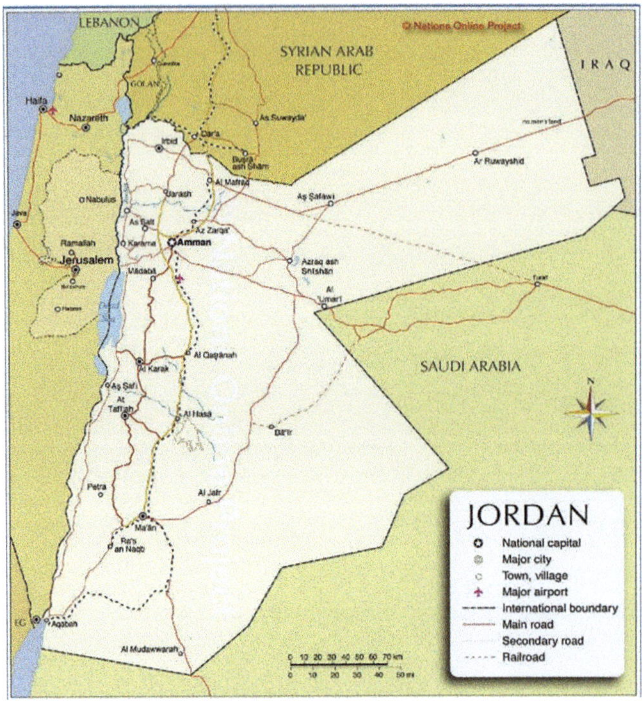

Physical map of Jordan

North Region
1. Irbid
2. Ajloun
3. Jerash
4. Mafraq

Central Region
5. Balqa
6. Amman
7. Zarqa
8. Madaba

South Region
9. Karak
10. Tafila
11. Ma'an
12. Aqaba

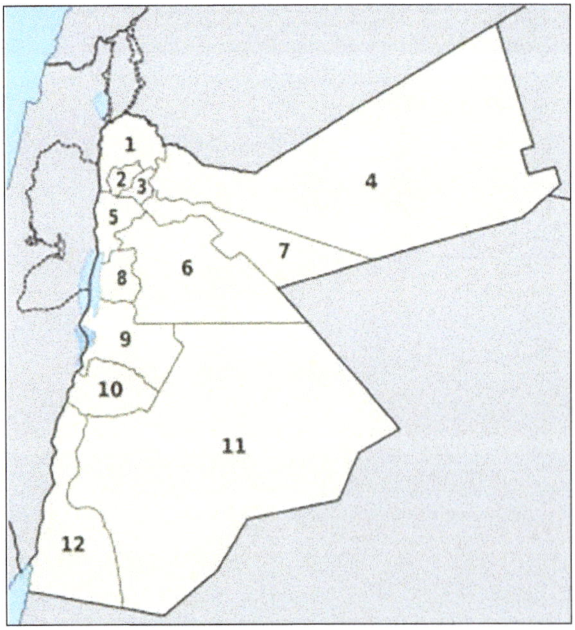

Governorates of Jordan

Herbaria consulted and cited in checklist

AUB — Post Herbarium, American University of Beirut, Lebanon.
B — Botanischer Garten und Botanisches Museum Berlin-Dahlem, Berlin.
E — Royal Botanic Garden, Edinburgh, Scotland, UK. [Barcode E00-]
HUJ — Jerusalem University Herbarium, Israel.
K — Herbarium Royal Botanic Gardens Kew, UK. [Barcode K000-]
MSB-K — Millennium Seed Bank, Kew (Voucher specimens with seed stored at MSB herbarium and K).
TUB — Herbarium Tubingense, Eberhard Karls Universität Tübingen, Tübingen.

Herbaria in Jordan (not in Index Herbariorum)

BRDC-H — Badia Research and Development Center Herbarium, Amman.
FOR-H — Herbarium of the Forestry Department.
IBBU-H — Ibn Al-Bitar Herbarium, Al-Bayt University.
NCARE — Herbarium of the National Centre for Agriculture Research and Extension, Amman.
NHJ — Herbarium Royal Botanic Garden Jordan.
RSCN-H — Herbarium of the Royal Society of the Conservation of Nature (incl. Dibeen Forest Reserve Herbarium, Ajloun Forest Reserve Herbarium).
YAR-H — Yarmouk University Herbarium, Irbid.

Duplicates of collections at the Herbarium of the Department of Botany, University of Jordan, Amman (AMM), deposited at other national and international herbaria have been studied; herbarium material at AMM cited in publications is also integrated into the checklist.

Annotated checklist

FERNS AND FERN ALLIES

Equisetaceae

Equisetum ramosissimum Desf., Fl. Atlant. 2: 398. 1799.
Petra, *Frey & Kürschner* 80–898 (TUB).

Aspleniaceae

Asplenium ceterach L., Sp. Pl. 2: 1080. 1753.
Ref. Aqaba, Danin (2004).

Asplenium sagittatum (DC.) Bunge, Bull. Mens. Soc. Linn. Lyon 21: 84. 1952.
Ref. Aqaba, Danin (2004).

Ceterach officinarum Willd., Anleit.Selbststud. Bot. 578. 1804.
Wadi Mujib, *Frey & Kürschner* 80–442 (TUB); Tal Al-Rumman, *Taifour & Aldamen* (NHJ 2010.0010).

Dryopteris raddeana (Fomin) Fomin, Věstn. Tiflissk. Bot. Sada 20: 50. 1911.
Syn. *Dryopteris pallida* Fomin (1910).
Ref. Ajloun, Danin (2004).

Pteridaceae

Adiantum capillus-veneris L., Sp. Pl. 2: 1096. 1753.
Tal Al-Rumman, *Taifour & Aldamen* (NHJ 2010.0035).

Anogramma leptophylla (L.) Link, Fil. Spec. 137. 1841.
Ref. Ajloun, Danin (2004).

Cheilanthes acrostica (Balb.) Tod., Giorn. Sci. Nat. Econ. Palermo 1(10): 215. 1866.
Syn. *Pteris acrostica* Balb. (1801).
Ref. Aqaba, Danin (2004).

Cheilanthes fragrans (L.) Swartz, Synopsis Filicum 127. 1806.
Wadi Araba, *Frey & Kürschner* 86–249 (TUB).

Cheilanthes pteridioides C.Chr., Index Filic. 178. 1905.
Taifour & Aldamen (NHJ 2009.0094)**;** Dibeen Forest Reserve; Boulos, 3–4 km West Rum Rest House (Ref. 1975).

Cheilanthes vellea (Aiton) F.Muell., Fragm. (Mueller) 5(36): 123. 1866.
Ref. 3–4 km West Rum Rest House, *Boulos et al.* (1975).

Cosentinia vellea (Aiton) Tod., Giorn. Sci. Nat. Econ. Palermo 1. 220 (seors. 15). 1866.
Syn. *Acrostichum velleum* Aiton (1789).
Ref. Ma'an, Danin (2004).

GYMNOSPERMS

Cupressaceae

Cupressus sempervirens var. **horizontalis** L., Sp. Pl. 2: 1002. 1753.
Dana (NHJ 2013.0631).

Juniperus phoenicea subsp. **phoenicea** L., Sp. Pl. 2: 1040 .1753.
N of Petra, *Frey & Kürschner* 80–817 (TUB), Dana Reserve, Barra (NHJ 2013.0477).

Ephedraceae

Ephedra alata Decne., Ann. Sci. Nat. (Paris) 2: 239. 1824.
Abulaila & Tehabshem 2005JOR38 (K000459041); Tal Al-Rumman, *Al-Eisawi* (NHJ 2006.0186).

Ephedra aphylla Forssk., Fl. Aegypt.-Arab.: 170 1775.
Near Tafila, *Frey & Kürschner* 80–797 (TUB); Wadi Musa, *Ananbeh et al.* (NHJ 2013.0644).

Ephedra foeminea Forssk., Fl. Aegypt.-Arab.: 219. 1775.
Jabal Ataita towards Dana, *Frey & Kürschner* 80–797 (TUB).

Ephedra foliata Boiss. ex C.A.Mey., Monogr. Ephedra (Mem. Acad. Sci. St. Petersb. 6: 107. 1846.
Syn. *Ephedra ciliata* Fisch. & C.A.Mey. (1845); *E. peduncularis* Boiss. (1884).
Wadi Mudeisisat S of Azraq, *Frey & Kürschner* 80–373 (TUB); Tal Al-Rumman, *Shishani* (NHJ 2011.0117).

Ephedra fragilis Desf., Fl. Atlant. 2: 372. 1799.
Petra, *Dinsmore* (K).

Ephedra pachyclada Boiss., Fl. Orient. 5: 713. 1884.
subsp. **sinaica** (Riedl.) Frietag & Maier-st., Edinb. J. Bot. 49: 92. 1992.
Dana area, *Walid* 8068 (K000459143); *Dinsmore* 13717 (K000459230); Tafila, Agricultural Station near Dana, *Abu-Laila & Slageren* 151979 (K000459142).

Pinaceae

Pinus halepensis Mill., Gard. Dict., ed. 8. n. 8. 1768.
Suweleh towards Jerash, *Frey & Kürschner* 80–73 (TUB); Dibeen Forest, *RBG-Ire & RBG-Jo* (NHJ 2008.0066).

ANGIOSPERMS

Acanthaceae

Acanthus hirsutus (Boiss.) Brummitt subsp. **syriacus**, Notes Roy. Bot. Gard. Edinburgh 46(1): 8. 1989.
Syn. *Acanthus syriacus* Boiss. (1849).
Amman, Wadi Al Seer (MSB-K 228060); Tal Al-Rumman RBG, Sector 58, *Taifour & Al Khawaldah* (NHJ 2013.0326).

Blepharis ciliaris (L.) B.L.Burtt, Notes Roy. Bot. Gard. Edinburgh 22: 94. 1956.
Syn. *Ruellia ciliaris* L. (1767); *Acanthus edulis* Forssk. (1775).
Dead Sea area, near Zarqa Ma'in, *Frey & Kürschner* 80–988 (TUB).

Blepharis attenuata Napper, Israel J. Bot. 21(3): 164. 1972.
Dead Sea highway, before Zara Springs (hot springs) (MSB-K 227801); Wadi Yutum, *Taifour et al.* (NHJ 2013.0705).

Aizoaceae

Aizoanthemum hispanicum (L.) H.E.K.Hartmann, Illustr. Handb. Succ. Pl.: Aizoaceae A-E: 29. 2001.
Syn. *Aizoon hispanicum* L. (1753).
Umm El-Jimal antiquities, *Royal Botanic Garden Team* (NHJ 2008.0010).

Aizoon canariense L., Sp. Pl.: 488. 1753.
Wadi Rum, *El-Oqlah et al.* (NHJ 2013.0607).

Mesembryanthemum forsskahlii Hochst. ex Boiss., Fl. Orient. 2: 765. 1872.
Syn.*Opophytum forsskahlii* N.E.Br. (1928).
Townsend 65/434 29441 (K).

Mesembryanthemum nodiflorum L., Sp. Pl.: 481. 1753.
Syn. *Gasoul nodiflorum* (L.) Rothm. (1941); *Cryophytum nodiflorum* (L.) L.Boulos (1927).
Azraq Oasis, *Tackholm et al.* 9083 (K000286065); Tal Al-Rumman, *Al-Eisawi* (NHJ 2006.0163).

Tetragonia tetragonioides (Pall.) Kuntze, Revis. Gen. Pl. 1: 264. 1891.
Native of New Zealand, cultivated and naturalised in many warm temperate regions including Jordan.

New Zealand spinach.
Taifour (in obs.). No specimens seen.

Trianthema portulacastrum L., Sp. Pl.: 223. 1753.
Araba valley, *Danin* [http://flora.org.il/en/plants/TRIPOR/].

Zaleya pentandra (L.) C.Jeffrey, Kew Bull. 14: 238. 1960.
Syn.*Trianthema pentrandra* L. (1767).
Aqaba, *Boulos & Lahham* 5408 (AMM, BM, E).

Adoxaceae

Sambucus ebulus L., Sp. Pl.: Sp. Pl. 269. 1753.
Wadi Ajloun, RBG Jordan Team 2016.

Sambucus nigra L., Sp. Pl.: 269. 1753.
Ajloun, Wadi Ajloun, 17.vi.1981 (YAR-H).
Naturalised along streams.

Alismataceae

Alisma lanceolatum With., Bot. Arr. Brit. Pl. ed. 3 2: 362. 1796.
Ref. RSCN database; Ajloun, Danin (2004).

Alisma plantago-aquatica L., Sp. Pl. 342. 1753.
Ref. RSCN database.

Damasonium alisma Mill., Gard. Dict. ed. 8, 1. 1768.
Syn. *Alisma damasonium* L. *(*1753.); *Damasonium alsima* Mill. var. *compactum* (M.Mich.) Dur. et Schinz (1881*)*; *Damasonium alsima* Mill. subsp. *bourgaei* (Coss.) Maire (1890).
Upper Jordan Valley [http://flora.org.il/en/plants/DAMALI/].

Damasonium polyspermum Coss., Notes Pl. Crit. 47. 1849.
Recorded as very rare. [http://flora.org.il/en/plants/DAMPOL/].

Amaranthaceae

In the APG III (2009) & APG IV (2016) classification, the Amaranthaceae includes the plants formerly treated in the family Chenopodiaceae. The monophyly of the broadly defined Amaranthaceae has been strongly supported by both morphological and phylogenetic analyses. The main differences between Amaranthaceae (*sensu stricto*) and Chenopodiaceae (*sensu stricto*) are the membraneous petals and stamens often united in a ring structure in Amaranthaceae. The taxonomy of many Chenopodiaceae (*sensu stricto*) has been changed in the last decade; the circumscription of each genus is given in accordance to the results of detailed molecular investigations.

Achyranthes aspera L., Sp. Pl. 204. 1753.
Dead Sea, 12.vi.1884 (AUB).

Aerva javanica (Burm.f.) Juss. ex Schult., Ann. Mus. Hist. Nat. 11: 131. 1808.
Syn. *Iresine javanica* Burm.f. (*1768)*; *Aerva tomentosa* Forssk. (1775); *Aerva persica* (Burm.f.) Merr. (1921).
Ma'an, 5 km NW of Rum, *Frey & Kürschner* 5035 (E00161899, TUB); Wadi Araba, *Al Mfadi et al.* (NHJ 2013.0519, K)

Agathophora alopecuroides (Delile) Bunge, Mém. Acad. Sci. Pétersb. ser. 7, 4, 11: 92. 1862.
Syn. *Salsola alopecuroides* Delile (1813); *Agathophora postii* (Eig) Botsch. (1977); *Halogeton alopecuroides* (Delile) Moq. (1840).
Wadi Rum, *Boulos & Lahham* 7737 (AMM, BM, E).

Alternanthera sessilis (L.) DC., Cat. Pl. Horti Monsp. 77. 1813.
Syn. *Suaeda sessilis* L. (1753).
Upper Jordan Valley (YAR-H).

Amaranthus albus L., Syst. Nat. ed. 10 2: 1268. 1759.
Aqaba, Wadi Rum, Site no 34. (MSB-K 497826).

Amaranthus arenicola I.M.Johnst., J. Arnold Arbor. 29: 193. 1948.
Along Zarqa River, 2004, *Al-Dwiri* s.n. (AMM).
Introduced.

Amaranthus blitoides S.Watson, Proc. Amer. Acad. Arts 12: 273. 1877.
Salt, on the way to Wadi Shuaib (MSB-K 185730); Irbid, 03/6/2002, Jordan Univ. Science & Tech. (YAR-H).

Amaranthus blitum L., Sp. Pl. 990. 1753.
Fifa, Wadi Araba, Aqaba (Moab and Edom), *Danin* [http://flora.org.il/en/plants/AMABLT/].

Amaranthus graecizans L., Sp. Pl. 2: 990. 1753.
Jordan Valley (RSCN database).

Amaranthus hybridus L., Sp. Pl. 2: 990. 1753.
Syn. *Amaranthus hypochondriacus* L. (1753); *Amaranthus chlorostachys* Willd. (1790).
Irbid, 3/6/2002, Jordan Univ. Science & Tech. (YAR-H).

Several subspecies are recognized of which one occurs in Jordan separated by the following key.

Larger bracteoles of female flowers mostly 1.5–2 × as long as perianth. Style bases and upper part of lid of fruit swollen, so that the fruit has an inflated beak. . . . . subsp. **hybridus**

Larger bracteoles of female flowers mostly 1–1.5 × as long as perianth. Style bases and upper part of lid of fruit not swollen, so that the fruit has a smooth, narrow beak. . subsp. **cruentus**

Amaranthus hybridus L. subsp. **cruentus** (L.) Thell., Fl. Advent. Montpellier 205. 1912.
Syn. *Amaranthus cruentus* L. (1759); *Amaranthus paniculatus* L. (1763); *Amaranthus hybridus* L. var. *paniculatus* (L.) Thell. (1912).
Jordan valley (RSCN database).

Amaranthus palmeri S.Watson, Proc. Amer. Acad. Arts 12: 274. 1877.
Dead Sea area, *Danin* (http://flora.org.il/en/plants/AMAPAL/).

Amaranthus retroflexus L., Sp. Pl. 2: 991. 1753.
Syn. *A. retroflexus* var. *delilei* (Richat & Loret) Thell. (1907).
Salt, Bellauna village (MSB-K 114356).

Amaranthus spinosus L., Sp. Pl. 991. 1753.
Along Zarqa River, 2004, *Al-Dwiri* s.n. (AMM).

Amaranthus tuberculatus (Moq.) Sauer, Madroño 13: 18. 1955.
Syn. *Acnida tuberculata* Moq. (1849); *Amaranthus rudis* J.D.Sauer (1972).
Jordan valley (YAR-H).

Amaranthus viridis L., Sp. Pl., ed. 2. 2: 1405. 1763.
Syn. *Amaranthus gracilis* Desf. (1804).
Amman, Desert road to Thibham-Umm Al Rasas. (MSB-K 186036); Irbid, 14/11/1981, El Ghor (YAR-H).

Anabasis articulata (Forssk.) Moq., DC., Prodr. 13 (2): 212. 1849.
Syn. *Salsola articulata* Forssk. *(*1775).
Shaumari Wildlife Reserve, 1/12/1977, *Clarke* (E00451879).

Anabasis setifera Moq., Chenop. Monogr. enum. 164. 1840.
Syn. *Seidlitzia lanigera* Post (1896).
Tafila, Shaumari Wildlife Reserve, *Clarke* (E00676796); Between Zarqa road to Qa' AlKhanna, *Addison* (NHJ 2008.0011).

Anabasis syriaca Iljin, Bot. Mater. Gerb. Bot. Inst. Komarova Akad. Nauk S.S.S.R. 20: 138. 1960.
Tafila, Jurf-ud-Darawish, *Dinsmore* 10983 (E00676792); Northeastern Badia Daer, *Al Shudiefat et al.* (NHJ 2013.0012).
Anabasis lacnantha and *Anabasis oropediorum* are similar to *A. articulta* and may be found in Jordan — they are distinguished by the characters given below. No specimens seen from Jordan and some material possibly misidentified for *A. articulata*.
[*Anabasis lachnantha* Aellen & Rech.f., Anz. Math.-Nat. Kl. Österr. Akad. Wiss. 98: 26 1961.]
Small shrub to 30 cm. Leaves minute, rounded, divergent from the stem. Flowers solitary, in short spikes at end of branches, arisng in axils with white tufts of hairs. Fruiting perianth with pinkish ovate-orbicular sinuate wings.

Found in Kuwait, Saudi Arabia, Iraq, S Iran.

Similar to *A. articulata* in its minute scale-like leaves, but separated by the following characters:

Annual stems greyish or green; hairs shorter than perianth . . . . 1. *A. articulata*
Annual stems white; hairs in the axils of bracts of equal length as perianth . . .2. *A. lachnantha*

Anabasis oropediorum Maire, Bull. Soc. Hist. Nat. Afr. Nord. 29: 447 (1938).
Syn. *Anabasis articulata* (Forssk.) Moq. subsp. *oropediorum* (Maire) Ozenda, Fl. Sahara 235 1958.
A. oropediorum is very close to *A. articulata* resembling the latter in most characters. In the field it differs by being abundantly browsed down to woody trunk, and with prominent narrow triangular, pointed free scales at the tip of vegetative branches (as opposed to the rounded tip of the vegetative branches of typical *A. articulata* where no scales are seen)].

Arthrocnemum macrostachyum (Moric.) K.Koch, Hort. Dendrol. 96. 1853.
Syn. *Salicornia macrostachya* Moric. *(1820); Arthrocnemum glaucum* (Moq.) Ung.- Sternb.(1876).
Amman, Dead Sea, opposite Amman Beach Wadi, site no 28 (MSB-K 497516).

Several species of **Atriplex** have been listed for Jordan which may be a result of misidentifications and should be referable to other species: *Atriplex sagittata* is easily recognized due to its folded leaves, but we have not seen any specimens from Jordan. *A. hastata* L. (now *A. prostrata* Boucher ex DC.) is morphologically close to *A. micrantha* and *A. davisii*. The southernmost distribution of *A. prostrata* is the Eastern Mediterranean (Syria and Lebanon). It is probable that material identified as *A. micrantha* and *A. davisii* (and *A. hastata*) belong to *A. prostrata*.

Atriplex davisii Aellen, Notes Roy. Bot. Gard. Edinburgh 28: 30. 1967.
Branched annual with spreading stems. Leaves green or greyish below, often shiny, triangular-hastate, entire or weakly toothed. Both male and female flowers intermixed, in clusters forming loose, usually leafless inflorescence. Bracteoles in fruit rounded, small, smooth or mostly with two dot-like appendages in upper part.
In habit very similar to **A. prostrata** Boucher ex DC., but the latter species has triangular bracteoles united in their basal part. In Turkey, Eastern Mediterranean to Sinai, Cyprus, Greece, Iran, Afghanistan. No specimens seen.

Atriplex dimorphostegia Kar. & Kir., Bull. Soc. Imp. Naturalistes Moscou. 15: 438. 1842.
Wadi Rum, *Gillett* 16037 (K000286309).

Atriplex glauca L., Sp. Pl. ed. 2: 1493. 1763.
Syn. *Atriplex stylosa* Viv. (1836).
Ma'an, Petra, Karak (Edom and Moab), *Danin* (http://flora.org.il/en/plants/ATRGLA/).

Atriplex halimus L., Sp. Pl. 1052. 1753.
Wadi Essier on the way to Bahath & Wadi Al-Shitta (MSB-K 244606); Tabaget Fahel, 9.v.1967, *Edgecombe* (AUB).

Atriplex holocarpa F.Muell., Rep. Pl. Babbage's Exped. 19. 1859.
Syn. *Atriplex spongiosa* F.Muell. (1858); *Senniella spongiosa* (F.Muell.) Aellen (1938) var. *holocarpa* (F. Muell.) Aellen (1938).
Ma'an, toward Wadi Yatum, 2 km after Tax Point (MSB-K 256933).

Atriplex leucoclada Boiss., Diagn. Pl. Orient. 12: 95. 1853.
Syn. *Atriplex inamoena* Aellen (1939).
Two varieties are recognized: **A. leucoclada** var. **inamoena** with small, ovatecordate leaves up to 1.5 cm long, and a fruiting perianth with 5 apical lobes. The second variety, **A. leucoclada** var. **turcomanica** (Moq.) Zohary has larger, triangular-deltoid leaves, up to 3.5 cm, and a fruiting perianth that is quadrangular to campanulate.
Tal Al-Rumman, *Taifour & Aldamen* (NHJ 2010.0034).

Altriplex micrantha Ledeb., Icon. Pl. Fl. Ross. 1: 11. 1829.
Syn. *Atriplex heterosperma* Bunge (1851).
Ma'an, *Danin* (http://flora.org.il/en/plants/ATRMIR/); Wadi Araba and Aqaba, Danin 2005; RSCN database.

Atriplex muelleri Benth., Fl. Austral. 5: 175. 1870.
Wadi Araba and Aqaba, Danin 2005; RSCN database.

Atriplex nummularia Lindl., J. Exped. Trop. Australia 64. 1848.
Syn. *Atriplex asphaltitis* Kasapligil (1966).

This shrubby species differs from *A. halimus* in its dioecious habit, leafy female inflorescences and in having appendages on fruit bracts. ?Cultivated.

Atriplex portulacoides L., Sp. Pl. 2: 1053. 1753.
Syn. *Halimione portulacoides* (L.) Aellen (1938).
Ref. Karak, Moab, Danin (2004).

Atriplex prostrata Boucher ex DC., Fl. Franc. 3, 3: 387. 1805.
Syn. *Atriplex hastata* auctt.
Ma'an, *Danin* (http://flora.org.il/en/plants/ATRPRS/).

Atriplex rosea L., Sp. Pl. ed. 2: 1493. 1763.
From Al Waha circle toward Al Madeevel Atebieh Street, 5 km before the bridge (MSB-K 341479); Tal Al-Rumman, *Al-Eisawi* (NHJ 2006.0150).

Atriplex semibaccata R.Br., Prodr. Fl. Nov. Holl. 406. 1810.
Ref. Al-Eisawi 83 (2013); *Danin* (http://flora.org.il/en/plants/ATRSEM/).

Atriplex suberecta I.Verd., Bothalia 6: 418. 1954.
Ma'an, *Danin* (http://flora.org.il/en/plants/ATRSUB/).

Atriplex tatarica L., Sp. Pl. 1053. 1753.
Syn. *Atriplex lasiantha* Boiss. (1853); *A. tataricum* L. var. *virgatum* Boiss. (1879.) p.p.; *A. autrani* Post (1896); *A. mesopotamica* Eig (nomen in herb. HUJ).
Jordan, Ghour Safi (AUB); Karak, 28.x.1926 (28261-HUJ).

Bassia arabica (Boiss.) Maire & Weiller, Fl. Afr. Nord 8: 54. 1962.
Syn. *Chenolea arabica* Boiss. (1853); *Chenoleoides arabica* (Boiss.) Botsch. (1976).
Dead Sea valley, *Danin* (http://flora.org.il/en/plants/BASARA/).

Bassia eriophora (Schrad.) Aschers.in Schweinf., Beitr. Fl. Aethiop.: 187. 1867.
Syn. *Kochia eriophora* Schrad. (1809); *Bassia eriantha* (Fisch. & C.A.Mey.) Kuntze (1891); *Bassia latifolia* (Fresen.) Asch. & Schweinf. (1887).
Jordan, *Frey & Kürschner* 5428 (E00596890, TUB); Karak–Ghor Safi road, 1.5 km after the potash. (MSB-K 228107); 25 km S of Ma'an, *Frey & Kürschner* VO 5428 (E).
Recently reported from Jordan by Khoshravesh & Akhani (2013).

Bassia indica (Wight) A.J.Scott, Feddes Repert. 89(2–3): 108. 1978.
Syn. *Kochia indica* Wight (1852).
Dead Sea valley, *Danin* (http://flora.org.il/en/plants/BASIND/).

Bassia muricata (L.) Asch. & Schweinf., Schweinf., Beitr. Fl. Aethiop.: 187. 1867.
Syn. *Salsola muricata* L. (1767); *Kochia muricata* (L.) Schrad. (1809).
Wadi Rum *Gillett* 16068 (K000286313); Aqaba, *Boulos & Lahham* 5375 (AMM, BM, E).

Bassia pilosa (Fisch. et C.A. Mey.) Freitag & G. Kadereit in Taxon 60(1): 73. 2011.
Syn. *Panderia pilosa* Fisch. & C.A. Mey. (1835).
Azraq Druz [as Duruz], *Townsend* 217 (K000286324).

Bassia scoparia (L.) A.J.Scott, Feddes Repert. 89: 108. 1978.
Syn. *Chenopodium scoparium* L. (1753.)
Ref. Al-Eisawi 84 (2013).
Grown in gardens for the beautiful reddish colour which it develops on senescence – hence its English name of "Burning Bush". Cultivated elsewhere including neighboring territories (Arabia, Kuwait, Palestine).

Beta maritima L., Sp. Pl. 322. 1763.
Various forms and varieties are recognized, but most forms in cultivation belong to *B. vulgaris* var. *vulgaris*, the beetroot, that has originated from *B. maritima*.
Shuna (MSB-K 114105); Tabaget Fahel, 26.iv.1967, *Edgecombe* (AUB).

Bienertia sinuspersici Akhani in Syst. Bot. 30 (2): 291. 2005.
Syn. *Bienertia cycloptera* auct. non *Bunge* ex *Boiss.*; *Suaeda guestii* Aellen in sched. (G).
Ref. Al-Eisawi 84 (2013) [as *B. cycloptera*].

Caroxylon cyclophyllum (Baker)Akhani & Roalson in Int. J. Pl. Sci. 168(6): 948. 2007.
Syn. *Salsola cyclophylla* Baker (1894).
Wadi Araba, *Danin* (http://flora.org.il/en/plants/SALCYC/).

Caroxylon divaricatum Moq., Prodr. 13(2): 177. 1849.
Syn. *Salsola longifolia* Forssk. (1775).
Ref. Al-Eisawi 86 (2013) [as *S. longifolia*].

Caroxylon gaetulum (Maire) Akhani & Roalson, Int. J. Pl. Sci. 168: 947. 2007.
Syn. *Salsola imbricata* Forssk. subsp. *gaetula* (Maire) Boulos (1995); *Salsola gaetula* (Maire) Botsch. (1975); *Salsola baryosma* (Schult.) Dandy subsp. *gaetula* (Maire) Freitag (1989).
Aqaba and Araba valley, *Danin* (http://flora.org.il/en/plants/SALGAE/).

Caroxylon imbricatum (Forssk.) Akhani & Roalson, Int. J. Pl. Sci. 168: 947. 2007.
Syn. *Salsola imbricata* Forssk. (1775).
Aqaba and Araba valley, *Danin* (http://flora.org.il/en/plants/SALIMB/).

Caroxylon incanescens (C.A.Mey.) Akhani & Roalson, Int. J. Pl. Sci. 168: 947. 2007.
Syn. *Salsola incanescens* C.A. Mey. (1833).
Dead Sea, 28.VII.1890 (AUB); Aqaba, *Danin* (http://flora.org.il/en/plants/SALINC/).

Caroxylon inerme (Forssk.) Akhani & Roalson in Int. J. Pl. Sci. 168(6): 947. 2007.
Syn. *Salsola inermis* Forssk. (1775).
Aqaba area, *Danin* (http://flora.org.il/en/plants/SALINE/).

Caroxylon jordanicola (Eig) Akhani & Roalson in Int. J. Pl. Sci. 168: 948. 2007.
Syn. *Salsola jordanicola* Eig (1945).
Ma'an and Dead Sea valley (http://flora.org.il/en/plants/SALJOR/).

Caroxylon tetrandrum (Forssk.) Akhani & Roalson in Int. J. Pl. Sci. 168: 948. 2007.
Syn. *Salsola tetrandra* Forssk. (1775).
Ma'an, *Danin* (http://flora.org.il/en/plants/).

Caroxylon vermiculatum (L.) Akhani & Roalson in Int. J. Pl. Sci. 168: 948. 2007.
Syn. *Salsola vermiculata* L. (1753); *Salsola villosa* Del. ex Roem. & Schult. (1820); *Salsola vermiculata* L. var. *villosa* (Delile ex Roem. & Schult.) Moq. (1849).
Karak, 10 km after Karak toward Tafila (MSB-K 341516).

Chenopodiastrum murale (L.) S.Fuentes, Uotila & Borsch, Willdenowia 42 (1): 14. 2012.
Syn. *Chenopodium murale* L. (1753).
Tabaget Fahel, 20.IV.1967 (AUB–224); Aqaba, *Boulos & Lahham* 5376 (AMM, BM, E).

Chenopodium album L., Sp. Pl. 219. 1753.
Ghor Safe, Ghor Hadipha at Village (MSB-K 113821); Jordan, Balqa Governorate, Dead Sea, Zara, *Nuaimat, Taifour, Shishani, Abuhammour, Al Mfadi & Al Khateeb* (NHJ 2012.0010). A polymorphic species with many ecological forms.

Chenopodium ficifolium Sm.Fl. Brit. 1: 276. 1800.
Ref. Lower Jordan Valley, Danin (2004).

Chenopodium opulifolium Schrad., Cat. Pl. 6. 1814.
Ref. Amman, Danin (2004).

Chenopodium rubrum L., Sp. Pl.: 218. 1753.
Along Zarqa River, 2004, *Al-Dwiri* s.n. (AMM).

Chenopodium vulvaria L., Sp. Pl.: 220. 1753.
Tafila Province, 2014, *Oran* (AMM).

Climacoptera iraqensis Botsch., Novosti Sist. Vyssh. Rast. 19: 76. 1982.
Syn. *Salsola crassa*auctt. non M.Bieb. (1811).
Ref. Ma'an, Danin (2004).

Cornulaca setifera (DC.) Moq., Prodr. 13(2): 218. 1849.
Syn. *Astragalus setiferus* DC. (1825).
Jebel Guzlan, Rum, *Borzatti & Barsotti* (AMM).

Dysphania ambrosioides (L.) Mosyakin & Clemants, Ukrainsk. Bot. Zhurn. 59(4): 382. 2002.
Syn. *Chenopodium ambrosioides* L. (1753).
Jarash, before the nursery shops by 0.5 km (MSB-K 243540); **a**long Zarqa River, 2004, *Al-Dwiri* s.n. (AMM).

Halimocnemis pilifera Moq. in Hist. & Mém. Acad. Roy. Sci. Toulouse5 (1): 181. 1839.
Syn. *Halimocnemis pilosa* Moq. (1840); *Halimocnemis gibbosa* Wol. ex Stapf (1886); *Halotis pilosa* (Moq.) Iljin (1936); *Halotis pilifera* (Moq.) Botsch. (1971).
Ref. Ma'an (Moab), Danin (2004).

Halocnemum strobilaceum (Pall.) M.Bieb., Fl. Taur.-Caucas. 3: 3. 1819.
Syn. *Salicornia strobilacea* Pall. *(1771).*
Townsend 65/100 7148 (K); Wadi Araba, *Danin* http://flora.org.il/en/plants/HALSTR.

Halopeplis amplexicaulis (Vahl) Ung.-Sternb. ex Ces., Pass. & Gibelli, Comp. Fl. Ital. 271. 1874.
Syn. *Salicornia amplexicaulis* Vahl (1791).
Azraq, Shishan, *Townsend* 99 (K000286251).

Halothamnus hierochunticus (Bornm.) Botsch., Novosti Sist. Vyssh. Rast 18: 156. 1981.
Syn. *Salsola hierochuntica* Bornm. (1912).
Karak, from Karak city toward Jordan valley, 200 m after military station (MSB-K 228255).

Halothamnus lancifolius (Boiss.) Kothe-Heinr., Biblioth. Bot. 143: 88. 1993.
Syn. *Salsola lancifolia* Boiss. (1879); *Aellenia lancifolia* (Boiss.) Ulbrich (1934); *Halothamnus acutifolius* (Moq.) Botsch. (1981).
Ref. Ma'an, Amman, Danin (2004).

Halothamnus hierochunticus(Bornm.) Botsch., Novit. Syst. Pl. Vasc. 18: 156. 1981.
Syn. *Salsola hierochuntica* Bornm. (1912); *Aellenia autranii* (Post) Zohary (1966) p.p.
Type: Israel, Maris Mortui, ad viam ex Khan-Hadrur ad Jericho, in agris, 15.ix.1905, *J. E. Dinsmore* 1001 (holo. JE, iso. W).
Ref. Amman & Ajloun, Danin (2004).

Haloxylon negevensis (Iljin & Zohary) Boulos, Fl. Egypt Checklist: 26. 1995.
Syn. *Hammada negevensis* Iljin &Zohary (1964).
Araba valley, *Danin* (http://flora.org.il/en/plants/HALNEG/).

Haloxylon persicum Bunge ex Boiss. & Buhse, Nouv. Mém. Soc. Imp. Naturalistes Moscou 12: 189. 1860.
Syn. *Haloxylon ammodendron* auctt. non (C.A. Mey.) Bunge.
Wadi Sunfahah, *Ananbeh, Al Khawaldah & Alshwaer* (NHJ 2013.0638).

Hammada articulata (Moq.) O. Bolòs & Vigo, Butl.Inst. Catalana Hist. Nat., sec. bot. 38(1): 89. 1974.
Syn. *Caroxylon articulatum* Moq. in DC. (1849); *Salsola articulata* Cav. (1775).

Hammada eigii Iljin in Novit. Syst. Pl.Vasc. 1964: 72. 1964.
Syn. *Haloxylon articulatum* Bunge ssp. *ramosissimum* (Benth. & Hook. f.) Eig (1945); *Haloxylon ramosissimum* Benth. & Hook.f. (1880) nomen.; *Hammada ramosissima* (Benth. & Hook.f.) Iljin (1948), comb. illegit.
Amman, Karak, Mujib, (Moab and Ammon), *Danin* (http://flora.org.il/en/plants/HALEIG/).

Hammada salicornica (Moq.) Iljin, Bot. Zhurn. 33: 583. 1948.
Syn. *Caroxylon salicornicum* Moq. in DC. (1849); *Haloxylon salicornicum* (Moq.) Bunge ex Boiss. (1879); *Hammada elegans* (Bunge) Botsch. (1964).
Ma'an, Araba valley; common (RSCN database).

Hammada scoparia (Pomel) Iljin, Bot. Zhurn. (Moscow & Leningrad) 33: 583. 1948.
Syn. *Haloxylon scoparium* Pomel (1875).
Azraq, near Shomari, *Taifour et al.* (NHJ 2013.0729).

Girgensohnia oppositiflora (Pall.) Fenzl in Ledeb., Fl. Ross. 3: 835. 1851.
Syn. *Salsola oppositiflora* Pall. (1773).
Araba valley.

Kali tragus (L.) Scop., Fl. Carniol. Ed. 2, 1: 175. 1771.
Syn. *Salsola kali* R.Br.subsp. *tragus* (L.) Celak. (1871).
Regarded for Palestine and Jordan as part of the *Salsola kali* aggr. (Uotila 2011).
Wadi Rum, 5 km after main junction of desert highway towards Rum. Site no 34. (MSB-K 497941).

Lagenantha cycloptera (Stapf) M.G.Gilbert & Friis, Nord. J. Bot. 11: 315.
Syn. *Salsola cycloptera* Stapf (1903).
Ref. Ma'an, Danin (2004).

Noaea mucronata (Forssk.) Asch. & Schweinf., Ill. Fl. Égypte 131. 1887.
Syn. *Salsola mucronata* Forssk. (1775); *Anabasis spinosissima* L.f. (1781).
Irbid, 26.ix.1967, *Kaleem Hajj* s.n. (AUB).

Oxybasis chenopodioides (L.) S. Fuentes, Uotila et Borsch, Willdenowia 42(1): 15. 2012.
Syn. *Oxybasis minutiflora* Kar. et Kir. (1841); *Chenopodium chenopodioides* (L.) Aellen (1933); *Blitum chenopodioides* L. (1771); *Chenopodium rubrum* (auct. non L.).
Recorded as rare by Danin (http://flora.org.il/en/plants/CHECHE/).

Salsola florida (M.Bieb.) Poir, Encycl. Suppl. 5: 191. 1817.
Syn. *Anabasis florida* M.Bieb. (1808); *Seidlitzia florida* (M.Bieb.) Bunge ex Boiss. (1879).
Shaumari Reserve, *Clarke* s.n. (K000286253).

Salsola rosmarinus (Bunge ex Boiss.) Eig, Palestine J. Bot., Jerusalem Ser. 3: 132. 1945.
Syn. *Suaeda rosmarinus* Ehrenb. ex Boiss. (1879); *Seidlitzia rosmarinus* Bunge ex Boiss. (1879);
Type: (Gulf of Akaba): In insula Maksuk Arabiae, *C.C. Ehrenberg* 123 (syntypes in G, K).
Amra Wadi Bottom, *Taifour, Al Mfadi & Bani Said* (NHJ 2013.0050); Azraq Oasis, *Tackholm, El-Hadidi, Lahham, & Boulos* 9089 (K000286257).

Salsola schweinfurthii Solms-Laub., Bot. Zeit. 59: 173. 1901.
Syn. *Darniella schweinfurthii* (Solms-Laub.) Brullo (1984); *Caroxylon schweinfurthii* (Solms-Laub.) Akhani & Roalson (2007).
Ref. Ma'an & Amman, Danin (2004).

Salsola sinus-persica Akhani, Pak. J. Bot. 40(4): 1638. 2008.
Dead Sea populations in Palestine show that the *Salicornia* there belong to the *S. persica* clade. Correct species would be *S. sinus-persica*.

Sarcocornia fruticosa (L.) A.J.Scott, Bot. J. Linn. Soc. 72: 367. 1978.
Syn. *Salicornia europaea* L. var. *fruticosa* L. (1753).
Al Karak, *Frey & Küschner* 5638 (E00476124, TUB).

Sarcocornia perennis (Mill.) A.J.Scott, Bot. J. Linn. Soc. 72: 367. 1978.
Lower Jordan valley (http://flora.org.il/en/plants/SARPER/).

Suaeda aegyptiaca (Hasselq.) Zohary, J. Linn. Soc., Bot. 55: 635. 1957.
Syn. *Chenopodium aegyptiacum* Hasselq. (1757); Forssk. (1775) nomen; *Suaeda hortensis* Forssk. ex J.F.Gmel., (1776); *Suaeda baccata* Forssk. ex J. F. Gmel. (1791); *Schanginia baccata* (Forssk.) Moq. (1840); *Suaeda maris-mortui* Post (1896).
Amman, Dead Sea. 2 km after the hot water towards Aqaba, Site no 18 (MSB-K 496807).

Suaeda asphaltica Boiss., Fl. Orient. 4(2): 938. 1879.
Karak, Mujib (Moab), *Danin* (http://flora.org.il/en/plants/SUAASP/).

Suaeda fruticosa Forssk. ex J.F. Gmel., Syst. Nat. ed. 13, 2: 503 179; Forssk., Fl. Aegypt.-Arab. 70 1775 non "*S. fruticosa* (L.) Forssk." auct. mult. nec *Chenopodium fruticosum* L. (1753).
Dead Sea valley (RSCN database).

Suaeda monoica Forssk. ex J.F.Gmel., Onomat. Bot. Compl. 8: 798. 1776.
Nr. Aqaba, *Boulos & Lahham* 5386 (AMM, BM, E).

Suaeda palaestina Eig & Zohary, Palestine J. Bot., Jerusalem Ser. 3: 126. 1945.
Dead Sea Valley, *Danin* (http://flora.org.il/en/plants/SUAPAL/).

Suaeda vermiculata Forssk. ex J.F.Gmel., Onomat. Bot. Compl. 8: 798. 1776.
Syn. *Suaeda mesopotamica* Eig (1945).
Aqaba (RSCN database).

Traganum nudatum Delile, Descr. Egypte, Hist. Nat. 204. 1813.
Irbid, 16.viii.1962, *Kaleem Hajj* s.n. (AUB).

Amaryllidaceae

Allium ampeloprasum L., Sp. Pl.: 294. 1753.
Syn. *Allium porrum* L. (1753).
Amman, Al Saravo area, opposite Princess Alia horse farm (MSB-K 243481); Northeastern Badia, Aretein *Al Shudiefat & Khalidi* (NHJ 2013.0011).

Allium artemisietorum Eig & Feinbrun, Palestine J. Bot., Jerusalem Ser. 3: 18. 1943.
Jordan, Site 52, Jordan University Research Station, *Mwagar* (MSB-K 395760).

Allium ascalonicum L., Fl. Palaest. 117. 1756.
Syn. *Allium hierochuntinum* Boiss. (1882).
Amman, Al-Geezeh, Daba'a Village, inside Daba'a Agricultural Research Station, site no 12 (MSB-K 497860).

Allium aschersonianum Barbey, Herbor. Levant: 163. 1882.
Deir Alla, *Simpson* 53388 (K000318350).

Allium carmeli Boiss., Diagn. Pl. Orient. 13: 28. 1854.
Tal Al-Rumman RBG, Sector No. 3, *El-Oqlah, Taifour & Al Khawaldah* (NHJ 2013.0216).

Allium curtum Boiss. & Gaill., Diagn. Pl. Orient. 2, 4: 116. 1859.
Ref. Wadi Ibn Hammad, Wadi Ibn Hammad SCA baseline survey 2011.

Allium daninianum Brullo, Pavone & Salmeri, Willdenowia 26: 239. 1996.
Wadi Wala, 7.5.1974, *Boulos* 7182(K); Gebel Uxeigi (Aqaba), *Hwavis* 9211(K).

Allium desertorum Forssk., Fl. Aegypt.-Arab.: 72. 1775.
Al Mafraq, road between Mafrag & H 5, *Hunting Aero Survey* 215A (E00338435); Azraq Wadi Bottom (South west the Castle Amra) *El-Oqlah* (NHJ 2013.0566).

Allium dictyoprasum C.A.Mey. ex Kunth, Enum. Pl. 4: 390. 1843.
Amman, Muwagar, University of Jordan Research station in Muwagar, (NCARE 2006); Danin (http://flora.org.il/en/plants/ALLDIC/).

Allium erdelii Zucc., Abh. Math.-Phys. Cl. Königl. Bayer. Akad. Wiss. 3: 236. 1843.
Jerash, *Davis* 9643 (E00344738); Tal Al-Rumman RBG, Sector No. 71, *Taifour* (NHJ 2013.0200).

Allium lachnophyllum Paine, Palestine Explor. Soc. 3: 125. 1875.
Ref. Al-Eisawi 121 (2013).

Allium libani Boiss., Diagn. Pl. Orient. 13: 26. 1854.
Ref. Al-Eisawi 121 (2013).

Allium neapolitanum Cirillo, Pl. Rar. Neapol. 1: 13. 1788.
Ma'an, Wadi Musa, near Petra, *Boulos & Jallad* 8120 (E); Tal Al-Rumman, *Al-Eisawi, Abu- Sondos, Abu Anza & Shishani* (NHJ 2006.0080).

Allium negevense Kollmann, Israel J. Bot. 18: 69. 1969.
Jarash, first turn toward Al Ketteh village, 500m after the turn (MSB-K 2003.49); Tal Al-Rumman, *Al-Eisawi* (NHJ 2006.0206).

Allium nigrum L., Sp. Pl. ed. 2: 430. 1762.
Ref. RSCN database; Al-Eisawi 121 (2013).

Allium orientale (Non Boiss.) Blakelock, Diagn. Pl. Orient. 13: 25. 1854.
Madaba, Jebel Siaghah, vii.1891 (AUB); Ajloun Forest Reserve, *Böcker* RB 902 (B100520882).

Allium pallens L., Sp. Pl. ed. 2: 427. 1762.
Syn. *Allium coppoleri* Tineo (1827); *Allium paniculatum* L. var. *pallens* (L.) Gren. & Godr. (1855); *Allium paniculatum* L. subsp. *pallens* (L.) Richt.; *Allium convallarioides* Grossh. (1924).
Tal Al-Rumman, *RBG-Ire & RBG-Jo* (NHJ 2008.0063).

Allium paniculatum L., Syst. Nat. ed. 10, 2: 978. 1759.
Dibeen Forest, *RBG-Ire & RBG-Jo* (NHJ 2008.0059).

Allium phanerantherum Boiss. & Hausskn., Fl. Orient. 5: 235. 1882.
Balqa, Tall Rumman reserve, left from junction, almost at end, site no 17 (MSB-K 395449).

Allium qasyunense Mouterde, Bull. Soc. Bot. France 100: 348. 1953.
Jordan, 300 m after site no. 12 (MSB-K 338271); Jerash, 1945, *Davis* (E).

Allium roseum L., Sp. Pl.: 296. 1753.
Ref. RSCN database.

Allium rothii Zucc., Abh. Math.-Phys. Cl. Königl. Bayer. Akad. Wiss. 3: 232. 1843.
30 km W Azraq, *Frey & Kürschner* 80–128 (E00344848, TUB).

Allium schubertii (Non Zucc.) Feinbrun, Abh. Math.-Phys. Cl. Königl. Bayer. Akad. Wiss. 3: 234. 1843.
Irbid, *Meyers, Fred & Dinsmore* G766 (E00344983).

Allium sinaiticum Boiss., Diagn. Pl. Orient. 13: 31. 1854.
Jebel Aneze, iv.1945 (7211-HUJ).

Allium sindjarensis Boiss. & Hausskn. ex Regel, Trudy Imp. S.-Peterburgsk. Bot. Sada 3(2): 121. 1875.
Ma'an, 5 km NW of Rum railway station, *Frey & Kürschner* (BSB).

Allium stamineum Boiss., Diagn. Pl. Orient., 2, 4: 119. 1859.
Madaba, Dhiban to Wadi Themed, *Davis* 9613 (E00344537); Zarqa, Al-Alouk, opposite fence of reserve on the way to Zarqa (MSB-K 176688); Dana Reserve, *RBG-Ire & RBG-Jo* (NHJ 2008.0109).

Allium trifoliatum Cirillo, Pl. Rar. Neapol. 2: 11. 1792.
Irbid, Umm Qeis, 25 km NW of Irbid, *Frey & Kürschner*; (YAR-H).

Allium truncatum (Feinbrun) Kollmann & Zohary, Fl. Palaestina 4: 90. 1986.
Dibeen National Park, 2005, *Mahklouf* s.n. (AMM).

Narcissus tazetta L., Diagn. Pl. Orient., 2, 4: 119. 1859.
Irbid, (FOR-H); Wadi Shuaib (RSCN database).

Pancratium sickenbergeri Asch. & Schweinf. ex Boiss., C.Barbey-Boissier & W.Barbey, Herb. Levant: 158. 1882.
Karak, Wadi Araba, *Frey & Kürschner* (BSB); Wadi Yutum (FOR-H); Jebel Rehebi, *Barsotti* (AMM).

Sternbergia clusiana (Ker-Gawl.) Spreng., Syst. Veg. 2: 57. 1825.
Syn. *Sternbergia stipitata* Boiss. & Hausskn. (1882); *Sternbergia macrantha* J.Gay ex Bak. (1888); *Sternbergia sparffordiana* Dinsm. (1928).
Mount Hor [Jabal an Nabi Harun], *Hart* 9 (K000286069); Jizeh, *Meyers, Fred & Dinsmore* M4834 (E00382124).

Sternbergia colchiciflora Waldst. & Kit., Descr. Icon. Pl. Hung. 2: 172. 1804.
Ajloun (RSCN database).

Sternbergia lutea (L.) Ker.-Gawl. ex Spreng., Syst. Veg. 2: 57. 1825.
Syn. *Amaryllis lutea* L. (1753); *Sternbergia aurantiaca* Dinsm. (1934).
Al Mafraq, Al Deganeih (YAR-H).

Vagaria parviflora (Desf. ex Delile) Herb., *Amaryllidaceae* 226. 1837.
Syn. *Pancratium parviflora* Desf. ex Delile (1815).
Ref. Al-Eisawi 31 (2013) as *Pancratium parviflora*.

Anacardiaceae

Pistacia atlantica Desf., Fl. Atlant. 2: 364. 1799.
P. mutica Fisch & Mey. (1838).
Jebel Attaris (Moab), *Davis* 8736 (E00436379); Tal Al-Rumman, *Taifour* (NHJ 2009.0101).

Pistacia khinjuk Stocks, Hooker's J. Bot. Kew Gard. Misc. 4: 143. 1852.
Syn. *Pistacia khinjuk* Stocks var. *populifera* Boiss.; *Pistacia khinjuk* Stocks var. *oblonga* Bornm.
Ma`an, 10 km N of Petra, near el-Hesheh, *Frey & Kürschner*; Amman (E00436439).

Pistacia lentiscus L., Sp. Pl. 2: 1026. 1753.
Jarash, Burma village, 9 km after Al Fawara Primary School, left roadside, (MSB-K 211978).
Pistacia lentiscus is native throughout the Mediterranean region, from Morocco and The Iberian Peninsula in the west through southern France and Turkey to Iraq and Iran in the east; also native to the Canary Islands.

Pistacia palaestina Boiss., Diagn. Pl. Orient. 9: 1. 1849.
Ajloun, *Davis* 9697 (E00436516]); Ajloun, *Taifour, Al Khawaldah & Alshwaer* (NHJ 2013.0664).
Zohary describes **Pistacia × saportae** Burnat (1896) as a hybrid between *P. lentiscus* and *P. palaestina*. This is an evergreen tree, leaves with 4–6 pairs of leaflets and with a winged rachis; inflorescence condensed; produces none or few viable seeds. Ref. Al-Eisawi (2013) recognizes this as a good species *P. saportae*.

Rhus coriaria L., Sp. Pl.: 265. 1753.
Ajloun, Anjara, Abu Nimer, (MSB-K 186427); Ajloun Road (Safsafa-Rajib) *Taifour et al.* (NHJ 2013.0665).

Searsia pentaphylla (Jacq.) F.A.Barkley, Prosp. Iraq Biol. 3: 57. 1962–1963 publ. 1965.
Syn. *Rhus pentaphylla* (Jacq.) Desf. (1798).

Searsia tripartita (Ucria) Moffet, Bothalia 37: 173. 2007.
Syn. *Rhamnus tripartitus* Ucria (1793); *Rhus tripartita* (Ucria) Grande (1918).
Jarash, Sad Elmalik Talal, 23/6/1981 (YAR-H); Tal Al-Rumman, *Taifour & Aldamen* (NHJ 2010.0029).

Apiaceae

Ainsworthia trachycarpa Boiss., Diagn. Pl. Orient. ser. 1, 10: 44. 1849.
Syn. *Ainsworthia cordata* (Jacq.) Boiss. (1844); *Tordylium cordatum* (Jacq.) Poir. subsp. *trachycarpum* (Boiss.) Holmboe; *Tordylium trachycarpum* (Boiss.) Holmboe (1914).
Dibeen Forest Reserve, 2004, *Mahklouf* s.n. (AMM); Jerash, Souf village, 4 Jun. 2003. (NCARE).

Ammi majus L., Sp. Pl.: 243. 1753.
Kafrein, NE of end of Dead Sea (MSB-K 113773).

Ammi visnaga (L.) Lam., Fl. Franç. 3: 462. 1778.
Syn. *Daucus visnaga* L. (1753); *Visnaga daucoides* Gaertn. (1779).
Upper Jordan Valley (YAR-H).

Anethum graveolens L., Sp. Pl.: 263. 1753.
Syn. *Peucedanum graveolens* (L.) C.B.Clarke (1879.) nom. illegit.
Al Mafraq, Al Mdawar (IBU-H).

Anisosciadium isosciadium Bornm., Repert. Spec. Nov. Regni Veg. 10: 468. 1912.
Jordan, *Frey & Kürschner* VO 5151(TUB).

Anisosciadium lanatum Boiss., Fl. Orient. Suppl.: 261. 1888.
Syn. *Echinosciadium arabicum* Zohary (1948).
Davis & Bokhari D56264 (E00042062).

Anthriscus lamprocarpa Boiss., Ann. Sci. Nat., Bot. 3, 2: 59. 1844.
Jarash, Souf, 5.v.1886 (AUB).

Apium graveolens L., Sp. Pl. 264. 1753.
Wadi Shuaib, *RBG-Ire & RBG-Jo* (NHJ 2008.0022).

Apium nodiflorum (L.) Lag., Amen. Nat. Españ. 1(2): 101. 1821.
Ajloun, Rajib, beside the bridge on the river (MSB-K 200497); Wadi Urjan (NHJ 2013.0555).

Artedia squamata L., Sp. Pl.: 242. 1753.
Zarqa, 2 km after Kamsha toward Alouk, left roadside, olive orchard, (MSB-K 176426); Tal Al-Rumman, *Taifour* (NHJ 2009.0049).

Astomaea seselifolia (A.DC.) Rauschert, Taxon 31: 563. 1982.
Syn. *Astoma seselifolium* DC. (1829).
Zarqa, 2 km after Kamsha toward Alouk, left roadside, olive orchard, (MSB-K 176448); Tal Al-Rumman, *Taifour* (NHJ 2013.0191).

Berula erecta (Huds.) Colvill, Contr. U.S. Natl. Herb. 4: 115. 1893.
Syn. *Sium erectum* Huds. (1762).
Al Mafraq, Jaber (IBBU-H).

Bifora testiculata (L.) Roth, Enum. Pl. Phaen. Germ. 1: 888. 1827.
Syn. *Coriandrum testiculatum* L. (1753).
Zarqa, (MSB-K 199681); Tal Al-Rumman, *Shishani & Alkoran* (NHJ 2011.0093).

Bilacunaria boissieri (Reut. & Hausskn. ex Boiss.) Pimenov & V.N.Tikhom., Feddes Repert. 94: 152. 1983.
Syn. *Hippomarathrum boissieri* Reut. & Hausskn. ex Boiss. (1872); *Cachrys boissieri* (Reut. & Hausskn. ex Boiss.) Hand (2011).
Ref. RSCN database; Al-Eisawi 32 (2013) as *Hippomarathrum boissieri.*

Bunium elegans (Fenzl) Freyn, Oesterr. Bot. Z. 42: 83. 1892.
Syn. *Carum elegans* Fenzl (1842).
Karak, olive grove opposite the university (MSB-K 277817); Salt, *Barbey* (AUB).

Bunium ferulaceum Sm., *Fl. Graec. Prodr. 1: 186. 1806.*
Karak (NCARE).

Bunium paucifolium DC., Prodr. 4: 117. 1830.
Ajloun, *Danin* (http://flora.org.il/en/plants/BUNPAU/).

Bupleurum brevicaule Schltdl., Linnaea 17: 124. 1843.
Jordan, *Balansa* 1233 (K000687168); Karak (MSB-K 277839).

Bupleurum gerardii All., Auct. Syn. Meth. Stirp. Hort. Regii Taur. 29. 1773.
Syn. *Bupleurum aristatum* Nábělek (1923) nom. illegit.; *Bupleurum gerardii* All. var. *trichopodioides* Wolff.
Jarash, Site 32. Souf (MSB-K 557928).

Bupleurum lancifolium Hornem., Hort. Bot. Hafn. 1: 267. 1813.
Irbid, 15 km northwest of Irbid near Hibraz village (MSB-K 151658); Tal Al-Rumman, *Taifour* (NHJ 2013.0281).

Bupleurum nodiflorum Sibth. & Sm., Fl. Graec. t. 260, 1821.
Jarash, Souf (MSB-K 497664).

Bupleurum odontites L., Sp. Pl.: 237. 1753.
Al Mafraq, Balama (YAR-H).

Bupleurum semicompositum L., Demonstr. Pl. 7. 1753.
Ref. Al-Eisawi 33 (2013).

Bupleurum subovatum Link ex Spreng., Sp. Umbell. 19. 1818.
Amman, nr. Al Yaduda, 15 May 1954 (FOR-H).

Chaetosciadium trichospermum (L.) Boiss., Fl. Orient. 2: 1078. 1872.
Syn. *Chaerophyllum trichospermum* (L.) Lam. (1785.); *Torilis trichosperma* (L.) Spreng. (1818); *Caucalis trichosperma* (L.) Delile (1840).
Amman, 40 km E of Amman, *Frey & Kürschner* VO 5050 (TUB); Tal Al-Rumman, *Taifour* (NHJ 2011.0091).

Conium maculatum L., Sp. Pl.: 243. 1753.
Irbid, Wadi Shuaib (YAR-H).

Daucus aureus Desf., Fl. Atlant. 1: 242, t. 61, 1798.
Al-Mafraq, road near Al Al-Bayet University (IBBU-H).

Daucus broteroi Ten., Syll. Pl. Fl. Neapol. App. 3: 4. 1830.
Tafila, Danin (2005); RSCN database.

Daucus carota L., Sp. Pl.: 242. 1753.
Syn. *Daucus glaber* Opiz ex Čelak. (1875).
Karak (MSB-K 558110); Salt (YAR-H).

Daucus durieua Lange, Prodr. Fl. Hispan. 3: 23. 1874.
Ajloun, 10 May 1927 (HUJ).

Daucus guttatus Sm., Fl. Graec. Prodr. 1: 184. 1806.
Salt (YAR-H).

Daucus jordanicus Post, J. Linn. Soc., Bot. 24. 431. 1888.
Wadi Rum, 16 Apr. 1908 (HUJ); Lower Jordan valley (RSCN database).

Daucus littoralis Sm., Fl. Graec. 3: 65. 1821.
Syn. *Daucus glaber* (Forssk.) Thell (1912).
About 40 km S of Ma'an, 1929 (HUJ).

Daucus subsessilis Boiss., Fl. Orient. Suppl. 272. 1888.
Tafila Province, 2014, *Oran* (AMM); Jerash (NCARE); Azraq (Ref. Guy Mountfort et al. - Azraq Wetland Reserve Herbarium, 2008).

Deverra tortuosa (Desf.) DC., Prodr. [A. P. de Candolle] 4: 143. 1830.
Syn. *Bubon tortuosum* Desf. (1798); *Pituranthos tortuosus* Benth. & Hook.f. ex Asch. & Schweinf. (1887).
Valley between Jebel E Brera and Umm Sabibel, *Borzatti & Barsotti* (AMM).

Deverra triradiata Hochst. ex Boiss., Fl. Orient. 2: 861. 1872.
Syn. *Pituranthos triradiatus* (Hochst.) Asch. & Schweinf. (1887).
Karak, Jebel Ataita, *Frey, Hilger & Kürschner* 80–882 (E, TUB).

Donnellsmithia juncea (Spreng.) Mathias & Constance, Contr. Univ. Michigan Herb. 11: 4. 1973.
Syn. *Peucedanum junceum* Humb. & Bonpl. ex Spreng. (1820).
Petra (YAR-H); Al-Eisawi 36 (2013) as *Peucedanum spreitzenhoferi* Dingl.

Ducrosia anethifolia (DC.) Boiss., Ann. Sci. Nat., Bot. sér. 3, 1: 341. 1844.
Syn. *Ducrosia olivieri* Boiss. & Buhse (1860).
Ref. RSCN database; Al-Eisawi 34 (2013).

Ducrosia flabellifolia Boiss., Ann. Sci. Nat., Bot. sér. 3, 1: 341. 1844.
Safawi, *Taifour et al.* (NHJ 2013.0731).

Eryngium creticum Lam., Encycl. 4(2): 754. 1798.
Jordan, (MSB-K 200590); Tal Al-Rumman (NHJ 2006.0315).

Eryngium falcatum F.Delaroche, Nouv. Bull. Sci. Soc. Philom. Paris 1: 88. 1807.
Tal Al-Rumman RBG, Sector No. 20 (NHJ 2013.0228).

Eryngium glomeratum Lam., Encycl. 4(2): 755. 1798.
Balqa, From Salt toward Jordan valley Arda main road, 10 km before Dair Alla (MSB-K 228004).

Eryngium maritimum L., Sp. Pl.: 233. 1753.
Irbid, Al Shounah (YAR-H).

Exoacantha heterophylla Labill., Icon. Pl. Syr. 1: 10. 1791.
Ajloun between Salt and Irbid, *Frey & Kürschner* (BSB).

Falcaria vulgaris Bernh., Syst. Verz. (Bernhardi) 176. 1800.
Syn. *Sium falcaria* L. (1753); *Bunium falcaria* (L.) M.Bieb. (1808); *Falcaria rivini* Host (1827); *Falcaria sioides* (Wibel) Asch. (1860).
Tal Al-Rumman RBG, Sector No. 28, *El-Oqlah & Al Mfadi* (NHJ 2013.0372).

Ferula biverticillata Thieb., Bull. Soc. Bot. France 82: 190. 1935.
Tafila, Gharandal (AUB).

Ferula blanchei Boiss., Fl. Orient. 2: 987. 1872.
Jordan, *Dinsmore*11309 (E00175646); Dana Reserve, *RBG-Ire & RBG-Jo* (NHJ 2008.0089).

Ferula communis L., Sp. Pl.: 246. 1753.
50 km S of Azraq, *Frey & Kürschner* 80–393 (TUB); Tal Al-Rumman, *Taifour* (NHJ 2013.0317).

Ferula negevensis Zohary, Fl. Palaestina 2: 460. 1972.
Ras Naqab, Danin, http://flora.org.il/en/plants/FERNEG.

Ferula ovina Boiss., Pl. Pers. Austr. 283. 1845.
Syn. *Peucedanum ovinum* Boiss. (1845); *Ferula barbeyi* (non Post.) H.Riedl.
Shobek to Gebel Areizi (Aqaba), *Davis* 9553 (E).

Ferula rutbaensis C.C.Towns., Kew Bull. 20: 77. 1966.
Inter Amman et Rutba, *Rechinger* 12872 (E).

Ferula sinaica Boiss., Diagn. Pl. Orient. 10: 40. 1849.
Near Quera, between Agaba and Ras Annagab, *Boulos & Jallad* 6239 (E).

Ferula tingitana L., Sp. Pl.: 247. 1753.
Tafila Province, 2014, *Oran* (AMM).

Foeniculum vulgare Mill., Gard. Dict., ed. 8. n. 1. 1768.
Syn. *Anethum foeniculum* L. (1753); *Foeniculum officinale* All. (1785); *Foeniculum piperitum* DC. (1826); *Foeniculum foeniculum* (L.) Karst. (1886).
Madaba (MSB-K 152002); Tal Al-Rumman, *Taifour & Aldamen* (NHJ 2010.0019).

Heptaptera anisoptera (DC.) Tutin, Feddes Repert. 74: 33. 1967.
Syn. *Prangos anisopetala* DC. (1830).
Tal Al-Rumman, Jarash (MSB-K 497697).

Hohenackeria exscapa (Steven) Grande, Bull. Orto Bot. Regia Univ. Napoli 8: 73. 1926.
Syn. *Fedia acaulis* Steven (1812).

Ref. Al-Eisawi 35 (2013).

Johrenia dichotoma DC., Coll. Mém. 5: 54. 1829.
Syn. *Krubera peregrina* (L.) Lowe (1864); *Capnophyllum peregrinum* (L.) Lange (1874); *Tordylium peregrinum* L. (1767).
Ref. Al-Eisawi 33 (2013) as *Capnophyllum peregrinum*.

Lagoecia cuminoides L., Sp. Pl.: 203. 1753.
Zarqa, on the way to Alouk (MSB-K 227856); Tal Al-Rumman, *Taifour, & Aldamen* (NHJ 2009.0118).

Lecokia cretica (Lam.) DC., Coll. Mém. 5: 67. 1829.
Syn. *Cachrys cretica* Lam. (1783).
Irbid, road from Zobia to Barkash (MSB-K 276924); Tal Al-Rumman, *Royal Botanic Garden Team* (NHJ 2011.0015).

Lisaea strigosa (Banks & Sol.) Eig, J. Bot. 75: 189. 1937.
Syn. *Caucalis strigosa* Banks & Sol. (1794).
Irbid, Zobia (YAR-H).

Malabaila secacul (Mill.) Boiss., Fl. Orient. 2: 1057. 1872.
Syn. *Tordylium secacul* Mill. (1768).
Ma'an, Al-Fujaij rangeland reserve, 3 km after Qadeseyeh (MSB-K 186346); Tal Al-Rumman, *Taifour* (NHJ 2009.0123).

Myrrhoides nodosa (L.) Cannon, Feddes Repert. 79: 65. 1968.
Syn. *Scandix nodosa* L. (1753).
Irbid, Malka & Wadi Shuaib (YAR-H).

Oenanthe pimpinelloides L., Sp. Pl.: 255. 1753.
Tal Al-Rumman RBG, Sector No. 69, *El-Oqlah, Taifour & Al Khawaldah* (NHJ 2013.0236).

Oenanthe prolifera L., Sp.Pl.: 254. 1753.
Salt, Apr. 1877 (AUB).

Oliveria decumbens Vent., Hort. Cels. 21, t. 21, 1801.
Syn. *Oliveria orientalis* DC. (1830); *Callistroma erubescens* Fenzl (1843).
Ref. (RSCN database); Al-Eisawi 36 (2013).

Opopanax hispidus (Friv.) Griseb., Spic. Fl. Rumel. 1: 378. 1843.
Syn. *Pastinaca opopanax* L. (1753); *Pastinaca hispida* Fenzl (1843); *Opopanax orientalis* Boiss. (1844).
Ref. Al-Eisawi 36 (2013).

Orlaya grandiflora (L.) Hoffm., Gen. Pl. Umbell. 58. 1814.
Syn. *Caucalis grandiflora* L. (1753); *Orlaya daucoides* (L.) Greuter (1967).
Dibeen Forest Reserve, May 2006, *Taifour & Othman* (FOR-H).

Pimpinella corymbosa Boiss., Ann. Sci. Nat., Bot. III, 1: 131. 1844.
Ajloun, Ajloun to Ishtafeina triangle (MSB-K 185888).

Pimpinella cretica Poir., Encycl. Suppl. 1: 684. 1811.
Amman, from Biaden Wadi Alseer toward Al Bassa (MSB-K 200224); Tal Al-Rumman, *Shishani & Alkoran* (NHJ 2011.0112).

Pimpinella eriocarpa Banks & Sol., Nat. Hist. Aleppo (ed. 2) 2: 249. 1794 (as "Eriocarpus").
Eastern upper slopes of the Jordan valley and Shoubak (NCARE).

Pimpinella olivieri Boiss., Ann. Sci. Nat., Bot. sér. 3, 1: 132. 1844.
Tafila Province, 2014, *Oran* (AMM).

Pimpinella peregrina L., Sp. Pl.: 264. 1753.
Ajloun, Anjara, on the way to Rajib, 500m after the junction (MSB-K 200464).

Pimpinella puberula (DC.) Boiss., Ann. Sci. Nat., Bot. 3(1): 129. 1844.
Syn. *Ptychotis puberula* DC. (1830).
Ref. RSCN database; Al-Eisawi 36 (2013).

Prangos ferulacea (L.) Lindl., Quart. J. Sci. Lit. Arts 10: 7. 1825.
Syn. *Laserpitium ferulaceum* L. (1762).
Ref. RSCN database; Al-Eisawi 36 (2013).

Pseudorlaya pumila (L.) Grande, Nuovo Giorn. Bot. Ital.n.s. 32: 86. 1925.
Syn. *Caucalis pumila* L. (1759).
Jebel El Muidir, 16 Mar. 1987, *Garbari & Giordani* (AMM).

Pycnocycla nodiflora Decne ex Boiss. Ann. Sci. Nat., Bot. III, 2: 88. 1844.
Jordan, *Danin* 981101 (E).

Pycnocycla saxatilis Danin, Hedge & Lamond, Willdenowia 30: 77. 2000.
Aqaba, Rum area, 10 km SSE of Wadi Rum resthouse, *Danin, Hedge & Lamond* (E).

Ridolfia segetum Moris, Ind. Sem. Hort. Taur. 43. 1841.
Al-Mafraq, Al-Dajniyah (IBBU-H); Dibeen Forest Reserve, 2004, *Mahklouf* (RSCN database).

Scaligeria hermonis Post, Bull. Herb. Boissier 1: 399. 1893.
Ajloun (RSCN database).

Scaligeria napiformis Grande, Bull. Orto Bot. Regia Univ. Napoli 4: 188. 1914.
Dana Reserve, *RBG-Ire & RBG-Jo* (NHJ 2008.0086).

Scandix australis L., Sp. Pl.: 257. 1753.
Syn. *Scandix falcata* Lond. (1811) nom. illegit.; *Scandix australis* subsp. *grandiflora* (L.) Thell. (1926).
Petra (YAR-H).

Scandix iberica M.Bieb., Fl. Taur.-Caucas. 1: 425. 1808.
Ma'an, North of Dana from main road to Rumman camp, near Ain Sahda, (MSB-K 151935); Ajloun, *Al Mfadi et al.* (NHJ 2013.0017).

Scandix palaestina Boiss., Fl. Orient. 2: 918. 1872.
Upper Jordan Valley (RSCN database).

Scandix pecten-veneris L., Sp. Pl.: 256. 1753.
Syn. *Scandix pecten-veneris* L. var. *brevirostris* Boiss.
Jarash, 500 m after first turn to Al Ketteh village (MSB-K 2003.50).

Scandix stellata Banks & Sol., Nat. Hist. Aleppo ed. 2, 2: 249. 1794.
Ma'an, Shoubak area, inside the king fenced farm in Al-Hisha hillside (MSB-K 277932); Dana Reserve, *RBG-Ire & RBG-Jo* (NHJ 2008.0092).

Scandix turgida Boiss., Fl. Orient. 2: 918. 1872.
Ref. RSCN database; Al-Eisawi 37 (2013).

Scandix verna O.Cohen., Israel J. Pl. Sci. 50(Suppl.): 25. 2002.
Ref. RSCN database.

Smyrniopsis cachroides Boiss., Fl. Orient. 2: 928. 1872.
Dibeen Forest Reserve, May, 2006, *Taifour* (RSCN) (Dibeen Forest Reserve Herbarium); Al Ramtha, Al Shajarah (YAR-H).

Smyrnium connatum Boiss. & Kotschy, Unger & Kotschy, Ins. Cypern 309. 1865.
Ajloun, 2 km before Safina coming from Ajloun after junction Safina-Ballas (MSB-K 176611).

Smyrnium olusatrum L., Sp. Pl.: 262. 1753.
Irbid, Al Mkheba Al-Foqa (IBBU-H).

Tordylium aegyptiacum (L.) Poir., Encycl. 7: 711. 1806.
Syn. *Hasselquistia aegyptiaca* L. (1755).
Irbid, Irbid-Mafraq rd. (MSB-K 114231).

Tordylium carmeli (Labill.) Al-Eisawi& Jury, Bot. J. Linn. Soc. 97(4): 392. 1988.
Syn. *Heracleum carmeli* Labill. (1812).
Tal Al-Rumman, *Al-Eisawi* (NHJ 2006.0237); Rumman, *Taifour & Aldamen* (NHJ 2009.0117).

Tordylium trachycarpum (Boiss.) Al-Eisawi, Mitt. Bot. Staatssamml. München 18: 179. 1982.
Tal Al-Rumman, *Taifour & Aldamen* (NHJ 2009.0124).

Torilis arvensis (Huds.) Link, Enum. Hort. Berol. Alt. 1: 265. 1821.
Syn. *Caucalis arvensis* Huds. (1762).
Dibeen Forest Reserve, *Taifour & Othman* (FOR-H); Eastern upper slopes of the Jordan valley (NCARE).

Torilis japonica (Houtt.) DC., Prodr. 4: 219. 1830.
Syn. *Caucalis japonia* Houtt. (1779).

Ajloun Castle (IBBU-H)

Torilis leptophylla (L.) Rchb.f., Icon. Fl. Germ. Helv. 21: 83, t. 169, 1866.
Syn. *Caucalis leptophylla* L. (1753).
Ajloun, Souf-Ibbein road, 1 km after the civil defence centre, forest on right side of road (MSB-K 276898); Tal Al-Rumman, *Taifour* (NHJ 2011.0017).

Torilis nodosa (L.) Gaertn., Fruct. Sem. Pl. 1: 82. t. 20. f. 6. 1788.
Syn. *Tordylium nodosum* L. (1753).
Tal Al-Rumman, *RBG-Ire & RBG-Jo* (NHJ 2008.0083).

Torilis tenella (Delile) Rchb.f., Icon. Fl. Germ. Helv. 21: 84. 1866.
Syn. *Caucalis tenella* Delile (1814).
Tal Al-Rumman, *Al-Eisawi* (NHJ 2006.0045).

Trachyspermum ammi (L.) Sprague, Bull. Misc. Inform. Kew 1929: 228. 1929.
Ar-Rumman, *Taifour & Aldamen* (NHJ 2009.0122).
Widely cultivated.

Turgenia latifolia (L.) Hoffm., Gen. Pl. Umbell. 59. 1814.
Syn. *Tordylium latifolium* L. (1753).
Ma'an, North of Dana from main road to Rumman camp, near Ain Sahda (MSB-K 151946).

Zosima absinthifolia (Vent.) Link, Enum. Hort. Berol. Alt. 1: 274. 1821 as "apsinthifolia".
Syn. *Zosima orientalis* Hoffm.(1814).
Dana Reserve, Rumana trail behind Rumana camp, *Al Awaji et al.* (NHJ 2013.0462).

Apocynaceae

Calotropis procera (Aiton) Dryan., Hortus Kew. 2: 78. 1811.
Syn. *Asclepias procera* Aiton (1789).
Fifa, *Nuaimat et al.* (NHJ 2013.0658).

Caralluma europaea (Guss.) N.E.Br., Gard. Chron. ser. 3, 12: 369. 1892.
Syn. *Boucerosia aaronis* Hart (1885); *Caralluma aaronis* (Hart) N.E.Br. (1892).
Wadi Rum, *Boulos et al.* 7750 (AMM, BM, E).

Caralluma sinaica (Decne.) A.Berger, Stapel. & Klein. 55. 1910.
Syn. *Boucerosia sinaica* Decne. (1844).
Balqa, Dead Sea area (HUJ); Dana (RSCN database).

Caralluma tuberculata N.E.Br., Bull. Misc. Inform. Kew 1895: 264. 1895.
Jordan, *Gillett* 16108 27316 (K).

Cionura erecta (L.) Griseb., Spic. Fl. Rumel. 2: 69. 1844.
Syn. *Marsdenia erecta* (L.) R.Br. (1810).
Ref. RSCN database; Al-Eisawi 38 (2013).

Cynanchum acutum L. subsp. **acutum**, Sp. Pl.: 212. 1753.
Syn. *Cynanchum monspeliacum* L. (1753).
Ghor-es-Safieh [As Safi] *Hart* 8 (K000286070).

Glossonema boveanum (Decne.) Decne., Ann. Sci. Nat., Bot. 2, 9: 335. 1838.
E Shores of Dead Sea Area, Wadi Araba (RSCN database).

Gomphocarpus fruticosus (L.) W.T.Aiton, Hort. Kew. ed. 2. 2: 80. 1811.
Ref. Al-Eisawi 38 (2013).

Gomphocarpus sinaicus Boiss., Diagn. Pl. Orient. 11: 80. 1849.
Ma'an, Petra, *Dinsmore* 13711 (E00646337); on the way to Dlaga village, *Taifour et al.* (NHJ 2013.0620).

Leptadenia pyrotechnica (Forssk.) Decne., Ann. Sci. Nat., Bot. 2, 9: 269. 1838.
Tafila, Site 57. Ghor-Al-Safi, 10 km south of NCARRT station toward Aqaba (MSB-K 395885); Ajloun Kraimeh road, *Taifour et al.* (NHJ 2013.0667).

Nerium oleander L., Sp. Pl.: 209. 1753.
Irbid, Irbid-Al Rafeed village from the village down toward Al Rafeed spring (MSB-K 227753); Tal Al-Rumman, Royal Botanic Garden Team (NHJ 2008.0017).

Oxystelma esculentum (L.f.) Schult., Mem. Wern. Soc. 1:40. 1810.
Ref. Al-Eisawi 38 (2013).

Oxystelma esculentum (L.f.) Schult. var. **alpini** (Decne.) N.E.Br., Fl. Trop. Afr. 4: 382. 1904.
Syn. *Oxystelma alpinii* Decne. (1844).
Dead Sea Area (RSCN database).

Pentatropis nivalis (J.F.Gmel.) D.V.Field & J.R.I.Wood, Kew Bull. 38: 215. 1983.
Amman, Dead Sea area, just at the hot water canal beside the pedestrian (MSB-K 419154).

Pergularia tomentosa L., Mant. Pl. 53. 1767.
Wadi Yutum, 12 km north of Aqaba, *Boulos, Jallad & Lahham* 7556 (E00676988); *Nuaimat et al.*, (NHJ 2013.0653).

Periploca aphylla Decne., Voy. Inde 4: 109. 1843.
North of Wadi Farah, *Dinsmore* 9221 (E00676927).

Periploca graeca L., Sp. Pl.: 211. 1753.
Karak (RSCN database).

Periploca laevigata (Labill.) subsp. **angustifolia** (Labill.) Markgr., Bot. J. Linn. Soc. 64: 375. 1971.
Syn. *Periploca angustifolia* Labill. (1791).
Wadi Rum, *Boulos et al.* 7653 (AMM, BM, E).

Vinca herbacea Waldst. & Kit., Descr. Icon. Pl. Hung. 1: 8. 1799.
Jordan, *Eig* 1922 (E00055982).

Araceae

Arisarum vulgare Targ.Tozz., Ann. Mus. Imp. Fis. Firenze 2(2): 67. 1810.
Ajloun Forest Reserve, RSCN database.

Arum dioscoridis Sm., Fl. Graec. Prodr. 2: 245. 1816.
Madaba, Ain Hesban, *Meyers & Dinsmore* M662 (E00325674).

Arum elongatum Steven, Bull. Soc. Imp. Naturalistes Moscou 30(1): 67. 1857.
Ma'an, Petra, *Boulos & Jallad* 6245 (E00330006).

Arum hygrophilum Boiss., Diagn. Pl. Orient. 13: 8. 1854.
Jerash, *Meyers & Dinsmore* G643 (E00330024); Tal Al-Rumman, *Taifour & Aldamen* (NHJ 2009.0096).

Arum palaestinum Boiss., Diagn. Pl. Orient. 13: 6. 1854.
Wadi Sir, *Meyers & Dinsmore* M663 (E00330038); Tal Al-Rumman, *Taifour & Aldamen* (NHJ 2010.0020).

Arum rupicola Boiss., Diagn. Pl. Orient. 13: 7. 1854.
Syn. *Arum conophalloides* Kotschy ex Schott, has been reported from Zibb Atuf by Holzapfel.
It is not mentioned by Danin (http://flora.org.il/en/plants/). The record needs confirmation.

Biarum angustatum (Hook.f.) N.E.Br., J. Linn. Soc., Bot. 18: 255. 1880.
Irbid, Aqraba Wadi, *Taifour et al.* (NHJ 2013.0712).

Biarum eximium (Schott & Kotschy) Engl., Monogr. Phan. 2: 576. 1879.
Ma'an, Petra, *Dinsmore* 11725 (E00330074).

Biarum olivieri Blume, Rumphia 1: 115 .1836.
Ma`an, Dabbit Hanot, 2015, *Taifour* (NHJ).

Biarum pyrami (Schott.) Engler, *Monogr. Phan. 2: 576. 1879.*
Jebel Gill, Ras en Naqb, *Borzatti & Barsotti* s.n. (AMM).

Eminium spiculatum (Blume) Schott, Syn. Aroid. 17. 1856.
Um Al Qittayn, *Al Khalidi* (NHJ 2011.0019).

Lemna gibba L., Sp. Pl.: 970. 1753.
Jordan, *Townsend* (65/325 27799 K).

Lemna minor L., Sp. Pl.: 970 1753.
Ref. RSCN database; RBG Jordan checklist (unpubl.)

Araliaceae

Hedera helix L., Sp. Pl.: 202. 1753.
Petra, *Danin* (http://flora.org.il/en/plants/HEDHEL/).

Arecaceae

Hyphaene thebaica (L.) Mart., Hist. Nat. Palm. 3: 226. 1838.
Aqaba, cornish road, in middle of road island (MSB-K 338190).

Phoenix dactylifera L., Sp. Pl. 2: 1188. 1753.
Balqa, Alzara (MSB-K 211945); Wadi Sunfahah (NHJ 2013.0639).
Cultivated.

Aristolochiaceae

Aristolochia billardieri Jaub. & Spach, Ill. Pl. Orient. 1: 175. 1844.
Ajloun, Anjara, 25/5/1981 (YAR-H).

Aristolochia bottae Jaub. & Spach, Ill. Pl. Orient. 1: 173. 1844.
Syn. *Aristolochia maurorum* L. (1763).
Ajloun, Ibliin, *El-Oqlah* (NHJ 2012.0162); Jarash, Digol forest, 04/10/1983, (YAR-H).

Aristolochia paecilantha Boiss., Diagn. Pl. Orient. 12: 104. 1853.
Syn. *Aristolochia scrabidula* Boiss. *(*1853).
Irbid, Yarmmuk village, 24/3/1983 (YAR-H).

Aristolochia parviflora Sm., Fl. Graec. Prodr. 2: 22. 1816.
Irbid, 04/11/1984; Salt, Alfhais Al Khateeb (NHJ 2013.0155).

Aristolochia sempervirens L., Sp. Pl.: 961. 1753.
Ajloun, Anjara, 25/5/1981 (YAR-H).

Asparagaceae

Asparagus acutifolius L., Sp. Pl.: 314. 1753.
Petra, vii.1891 (AUB).

Asparagus aphyllus L., Sp. Pl.: 314. 1753.
Below Ain Laqhar, *Hunting Aero Survey* 151A (E00345101); Tal Al-Rumman, *Taifour* (NHJ 2009.0072).

Asparagus horridus L., Syst. Veg. ed. 13, 274. 1774.
Syn. *Asparagus stipularis* Forssk. (1775).
Wadi Zarqa, *Simpson* 53049 (K000318514).

Asparagus palaestinus Baker, J. Linn. Soc., Bot. 14: 602. 1875.
Lower Jordan Valley, *Eig, Zohary & Feinbrun* (E).

Bellevalia ciliata (Cirillo) T.Nees, Gen. Fl. Germ. 4: t. 68. 1834.
Tafila Province, 2014, *Oran* (AMM).

Bellevalia densiflora Boiss., Diagn. Pl. Orient. 7: 109. 1846.
Karak, Merj, 21 Apr.1965 (AUB).

Bellevalia desertorum Eig & Feinbrun, Beih. Bot. Centralbl. 49(2): 666. 1932.
Dead Sea, 20 km N of Sofi, north face of Wadi, *Jones* (E00340557).

Bellevalia glauca (Lindl.) Kunth, Enum. Pl. 4: 309. 1843.
Az Zarqa, Shaumari Wildlife Reserve, *Clarke* (E00340707).

Bellevalia eigii Feinbrun, Palestine J. Bot., Jerusalem Ser. 1: 357. 1939.
Irbid, 2 km S of Al-Ramtha (IBBU-H); Ma'an, Juhaira village, nr. Shoubak, 11 Jun 1996 (NCARE).

Bellevalia flexuosa Boiss., Diagn. Pl. Orient. ser. 1, 13: 36. 1854.
Tafila, Dana Rserve (MSB-K 416267); Tal Al-Rumman, *El-Oqlah* (NHJ 2012.0126).

Bellevalia kurdistanica Feinbrun, Palestine J. Bot., Jerusalem Ser. 1: 381. 1939.
Irbid, Al Ramtha (YAR-H).

Bellevalia longipes Post, Bull. Herb. Boissier 3: 165. 1895.
Al-Mafraq, Al-Bayet University (IBBU-H); Irbid, Ajloun, RSCN database.

Bellevalia macrobotrys Boiss., Diagn. Pl. Orient. 13: 35. 1854.
Tal Al-Rumman, *Al-Eisawi et al.* (NHJ 2006.0348).

Bellevalia mosheovii Feinbrun, Palestine J. Bot., Jerusalem Ser. 1: 380. 1939.
Petra, *Hepper* 3091 (K000315936).

Bellevalia stepporum Feinbrun, Palestine J. Bot., Jerusalem Ser. 1: 370. 1939.
Amman, Site 22, Muwaggar, University station at Muwaggar (MSB-K 557881); Tal Al-Rumman, *Al-Eisawi* (NHJ 2006.0401).

Bellevalia trifoliata (Ten.) Kunth, Enum. Pl. 4: 308. 1843.
Ajloun, Kufranja, *Al Mfadi et al.* (NHJ 2013.0028).

Bellevalia warburgii Feinbrun, Palestine J. Bot., Jerusalem Ser. 1: 355. 1939.
Balqa Governerate (NHJ); Dibeen Forest Reserve, 2004, *Mahklouf* (AMM).

Bellevalia zoharyi Feinbrun, Palestine J. Bot., Jerusalem Ser. 1: 372. 1939.
3 km W Azraq, *Frey & Kürschner* (E00340740).

Dipcadi erythraeum Webb & Berthel., Hist. Nat. Iles Canaries 2(3): 341. 1848.
Aqaba, wadi close to Saudi Arabian frontier 22 km south of Aqaba, *Boulos & Jallad* (E00345370).

Drimia maritima (L.) Stearn, Ann. Mus. Goulandris 4: 204. 1978.
Syn. *Urginea maritima* (L.) Baker (1873).
Madaba, Main town (MSB-K 211912); Tal Al-Rumman, *Shishani* (NHJ 2011.0122).

Drimia undulata Stearn, Ann. Mus. Goulandris 4: 208. 1978.
Syn. *Urginea undulata* (Desf.) Steinh. (1834).
Ma'an, *Danin* (http://flora.org.il/en/plants/URGUND/).

Hyacinthella nervosa (Bertol.) Chouard, Bull. Mus. Natl. Hist. Nat., sér. 2, 3: 178. 1931.
Syn. *Hyacinthus nervosus* Bertol. (1842).
Petra to Negle Gtashera (foot of Mt. Hot. E.), *Davis* (E00340837).

Hyacinthus orientalis L., Sp.Pl.: 317. 1753.
Tafila and Shoubak, Danin (2005); RSCN database

Leopoldia bicolor (Boiss.) Eig & Feinbrun, Palestine J. Bot., Jerusalem Ser. 4: 58. 1947.
Tal Al-Rumman RBG, Sector No. 23, *Al Mfadi* (NHJ 2013.0087).

Leopoldia comosa (L.) Parl., Giorn. Bot. Ital. 2(2): 160. 1847.
Syn. *Muscari comosum* (L.) Mill. (1768).
Petra, *Dinsmore* (E00345539).

Leopoldia eburenea Eig & Feinbrun, Palestine J. Bot., Jerusalem Ser. 4: 58. 1947.
Dibeen Forest Reserve May 2006, *Taifour & Othman* (RSCN – Dibeen For. Res. Herb).

Leopoldia longipes (Boiss.) Losinsk., Fl. URSS 4: 410. 1935.
Syn. *Muscari longipes* Boiss. (1854); *Muscari albicaule* Post (1890); *Muscari aaronsohnii* Oppenh. & Beauverd (1931); *Leopoldia deserticola* (Rech.f.) Feinbrun (1986).
Ma'an, Site 16. Fujaij, just outside the station offices beside the interance gate (MSB-K 557836).

Leopoldia tenuiflora (Tausch) Heldr., Bull. Soc. Imp. Naturalistes Moscou 53(1): 70. 1878.
Syn. *Muscari tenuiflorum* Tausch (1841).
Amman, 40 km E Amman, *Frey & Kürschner* (E00345593); Tafila, Dana Reserve, Wadi Ahmer, *Al Awaji et al.* (NHJ 2013.0428).

Muscari commutatum Guss, Pl. Rar. 145. 1826.
Zarqa, on the way to Alouk, 2 km after sign of Kansha Alouk beside old oak tree (MSB-K 227889); Tal Al-Rumman, *Al-Eisawi et al.* (NHJ 2006.0091).

Muscari neglectum Guss. ex Ten., Syll. Pl. Fl. Neapol. App. 5: 13. 1842.
Amman, *Buxton* 1 (K000318279).

Muscari parviflorum Desf., Fl. Atlant. 1: 309. 1798.
Near Umm Qeis, NW Irbid, 1973, *Boulos & Hauke* (E).

Muscari pulchellum Heldr. et Sart., Diagn. Pl. Orient. ser. 2, 4: 109. 1859.
Tal Al-Rumman, *Taifour* (NHJ 2013.0139).

Ornithogalum arabicum L., Sp. Pl.: 307. 1753.
Ref. Al-Eisawi 125 (2013).

Ornithogalum collinum Guss., Index Seminum (NAP) 1825: s.p. 1825.
N and C Naqab, Wadi 'Araba, Karak (RSCN database).

Ornithogalum divergens Boreau, Notes Pl. Franç. 3: 15. 1847.
AlMafraq, Al Bayet University (IBBU-H).

Ornithogalum fimbriatum Willd., Neue Schriften Ges. Naturf. Freunde Berlin 3: 420. 1801.
Medaba, 24 Apr. 1911, *Meyers & Dinsmore* (E).

Ornithogalum fuscescens Boiss. & Faill., Diagn. Pl. Orient. 2, 4: 107. 1859.
Ref. (RSCN database).

Ornithogalum lanceolatum Labill. Icon. Pl. Syr. 5: 11. 1812.
Gharaibeh (IBBU-H).

Ornithogalum montanum Cirillo, Fl. Napol. 1: 176. 1811.
Petra, *C. Holzapfel* (1983).

Ornithogalum narbonense L., Cent. Pl. 2: 15. 1756.
Amman, Wadi Al Seer, Al Bahath, coming from Wadi El Sheta (MSB-K 276474); Tal Al-Rumman, *Taifour* (NHJ 2010.0053).

Ornithogalum neurostegium Boiss., Fl. Orient. 5: 222. 1882.
Karak, Mu'tah-Ay road, side road to the right toward Kherbet Chferaz, first turn to the left back of ruined building (MSB-K 418364); Tafila, Dana Reserve, Wadi Ahmer, *Al Awaji et al.* (NHJ 2013.0426).

Ornithogalum neurostegium Boiss. & Blanche subsp. **eigii** (Feinbrun) Feinbrun, Fl. Palaestina 4: 55. 1986.
Syn. *Ornithogalum eigii* Feinbrun (1941).
Ajloun Castle, 01 May 1945, *Davis* 9662 (E00349059).

Ornithogalum polyphyllum Jacq., Icon. Pl. Rar. 2: 19, t. 430, 1794.
Ajloun Castle (IBBU-H).

Ornithogalum trichophyllum Boiss., Diagn. Pl. Orient. 2, 4: 108. 1859.
Wadi Sir, *Dinsmore* (E00349721).

Ornithogalum umbellatum L., Sp. Pl.: 307. 1753.
Salt, iv.1886, *Barbey* (AUB).

Prospero autumnale (L.) Speta, Veröff. Int. Clusius-Forschungsges. Güssing 5: 4. 1982.
Syn. *Scilla autumnalis* L. (1753).
Irbid, nr. Umm Qeis, NW Irbid, *Boulos & Hauke* (E00349204).

Prospero hanburyi (Baker) Speta, Veröff. Int. Clusius-Forschungsges. Güssing 5: 13. 1982.
Syn. *Scilla hanburyi* Baker (1873).
Ref. Al-Eisawi 126 (2013) as *Scilla hanburyi*.

Ruscus aculeatus L., Sp. Pl.: 1041. 1753.
RSCN database; 2011, *Taifour* (Wadi Ibn Hammad Herbarium).

Scilla hyacinthoides L., Syst. Nat. ed. 12 2: 243. 1767.
Madaba, Jebel Siaghah, iv.1885 (AUB).

Asphodelaceae

Aloe vera (L.) Burm.f., Fl. Indica: 83. 1768.
Petra, Wadi Musa (YAR-H). Also cultivated.

Asphodeline brevicaulis (Bertol.) J.Gay ex Baker var. **edumea** (Zohary) Tuzlaci, Notes Roy. Bot. Gard. Edinb. 41: 85. 1983.
Syn. *Asphodeline edumea* (Zohary) Mouterde (1966).
Shobek to Jebel Aneizi, *Davis* 9646 (E00337931).

Asphodeline lutea (L.) Rchb., Fl. Germ. Excurs.: 116. 1830.
Syn. *Asphodelus luteus* L. (1753).
S of Ras El-Naqb, *Boulos, Al-Eisawi & Jallad* 5992 (E00337958).

Asphodeline recurva Post, Bull. Herb. Boissier 3: 166. 1895.
Irbid, iv.1940 (AUB); 21 km SW of Ma'an, 30.3.1936 (8361-HUJ).

Asphodelus fistulosus L., Sp. Pl.: 309. 1753.
Jerash, *Meyers, Fred & Dinsmore* G 450 (E00337676).

Asphodelus ramosus L., Sp. Pl.: 310. 1753.
Syn. *Asphodelus aestivus* Brot. (1804); *Asphodelus microcarpus* Salzm. & Viv.
Tal Al-Rumman, *Al-Eisawi, Abu Sondos, Abu Anza & Shishani* (NHJ 2006.0104).
Yarmouk protectred area (YAR-H).

Asphodelus refractus Boiss., Diagn. Pl. Orient. 13: 23. 1854.
Wadi Rum, *Frey & Kürschner* 80–586 (E00337699).

Asphodelus tenuifolius Cav., Anales Ci. Nat. 3: 46. 1801.
Madaba, from the cross of Dead Sea - Amman - Madaba toward Madaba up side (MSB-K 227812); nr. Aqaba, *Boulos & Lahham* 5416 (AMM, BM, E).

Asphodelus viscidulus Boiss., Diagn. Pl. Orient. 7: 118. 1846.
Tal Al-Rumman, *RBG-Ire & RBG-Jo* (NHJ 2008.0073).

Eremurus spectabilis M.Bieb., Cent. Pl. Ross. Merid.: 2. 1810.
E of Medaba, *Meyers, Fred & Dinsmore* M1664 (E00338118).

Asteraceae

Aaronsohnia factorovskyi Warb. & Eig, Leafl. Agric. Exp. Sta. Zionist Organ. Inst. 6: 40. 1927.
Wadi Rum, *Davis* 9106 (E00588392); Dana, *Taifour* (NHJ 2013.0418).

Acantholepis orientalis Less., Linnaea 6: 88. 1831.
Azraq, Wadi Ratam, Azraq *C.C. Townsend* (K).

Achillea aleppica DC., Prodr. 7(1): 296. 1838.
Jerash, Thagrat Asfour, *Taifour et al.* (NHJ 2013.0668).

Achillea arabica Kotschy, 255. 1866.
Syn. *Achillea biebersteinii* Hub.-Mor. (1975); *A. micrantha* Willd. (1789).
Mafraq, Aldogmsah, *Alshudiefat & Khalidi* s.n. (NHJ); Ajloun castle, *Davis* 9661 (E00582356); Ajloun, 15 km north of Ajloun to Irbid at 3 km after Ibbin junction at Ras Munif (MSB-K 151706); Ajloun district, near Orjan village, BOT ERA project (NHJ 2013.0539).

Achillea cretica L., Sp. Pl.: 899. 1753.
Syn. *Achillea santolina* L. (1753).
Ma'an, Shoubak, Wadi Musa to Shoubak before Hisha on the right roadside (MSB-K 497882).

Achillea falcata L., Sp. Pl.: 897. 1753.
Wadi Heidan (Moab), *Davis* 9189 (E00577356); Amman, Khalda, from the Al Waha Circle toward Sweleh, (MSB-K 200291).

Achillea fragrantissima (Forssk.) Sch.Bip., Flora 38: 13. 1855.
Syn. *Santolina fragrantissima* Forssk. (1775).
Ma'an, 0 km N von Petra, *Frey& Kürschner* (E00577407); Zarqa-Azraq road, *Taifour, Al Khawaldah & Alshwaer* (NHJ 2013.0678).

Achillea membranacea (Labill.) DC., Prodr. 6: 32. 1838.
Syn. *Anthemis membranacea* Labill. (1809).
Amman, Site 11, Dabala site on the desert highway (MSB-K 558028).

Achillea santolinoides subsp. **wilhelmsii** (K.Koch) Greuter 39. 2003.
Syn. *Achillea wilhelmsii* K.Koch (1851).
Tafila, 20 km west of Jorf el Dorawish to Tafila (MSB-K 151810).

Achillea tenuifolia Lam., Encycl. 1(1): 26. 1783.
Ma'an, Wadi Musa, near Petra, *Boulos & Jallad* (E00577701); Dana, *Taifour* (NHJ 2013.0422).

Aetheorhiza bulbosa (L.) Cass., Dict. Sci. Nat., ed. 2. 48: 426. 1827.
Syn. *Leontodon bulbosus* L. (1753).
Ajloun, *Danin* (http://flora.org.il/en/plants/AETBUL/).

Ageratum conyzoides L., Sp. Pl.: 839. 1753.
Jordan valley & Wadi Shuaib, RSCN database.

Ambrosia confertifolia DC., Prodr. 5: 526. 1836.
Tafila, Al Rashadiah, 18 Feb. 1954 (FOR-H).
Exotic invasive.

Ambrosia maritima L., Sp. Pl.: 988. 1753.
Ref. Al-Eisawi 42 (2013).

Anthemis altissima L., Sp. Pl.: 893. 1753.
Ref. RSCN database; Al-Eisawi 42 (2013).

Anthemis bornmuelleri Stoj. & Acht., Notizbl. Bot. Gart. Berlin-Dahlem 13: 522. 1937.
Madaba (MSB-K 113902); Tal Al-Rumman, *Taifour* (NHJ 2008.0051).

Anthemis brachycarpa Eig, Palestine J. Bot., Jerusalem Ser. 1, 183. 1938.
Ajloun, Ain al Deek (YAR-H).

Anthemis chia L., Sp. Pl.: 894. 1753.
Wadi Arab (YAR-H).

Anthemis cornucopiae Boiss., Diagn. Pl. Orient. ser. 1, 11: 8. 1849.
Borqu (IBU-H); Irbid (YAR-H).

Anthemis cotula L., Sp. Pl.: 894. 1753.
Salt, iv.1886 (AUB).

Anthemis edumea Eig, Palestine J. Bot., Jerusalem Ser. 1: 173. 1938.
Shobela to Gebel Aneizi (Aqaba), *Davis* 9545 (E00581610).

Anthemis haussknechtii Boiss. & Reut., Fl. Orient. 3: 310. 1875.
Irbid, *Meyers & Dinsmore* G4392 (B).

Anthemis hebronica Boiss. & Kotschy, Diagn. Pl. Orient. ser. 2, 5: 108. 1856.
Irbid, Aqraba (YAR-H).

Anthemis hyalina DC., Prodr. 6: 4. 1838.
Madaba (MSB-K 114046).

Anthemis maris-mortui Eig, Palestine J. Bot., Jerusalem Ser. 1: 181. 1938.
Jebel Aneizi, *Davis* 9206 (E00581738).

Anthemis melampodina Delile, Fl. Egypte 268. t. 45, 1813.
Sof Naqb Ishtar, *Hunting Aero Survey* 49B (E00581766); Dana Reserve, Barra, BOT ERA project (NHJ 2013.0486).

Anthemis melampodina Delile subsp. **deserti** (Boiss.) Eig, 177. 1938.
Syn. *Anthemis deserti* Boiss. (1875).
Ghour Safi, iv.1895, *Fox* (AUB); 9 km NE of Aqaba, 1974, *Boulos et al.*

Anthemis nabataea Eig, Palestine J. Bot., Jerusalem Ser. 1: 184. 1938.
Wadi Al Hasa, nr. Qalat el Hasa, 28/3/1936 (28574-HUJ); Dibeen National Park, 2005, *Mahklouf* s.n. (AMM).

Anthemis palestina Reut., Fl. Orient. 3: 283. 1875.
Syn. *Anthemis melanolepis* Boiss. (1888).
Karak, Mu'tah-Ay road, side road to the right toward Kherbet Chferaz (MSB-K 418353); Tal Al-Rumman, *Taifour & Aldamen* (NHJ 2009.0011).

Anthemis parviflora Eig, Palestine J. Bot., Jerusalem Ser. 1, 198. 1938.
Irbid, Bani kanana, *El-Oqlah & Bani Said* (NHJ 2013.0135).

Anthemis pseudocotula Boiss., Diagn. Pl. Orient. ser. 1, 6: 86. 1846.
Wadi Themed to Ziga (Moab), *Davis* 9656 (E00581845); Tal Al-Rumman RBG, Sector No. 30, *Taifour & Al Khawaldah* (NHJ 2013.0280).

Anthemis rascheyana Boiss., Diagn. Pl. Orient. ser. 1, 11: 8. 1849.
Irbid, Al Ramtha (YAR-H); RSCN database.

Anthemis scariosa Banks & Sol., Nat. Hist. Aleppo, ed. 2, 2: 263. 1794.
Ref. Al-Eisawi 43 (2013).

Anthemis scrobicularis Yavin, Israel J. Bot. 21(3): 174. 1972.
Kurnub and Um Sahm, road to Mudawwara, *Hunting Aero Survey* (E00413603).

Anthemis tenuicarpa Eig, Palestine J. Bot., Jerusalem Ser. 1, 166. 1938.
Aqaba, 14 km S of Qal'at-el-Hasa, Eig, *Feinbrun & Zohary* (E00385738).

Anthemis tinctoria L., Sp. Pl.: 896. 1753.
Balqa, Road from Salt to Kufr Huda, 100m before turn to Kufr Huda at olive oil company (MSB-K 185822).

Anthemis zoharyana Eig, Palestine J. Bot., Jerusalem Ser. 1: 178. 1938.
Ref. Dead Sea area, RSCN database; Al-Eisawi 43 (2013).

Anvillea garcinii (Burm.f.) DC., Prodr. 5: 487. 1836.
Syn. *Anthemis garcinii* Burm.f. (1768).
Dead Sea area, *Böcker* RB 649 (B).

Artemisia arborescens L., Sp. Pl., ed. 2. 2: 1188. 1763.
Irbid, Ramtha-Mgier (YAR- H).

Artemisia jordanica Danin, Willdenowia 29(1–2): 147. 1999.
Endemic. Type: Jordan, Ma'an District, *Danin* 972901 (Holo. HUJ; iso. B, BM, E, K).

Artemisia judaica L., Syst. Nat., ed. 10. 2: 1208. 1759.
Aqaba, along the post to Aqaba, *Boulos et al.* 7593(E00574695); Mafraq Hazim, *Taifour, Al Mfadi&Bani Said* (NHJ 2013.0045).

Artemisia monosperma Delile, Fl. Egypte: 263. 1813.
Jebel Guzlan, *Borzatti & Barsotti* (AMM); Wadi Araba, RSCN database.

Artemisia sieberi Besser, Bull. Soc. Imp. Naturalistes Moscou 9: 80. 1836.
Syn. *Artemisia herba-alba* Asso (1779).
Amman, Azraq, *Frey & Kürschner* 4921 (E00573734); Dana Reserve, *RBG-Ire & RBG-Jo* (NHJ 2008.0075).

Aster subulatus (Michx.) Hort ex Michx., Fl. Bor.-Amer. 2: 111. 1803.
Along Zarqa River, 2004, *Al-Dwiri* s.n. (AMM).

Asteriscus aquaticus (L.) Less., Syn. Gen. Compos. 210. 1832.
Syn. *Buphthalmum aquaticum* L. (1753).
South of Wadi Farab, *Dinsmore*10219 (E00589283); Tal Al-Rumman, *Taifour & Aldamen* (NHJ 2009.0187).

Asteriscus graveolens Less., Syn. Gen. Compos. 210. 1832.
Aqaba, Wadi Araba, 15 km after Bee Mazkor rest house towards Amman (MSB-K 497435).

Atractylis cancellata L., Sp. Pl.: 830. 1753.
Balqa, Al-Rumman hills (MSB-K 340689); Tal Al-Rumman, *Taifour & Aldamen* (NHJ 2009.0055).

Atractylis carduus (Forssk.) C.Chr., Dansk Bot. Ark. iv. No. 3, 27. 1922.
Syn. *Centaurea carduus* Forssk. (1775).
Wadi Zerka Main, *Davis* 9409 (E00457548); nr. Aqaba, *Boulos & Lahham* 8533 (AMM, BM, E).

Atractylis mernephthae Asch., Schweinf & Letourn., Asch. & al. in Mém. Inst. Egypt. 2. 94. 1887.
Ref. Wadi Araba, Danin (2004).

Atractylis phaeolepis Pomel, Nouv. Mat. Fl. Atl.: 273. 1875.
Tal Al-Rumman RBG, Sector No. 47, *El-Oqla & Al Mfadi* (NHJ 2013.0328).

Atractylis prolifera Boiss., Diagn. Pl. Orient. ser. 1, 10: 96. 1849.
Above Jordan valley, *Trought* (E00457585); Tal Al-Rumman RBG, Sector No. 23, *El-Oqlah & Taifour* (NHJ 2013.0163).

Atractylis serratuloides Sieber ex Cass., Dict. Sci. Nat., ed. 2. [F. Cuvier] 50: 58. 1827.
Amman, 20 km W of Ma'in towards Zarqa/Dead Sea, *Frey & Kürschner* (BSB).

Bellis annua L., Sp. Pl.: 887. 1753.
Above Ain Fit, *Meyers & Dinsmore* (E).

Bellis perennis L., Sp. Pl.: 886. 1753.
Upper Jordan Valley, Ajloun (RSCN database).

Bellis sylvestris Cirillo, Pl. Rar. Neapol. ii. 22. t. 4. 1792.
Irbid, 06/03/1982 (YAR-H).

Brocchia cinerea (Delile) Vis., Pl. Aeg. et Nub. 35. 1836.
Syn. *Cotula cinerea* Delile (1814.).
Amman, Dead Sea, 21/04/1983 (YAR-H).

Calendula arvensis L., Sp. Pl., ed. 2. 2: 1303. 1763.
Midway Zarqa-Azraq road, *Boulos et al.* 7102 (E00525682); Tal Al-Rumman, *RBG-Ire & RBG-Jo* (NHJ 2008.0028)

Calendula pachysperma Zohary, Palestine J. Bot., Jerusalem Ser. 2: 172. 1941.
Ghor, Deir Alla (YAR-H).

Calendula palaestina Boiss., Diagn. Plant. Orient. Nov. ser. 1, 10. 1849.
Above Zarqa river, *Meyers & Dinsmore* (E00525647); Jarash, Site 5, Amman-Jerash highway (MSB-K 552705); Tal Al-Rumman, *Taifour* (NHJ 2013.0305).

Calendula tripterocarpa Rupr., Index Seminum [St. Petersburg], 1853 ex Bull. Phys.-Math. Acad. Petersb. 14: 231. 1856.
Tal Al-Rumman, *El-Oqlah* (NHJ 2013.0079).

Carduus argentatus L., Mant. Pl. Altera: 280. 1771.
Tal Al-Rumman, *Shishani & Alkoran* (NHJ 2011.0051).

Carduus argentatus L. subsp. **acicularis** (Bertol.) Meikle, Fl. Cyprus 2: 949, 1897. 1985.
Syn. *Carduus acicularis* Bertol. (1829).
Irbid, Almansourah, 03/04/1983 (YAR-H).

Carduus getulus Pomel, Nouv. Mat. Fl. Atl.: 275. 1875.
Shoubak, Al Fujaig agricultural reserve (MSB-K 199854); Tal Al-Rumman, *Taifour* (NHJ 2011.0046).

Carduus nigrescens Vill. subsp. **australis** (Nyman) Greuter, Willdenowia 33(1): 52. 2003.
Syn. *Carduus australis* Jord. (1849).
Ajloun, forest area, 12/04/1982 (YAR-H); Tal Al-Rumman, *Al-Eisawi* (NHJ 2006.0036).

Carduus pycnocephalus L. subsp. **arabicus** (Jacq. ex Murr) Nyman, Consp. Fl. Europ. 414. 1879.
Syn. *Carduus pycnocephalus* L. subsp. *cinereus* (M.Bieb.) Davis; *Carduus arabicus* Jacq. (1784).
Ma'an, Petra, Siq, *Davis* 9361(E00462272).

Carlina curetum Heldr. ex Halacsy, Fl. Orient. 3: 449. 1875.
Yarmouk protected area (YAR-H).

Carlina hispanica Lam., Encycl. 1(2): 624. 1785.
Syn. *Carlina libanotica* Boiss. (1849).
Tal Al-Rumman, *Al-Eisawi* (NHJ 2006.0201). Petra, *Post*, 27.vii.1907 (AUB).

Carlina lanata L., Sp. Pl.: 828 .1753.
Ajloun, 22 Sep. 1983, Anjerah (YAR-H).

Carlina racemosa L., Sp. Pl.: 829. 1753.
Along Zarqa River, 2004, *Al-Dwiri* s.n. (AMM).

Carthamus glaucus M.Bieb., Fl. Taur.-Caucas. 2: 284. 1808.
Amman, *Davis* 9703A (K000318484).

Carthamus nitidus Boiss., Fl. Orient. 3: 708. 1875.
Qadi Zarqa. Ma'an, 8.4.1930 (30139-HUJ).

Carthamus persicus Willd., Sp. Pl., ed. 4, 3(3): 1707. 1803.
Syn. *Carthamus palaestinus* Eig (1931).
Irbid, Wadi Arab (YAR-H).

Carthamus tenuis (Boiss. & Blanche) Bornm., Verh. Zool.-Bot. Ges. Wien, xlviii. 605. 1898.
Syn. *Kentrophyllum tenue* Boiss. & C.I.Blanche (1856).
Jordan, 16.viii.1962, *Kaleem Hajj* 89 (AUB); Irbid, Jord. Univ. Sci. Tech., 03/06/2002 (YAR-H).

Carthamus tinctorius L., Sp. Pl. 2: 830. 1753.
Jordan, 16.viii.1962, *Kaleem Hajj* 80 (AUB).

Catananche lutea L., Sp. Pl. 2: 812. 1753.
Amman, Site 28. Moubis-Alouk road (MSB-K 557250); Tal Al-Rumman, *Al-Eisawi* (NHJ 2006.0007).

Centaurea aegyptiaca L., Mant. Pl. 118. 1767.
Tafila, *El-Oqlah* (NHJ).

Centaurea ammocyanus Boiss., Diagn. Pl. Orient. ser. 1, 10: 109. 1849.
Wadi Rajil, 26 Apr. 1965 *Townsend* (K); Paine (K000794086); Al Mafraq, Al Bayt University (IBBU-H).

Centaurea ascalonica Bornm., Repert. Spec. Nov. Regni Veg. 24: 373. 1928.
Balqa, 3 km after Kufr-Huda on the way to A'eera & Yarga (MSB-K 200512).

Centaruea behen L., Sp. Pl.: 914. 1753.
Salt, Wadi Shuaib, 24/04/1983 (YAR-H).

Centaurea calcitrapa L., Sp. Pl. 2: 917. 1753.
Wadi Zerqa, *Hunting Aero Survey*,143A (E00472881).

Centaurea crocodylium L., Sp. Pl.: 919. 1753.
Dana, *Taifour* (NHJ 2013.0407).

Centaurea damascena Boiss., Diagn. Pl. Orient. ser. 1, 10: 112. 1849.
Petra, 26.vii.1907 (AUB).

Centaurea dumulosa Boiss., Diagn. Pl. Orient. ser. 1, 10: 111. 1849.
Amman, C Naqab (RSCN database).

Centaurea eryngioides Lam., Encycl. 1(2): 675 1785.
Rum, Jebel Ram, *Gillett* 16080 (K000315545).

Centaurea hyalolepis Boiss., Diagn. Pl. Orient. ser. 1, 6: 133. 1846.
Umm Qays, Gadara, *Tackholm, El-Hadidi, Lahham & Boulos* 8895 (K000315551); Tal Al-Rumman, *Al-Eisawi* (NHJ 2006.0026).

Centaurea iberica Trev. ex Spreng., Syst. Veg. (ed. 16) [Sprengel] 3: 406. 1826.
Tal Al-Rumman, *Al-Eisawi, Abu Sondos, Abu Anza & Shishani* (NHJ 2006.0105).

Centaurea lanulata Eig, Repert. Spec. Nov. Regni Veg. Beih. 63(1): 47, nomen. 1931.
Wadi Araba, Tafila road, 29/v/2007 (NCARE).

Centaurea laxa Boiss. & Hausskn., Fl. Orient. 3: 640 .1875.
Wadi Rum, *Boulos & Lahham* 7678 (AMM, BM, E).

Centaurea microcnicus Reese & Sam. in Rech.f., Ark. Bot. ser. 2, 1: 326. 1950.
Jordan, near camp on road to Kilwa, *Hunting Aero Survey* 92B (E00476590).

Centaurea pallescens Delile, Fl. Egypte 278. t. 49, 1814.
Ma'an, Rajef-Ras Nagab road (MSB-K 341273); Tal Al-Rumman, *RBG-Ire & RBG-Jo* (NHJ 2008.0041).

Centaurea postii Boiss., Fl. Orient. 3: 688. 1875.
Azraq, 20 Apr. 1965, *Townsend* (K).

Centaurea procurrens Sieber ex Spreng., Syst. Veg. (ed. 16) 3: 407. 1826.
Along Zarqa River, 2004, *Al-Dwiri* s.n. (AMM).

Centaurea pseudosinaica Czerep., Bot. Mater. Gerb. Bot. Inst. Komarova Akad. Nauk S.S.S.R. 20: 392. 1960.
Ma'an, 30 km N of Mudawwa, *Frey & Kürschner* (B); Eastern desert (RSCN database).

Centaurea rigida Banks & Sol., Enum. Pl. Suppl. 61. 1814.
Jarash, Alouk, 23/05/1983 (AUB); *Boulos* 9174 (B10021779).

Centaurea scoparia Sieber ex Spreng., Prodr. 6: 596. 1838.
3–4 km south of Ras en Naqb, *Boulos, Addis & Qumsiyeh* (K).

Centaurea sinaica DC., Prodr. 6: 592. 1838.
Tafila, Wadi Araba Tafila road (MSB-K 417758); Azraq Wadi Bottom (South west the Castle Amra) *El-Oqlah* (NHJ 2013.0580).

Centaurea solstitialis L., Sp. Pl. 2: 917. 1753.
Ref. RSCN database; Al-Eisawi 46 (2013).

Centaurea speciosa Boiss., Diagn. Pl. Orient. ser. 1, 10: 116. 1849.
Dibeen National Park, 2005, *Mahklouf* s.n. (AMM).

Centaurea verutum L., Cent. Pl. 1. 30. 1755.
Irbid, Maru agriculture station (MSB-K 200257); Irbid, Deir Elseneh, 13/05/1983 (YAR-H).

Chardinia orientalis (L.) Kuntze, Trudy Imp. S.-Peterburgsk. Bot. Sada 10: 201. 1887.
Syn. *Xeranthemum annuum* L. var. *orientale* L. (1753).
Jarash, Jarash-Souf way (MSB-K 413680).

Chiliadenus iphionoides (Boiss. & C.I.Blanche) Brullo, Webbia 34(1): 301. 1979.
Syn. *Varthemia iphionoides* Boiss. & C.I.Blanche (1856).
Ma'an, Petra, *Leonard* 7449 (E00587842).

Chiliadenus montanus (Vahl) Brullo, Webbia 34(1): 301. 1979.
Syn. *Chrysocoma montana* Vahl (1790).
Wadi Rum, 15 Ap. 1945, *Davis* (K); RSCN database.

Chlamydophora tridentata (Delile) Ehrenb. ex Less., Syn. Gen. Compos. 266. 1832.
Syn. *Balsamita tridentata* Delile (1814).
Lower Jordan valley, RSCN database.

Chondrilla juncea L., Sp. Pl. 2: 796. 1753.
Amman, University campus, *Boulos* 7299 (E00490451).

Chrysanthemum coronarium L., Sp. Pl. 2: 890. 1753.
Irbid, Yarmouk villege, 21/03/1983 (YAR-H); Tal Al-Rumman, *RBG-Ire & RBG-Jo* (NHJ 2008.0071).

Chrysanthemum segetum L., Sp. Pl. 2: 889. 1753.
Ref. Balqa, Danin (2004).

Cichorium endivia L., Sp. Pl. 2: 813. 1753.
Zarqa, 2 km after Kamsha toward Alouk, 17 Jun. 2006 (NCARE).

Cichorium intybus L., Sp Pl.: 813. 1753.
Fifa Reserve, Danin 2005; RSCN database.

Cichorium pumilum Observ. Bot. [Jacquin] 4: 3, t. 80, 1771.
Zarqa, 2 km after Kamsha toward Alouk (MSB-K 176437); Tal Al-Rumman, *Al-Eisawi* (NHJ 2006.0011).

Cirsium alatum (S.G.Gmel.) Bobrov subsp. **alatum**, Bot. Zhurn. 43(11): 1547. 1958.
Syn. *Serratula alata* S.G.Gmel. (possibly 1770–1774).
Azraq, RSCN database.

Cirsium arvense (L.) Scop., Fl. Carniol., ed. 2. 2: 126. 1772.
Azraq, RSCN database.

Cirsium creticum (Lam.) d'Urv. subsp. **gaillardotii** Boiss., Mém. Soc. Linn. Paris 1: 363. 1822.
Syn. *Cirsium gaillardotii* Boiss. (1856).
Al Mafraq, Bwadhah (IBBU-H).

Cirsium phyllocephalum Boiss. & C.I.Blanche, Diagn. Pl. Orient. ser. 2, 3: 39. 1856.
Ajloun, Eshtafeana, 26/08/1983 (YAR-H).

Cnicus benedictus L., Sp. Pl. 2: 826. 1753.
Pass between Petra & Mt. Hot, *Davis* 9197 (E00478651).

Coleostephus myconis (L.) Cass., Dict. Sci. Nat., ed. 2, 41: 43. 1826.
Syn. *Chrysanthemum myconis* L. (1753).
Tell el Kadi, *Meyers & Dinsmore* (E00586224).

Conyza aegyptiaca (L.) Aiton, Hortus Kew. (W. Aiton) 3: 183. 1789.
Syn. *Erigeron aegyptiacus* L. (1767).
Ref. RSCN database; Al-Eisawi 47 (2013).

Conyza canadensis (L.) Cronquist, Bull. Torrey Bot. Club 70: 632. 1943.
Syn. *Erigeron canadense* L. (1753).
Amman, Naor, Site 43 (MSB-K 39566); Jarash, Ain al Deek, 10/04/1983 (YAR-H).

Conyza stricta Willd., Numer. List [Wallich] n. 3044. 1831.
Jebel Rehebi, 23 Apr. 1984, *Borzatti & Barsotti* (AMM).

Cotula anthemoides L., Sp. Pl. 2: 891. 1753.
S of Jerash, Zarqa River (Al-Sayl) (IBBU-H).

Cousinia austrojordanica F.G.Davies & Boulos, Publ. Cairo Univ. Herb. 7–8: 287. 1977.
Jordan, *Jallad, Lahham & Hanazia* 590 (K000786346).

Cousinia hermonis Boiss., Diagn. Pl. Orient. ser. 1, 10: 102. 1849.
Syn. *Cousinia dayi* Post (1891).
Maan, 0 km N Petra Richtung El Haishah, *Frey & Kürschner* 80–805 (E00464718).

Cousinia moabitica Bornm. & Nábělek ex Bornm., Oesterr. Bot. Z. 63: 60. 1913.
Tafila, road from Dana reserve to Tafileh (MSB-K 277909).

Cousinia postiana C.Winkl., Trudy Imp. S.-Peterburgsk. Bot. Sada xii. 223. 1892.
Aqaba, RSCN database.

Crepis aculeata (DC.) Boiss., Fl. Orient. 3: 856. 1875.
Syn. *Barkhausia aculeata* DC. (1838).
Tell Ar Rumman, *Al Mfadi & Al Khateeb* (NHJ).

Crepis aspera L., Sp. Pl., ed. 2. 2: 1133. 1763.
Jerash ruins, *Davis* 9695 (E00491209); Tal Al-Rumman, *Taifour* (NHJ 2009.0009).

Crepis foetida L. subsp. **rhoeadifolia** (M.Bieb.) Celak., Sp. Pl. 2: 807. 1753.
Syn. *Crepis rhoeadifolia* M.Bieb. (1808).
Tal Al-Rumman, *Al-Eisawi* (NHJ 2006.0221).

Crepis hierosolymitana Boiss., Diagn. Pl. Orient. ser. 1, 11: 54. 1849.
Ajloun to Mar Elias, *Davis* 9457 (E00491386); Irbid, Baloqus, 04/05/1983 (YAR-H).

Crepis kotschyana (Boiss.) Boiss., Fl. Orient. 3: 852. 1875.
Syn. *Barkhausia kotschyana* Boiss. (1846).
Desertum Syriacum, inter Amman et Rutba, *Rechinger* 12910 (E00491414).

Crepis micrantha Czerep., Fl. URSS 29. 684. 1964.
Along Zarqa River, 2004, *Al-Dwiri* s.n. (AMM).

Crepis neglecta L., Mant. Pl. 107. 1767.
Ref. RSCN database; Al-Eisawi 47 (2013).

Crepis palaestina (Boiss.) Bornm., Beih. Bot. Centralbl. 31. 2. 236. 1914.
Syn. *Cymboseris palaestina* Boiss. (1849).
Jerash ruins, *Davis* 9680 (E00491476); Tal Al-Rumman, *Al-Eisawi* (NHJ 2006.0227).

Crepis pterothecoides Boiss., Fl. Orient. 3: 850. 1875.
Near Ail, S Jordan escarpment, *Hunting Aero Survey* (E).

Crepis reuteriana Boiss. & Heldr., Diagn. Pl. Orient. ser. 1, 11: 55. 1849.
Ajloun, 3 km E of Ishtafena (IBBU-H).

Crepis sancta (L.) Bornm. in Mitth. Thüring. Bot. Vereins 30. 79. 1913.
Midway between Amman and Madaba, *Boulos, Al-Eisawi & Jallad* 6069 (E00494049); Amman, on way to Bahhath (MSB-K 275086); Tal Al-Rumman, *Taifour* (NHJ 2013.0061).

Crepis senecioides Delile, Fl. Egypte 262. t. 42. f. 2. 1813.
Ma'an, Rajif Ras Al-Naqab road, site no 37 (MSB-K 497930); Tal Al-Rumman, *Al-Eisawi* (NHJ 2006.0024).

Crepis sprengeriana (L.) All., Fl. Pedem. 1: 221. 1785.
Syn. *Picris spergeriana* (L.) Chaix (1785).
Tal Al-Rumman, *Al-Eisawi* (NHJ 2006.0435).

Crepis syriaca (Bornm.) Babc. & Navashin, Gen. Crepis (Bibl. Genet. vi.) 75, in obs. 1930.
Syn. *Crepis alpina* L. var. *syriaca* Bornm.
Ajloun, foot of Ajloun Castle (MSB-K 176596).

Crupina crupinastrum (Moris) Vis., Fl. Dalmat. 2. 42. 1847.
Syn. *Centaurea crupinastrum* Moris (1852).

Ajloun Castle, *Davis* 9672 (E00469937); Zarqa, from sign of Alouk Kamsheh toward Alouk, after 2–3 km from Kamsheh (MSB-K 227915); Tal Al-Rumman, *Al-Eisawi* (NHJ 2006.0122).

Cyanus triumfettii subsp. **axillaris** (Čelak.) Štěpánek in Slavík & Štepánková, Kvetena Ceské Republ. 7. 455. 2004.
Syn. *Centaurea cyanoides* DC. (1838).
Salt, iv.1888 (AUB).

Cymbolaena griffithii (A.Gray) Wagenitz, Oesterr. Bot. Z. 119(4–5): 402. 1972.
Syn. *Stylocline griffithii* A.Gray (1873).
Wadi Musa territory, Al Taybeh & Um'Qaser area, 31 May 2004, *Abulaila, Tehabsem & van Slageren* (K).

Cynara auranitica Post, Bull. Herb. Boissier 1. 400. 1893.
Syn. Cynara syriaca Boiss. (1849).
Irbid, North Shuneh highway at traffic sign (MSB-K 186379); Irbid, Yarmouk University, 05/05/1982 (YAR-H).

Cynara cornigera Lindl., Fl. Graec. (Sibthorp). 10 (2): 74. 1840.
Ref. RSCN database; Al-Eisawi 48(2013).

Dittrichia graveolens (L.) Greuter, Exsicc. Genav. 4: 71. 1973.
Syn. *Erigeron graveolens* L. (1755).
Jarash, main road of Amman-Jarash under the bridge (MSB-K 212012).

Dittrichia viscosa (L.) Greuter, Exsicc. Genav. 4: 71. 1973.
Syn. *Erigeron viscosus* L. (1753).
Ma'an, Wadi-Musa, *Dinsmore*16288 (E00589107).

Doellia bovei (DC.) Anderb., Willdenowia 25(1): 21. 1995.
Syn. *Conyza bovei* DC. (1834); *Blumea abyssinica* Sch.Bip. ex A.Rich. (1848); *Erigeron bovei* (DC.) Boiss. (1875).
Wadi Hassa, Afra hot springs, BOT ERA Project (NHJ 2013.0503).

Echinops glaberrimus DC., Ann. Sci. Nat., Bot. sér. 2, 2: 260. 1834.
Jarash, Wadi Kofer Khal, 26/07/1983 (YAR-H).

Echinops philistaeus Feinburn &Zohary, Israel J. Bot. 25(12): 82. 1976.
Azraq Wadi Bottom (South west the Castle Amra), *El-Oqlah* (NHJ 2013.0559).

Echinops polyceras Boiss., Diagn. Pl. Orient. ser. 1, 10: 85. 1849.
Syn. *Echinops gaillardotii* Boiss. (1856).
Amman, *Davis*, Aug–45 (E00457759); Mafraq, Rehab, *Taifour* (NHJ 2013.0004).

Echinops pungens Trautv., Diss. Ech. 18. 1833.
Syn. Echinops adenocaulis Boiss. (1849).
Balqa, Oak forest (MSB-K 211864).

Echinops spinosissimus Turra, Farset. Nov. Gen. 13. 1765.
Ref. Al-Eisawi 49 (2013).

Echinops spinosus L., Mant. Pl.: 119. 1767.
On the Safawi-Mafeaq rd., 21 km from Safawi (BRDC-H).

Echinops tenuisectus Rech.f., Anz. Österr. Akad. Wiss., Math.-Naturwiss. Kl. ci. 16 (1964).
Ma'an, on the way to Shoubak, 2 km before Shoubak Castle (MSB-K 243805).

Echinops viscosus DC., Prodr. 6: 525. 1838.
Al Karak, Abu Hammur (FOR-H).

Eclipta prostrata (L.) L., Mant. Pl. 2: 286. 1771.
Syn. *Verbesina prostrata* L. (1753).
Ref. RSCN database; Al-Eisawi 49 (2013).

Erigeron bonariensis L., Sp. Pl. 2: 863. 1753.
Syn. *Conyza bonariensis* (L.) Cronquist (1943).
Amman, Al Bidden. Biaden wadi, 4 km from city centre, entrance of Um Assos village. Site 44, (MSB-K 395586).

Eupatorium cannabinum L., Sp. Pl. 2: 838. 1753.
Jarash, Badran farm, 17/04/1983 (YAR-H).

Filago argentea (Pomel) Chrtek & Holub, Preslia 35: 3. 1963.
Tal Al-Rumman RBG, Sector No. 93, *El-Oqlah, Taifour & Al Khawaldah* (NHJ 2013.0254).

Filago contracta (Boiss.) Chrtek & Holub, Preslia 53: 3. 1963.
Syn. *Evax contracta* Boiss. (1849).
Zarqa, 2 km after Kamsha (MSB-K 278294); Tal Al-Rumman, *Al-Eisawi* (NHJ 2006.0034); *Taifour & Aldamen* (NHJ 2009.0002).

Filago desertorum Pomel, Nouv. Mat. Fl. Atl.: 46. 1874.
Amman, 10 km N von Suweleh in Richtung Jerash, *Frey& Kürschner* 4856 (E00530464); Azraq Wadi Bottom (South west the Castle Amra), *El-Oqlah* (NHJ 2013.0576).

Filago eriocephala Guss., Pl. Rar. Sic. t. 69. 1832.
Ajloun, Kufuranjeh, 15/05/1983 (YAR-H); Tal Al-Rumman, *RBG-Ire & RBG-Jo* (NHJ 2008.0081).

Filago gallica (L.) L., Sp. Pl. 2: 857 (addenda) 1753.
Ajloun, Umelyanabia, 26/03/1983 (YAR-H); Tal Al-Rumman, *Taifour & Aldamen* (NHJ 2009.0032).

Filago inexpectata Wagenitz, Willdenowia 5: 414. 1969.
Wadi Rum, *Davis* 9023 (E00530558); Tal Al-Rumman, *Al-Eisawi* (NHJ 2006.0158).

Filago palaestina (Boiss.) Chrtek & Holub, Preslia 35: 3. 1963.
Syn. *Evax palaestina* Boiss. (1849).
Dhiban to Wadi Themed, *Davis* 9620 (E00530565); Irbid, Jordan Univ. Science & Technology, 20/05/2002 (YAR-H).

Filago pyramidata L., Sp. Pl. 2: 1199. 1753.
Syn. *Gnaphalium pyramidatum* (L.) Lam. (1788).
Madaba, 5 km after Madaba towards Mujib (MSB-K 276500); Tal Al-Rumman, *Taifour* (NHJ 2008.0070).

Garhadiolus hedypnois Jaub. & Spach, Ill. Pl. Orient. 3: 120. 1850.
Ref. Amman, Danin (2005); RSCN database.

Geropogon hybridus (L.) Sch.Bip., Webb & Berth. Phyt. Canar. 2. 472. 1850.
Syn. *Tragopogon hybridus* L. (1753).
Irbid, 15 km northwest of Irbid (MSB-K 151669); Tal Al-Rumman, *Taifour & Aldamen* (NHJ 2009.0183).

Glebionis coronaria (L.) Cass. ex Spach, Bot. Zhurn. (Moscow & Leningrad) 84(7): 117. 1999.
Syn. *Chrysanthemum coronarium* L. (1753).
Wadi Zerqa, *Hunting Aero Survey*145 (E00588261); Tal Al-Rumman, *Taifour & Aldamen* (NHJ 2009.0044).

Glebionis segetum (L.) Fourr., Ann. Soc. Linn. Lyon sér. 2, 17: 90. 1869.
Syn. *Chrysanthemum segetum* L. (1753).
Ref. Al-Eisawi 46 (2013) as *Chrysanthemum segetum.*

Gundelia tournefortii L., Sp. Pl. 2: 814. 1753.
Balqa, Kufr Houd, from Salt toward Aira (MSB-K 176378); Tal Al-Rumman, *Al-Eisawi* (NHJ 2006.0010).

Gymnarrhena micrantha Desf., Mém. Mus. Par. iv., 2. t. 1, 1818.
Poore, s.n. (K000315528); Wadi Araba, BOT ERA project (NHJ 2013.0516).

Hedypnois rhagadioloides (L.) F.W.Schmidt, Samml. Phys.-Oekon. Aufsatze i. 279. 1795.
Syn. *Hyoseris rhagadioloides* L. (1753).
Balqa, 4 km northwest of Salt to Zai at Tabat Al Awamila (MSB-K 151636); Tal Al-Rumman, *Taifour & Aldamen* (NHJ 2010.0032).

Helichrysum sanguineum (L.) Kostel., Allg. Med.-Pharm. Fl. 2. 709. 1833.
Syn. *Gnaphalium sanguineum* L. (1755).
Ajloun, pine wood south of Ajloun, *Hunting Aero Survey* 105A (E00538055); Irbid, Brgish, 13/04/1981 (YAR-H).

Helminthotheca echioides (L.) Holub, Folia Geobot. Phytotax. 8(2): 176. 1973.
Jarash, Jobbah, 23/05/1983 (YAR-H).

Hyoseris scabra L., Sp. Pl. 2: 809. 1753.
Tiberias, 28 Feb. 1912, *Meyers & Dinsmore* (E).

Ifloga spicata (Forssk.) Sch.Bip., Phyt. Canar. ii. 310. 1845.
Syn. *Chrysocoma spicata* Forssk. (1775).

Aqaba, 6 km north Aqaba, *Boulos, Al-Eisawi & Jallad* 6092 (E00541176); Tafila, Dana Reserve, Barra, BOT ERA project (NHJ 2013.0480).

Ifloga spicata (Forssk.) Sch.Bip. subsp. **albescens** Chrtek, Preslia 41: 243. 1969.
Syn. *Ifloga rueppellii* (Fresen.) Danin (1973).
Amman, Dead Sea area (MSB-K 275156); Tal Al-Rumman, *Taifour & Aldamen* (NHJ 2009.0067).

Inula viscosa (L.) Aiton, Hortus Kew. (W. Aiton) 3: 223. 1789.
Irbid, Ain Alturab, 17/11/1983 (YAR-H).

Iphiona maris-mortui Feinbrun, Israel J. Bot. 15: 22. 1966.
Balqa, Wadi Al Zara next to Dead Sea (MSB-K 176644).

Iphiona mucronata (Forssk.) Asch. & Schweinf., Asch. & Schweinf. in Mém. Inst. Égypt. 2. 86. 1887.
Syn. *Chrysocoma mucronata* Forssk. (1775).
Al Aqaba, 3–4 km west Rum Rest House, *Boulos, Jallad, Lahham & Abu Hmaidan* 7746 (E00587902); Dead sea, *Taifour, Al Khawaldah & Al Mogherbi* (NHJ 2013.0628).

Iphiona scabra DC. ex Decne., Ann. Sci. Nat., Bot. sér. 2, 2: 263. 1834.
Ma'an, 15 km S von Aqaba, *Frey & Kürschner* VO 5174 (E00587959).

Jurinea staehelinae (DC.) Boiss., Diagn. Pl. Orient. ser. 1, 6: 111. 1846.
Tafila, Fugaig Station, site no 30 (MSB-K 497871).

Klasea pusilla (Labill.) Greuter & Wagenitz, Willdenowia 33(1): 59. 2003.
Syn. *Cynara pusilla* Labill. (1809); *Rhaponticum pusillum* Boiss. (1875); *Serratula pusilla* (Labill.) Dittrich (1981).
Ma'an, Shoubak, NCARTT regional centre of Shoubak (MSB-K 341251); Tal Al-Rumman RBG, Sector No. 23, *El-Oqlah & Taifour* (NHJ 2013.0169).

Koelpinia linearis Pall., Reise iii. 755. 1776.
Karak, Midway between Qatrana and Jurf ed Darawish Station, *Boulos & Jallad* 7843 (E00479794); Irbid, Jordan Univ. of Science & Technology, 28/04/2002 (YAR-H).

Lactuca aculeata Boiss. & Kotschy ex Boiss., Fl. Orient. 3: 809. 1875.
Ref. Baqla, Danin (2004).

Lactuca orientalis (Boiss.) Boiss., Fl. Orient. 3: 819. 1875.
Syn. *Phaenopus orientalis* Boiss. (1893).
Karak, Jebel Ataita, *Frey & Kürschner* VO 5540 (E00030391); Karamah, *Taifour* (NHJ 2013.0364).

Lactuca saligna L., Sp. Pl. 2: 796. 1753.
Amman, Site 29, opposite University of Jordan (MSB-K 395807).

Lactuca serriola L., Cent. Pl. 2. 29. 1756.
Syn. *Lactuca scariola* L. (1763).
Ajloun, from Anjara village toward Ajloun centre (MSB-K 244330); Tal Al-Rumman, *Al- Eisawi, Abu-Sondos, Abu Anza & Shishani* (NHJ 2006.0098).

Lactuca tuberosa Jacq., Hort. Vindob. i. 18. 1770.
Irbid, Jordan Univ. of Science & Technology, 28/04/1999; Tal Al-Rumman, *Taifour* (NHJ 2013.0351).

Lactuca undulata Ledeb., Fl. Altaic. [Ledebour]. 4: 156. 1833.
Ayn Musa to Petra, 17.4.1929 (30444-HUJ); Ma'an, Fujaij range land station, *Abulaila, Tehabshem & van Slageren* (NCARE).

Lactuca viminea (L.) J. & C.Presl, Fl. Cech. 160. 1819.
Prenanthes viminea L. (1753).
Ref. Al-Eisawi 51 (2013).

Lapsana communis L. subsp. **pisidica** (Boiss. & Heldr.) Rech.f., Rech. f. in Akad. Wiss. Wien. Math.-Naturwiss. Kl., Denkschr. 105(1). 674. 1944.
Syn. *Lapsana pisidica* Boiss. & Heldr. (1849).
Ain Fit, *Dinsmore* (K).

Lasiopogon muscoides (Desf.) DC., Prodr. [A. P. de Candolle] 6: 246. 1838.
Syn. *Gnaphalium muscoides* Desf. (1799).
Jebel [Aniyi] (Aqaba) on basalt, *Davis* 9207 (E00541787); Tal Al-Rumman RBG, Sector No. 37, *El-Oqlah & Al Mfadi* (NHJ 2013.0351).

Lasiospermum brachyglossum DC., Prodr. [A. P. de Candolle] 6: 38. 1838.
Wadi Rum, d. 1996, *Al-Eisawi, von Lowenstern & Garbari* (AMM).

Launaea angustifolia (Desf.) O.Kuntze, Revis. Gen. Pl. 1: 351. 1898.
Syn. *Sonchus angustifolia* Desf. (1799).
Karak, Wadi Araba, *Frey & Kürschner* (BSB).

Launaea capitata (Spreng.) Dandy, F. W. Andrews, Fl. Pl. Sudan 3. 40. 1956.
Syn. *Sonchus capitatus* Spreng. (1826).
Nr. Aqaba, *Boulos & Lahham* 7409 (MM, BM, E).

Launaea fragilis (Asso) Pau, Bol. Soc. Aragonesa Ci. Nat. 16: 68. 1917.
Ref. Wadi Araba, Danin 2005; RSCN database.

Launaea massauensis (Fresen.) Sch.Bip. ex Kuntze, Revis. Gen. Pl. 1: 351. 1898.
Syn. *Heterachaena massauensis* Fresen. (1839).
Ref. RSCN database; Al-Eisawi 52 (2013).

Launaea mucronata (Forssk.) Muschl., Man. Fl. Egypt ii. 1057. 1912.
Syn. *Leontodon mucronatus* Forssk. (1775).
Wadi Rum, *Davis* 9299 (E00026199); 15 km east Ma'an, along the road to El-Jafr, *Boulos & Jallad* (AMM, E); Wadi Al Yabis (YAR-H).

Launaea mucronata Muschl. subsp. **cassiniana** (Jaub. & Spach) N.Kilian, Willdenowia 25(1): 277. 1995.
Syn. *Sonchus cassinianus* Jaub. & Spach (1848); *Launaea cassiniana* Muschl. (1912).
Ref. Al-Eisawi 52 (2013) as *Launaea cassiniana* Muschl.

Launaea nudicaulis (L.) Hook.f., Fl. Brit. India [J. D. Hooker] 3: 416. 1881.
Syn. *Chondrilla nudicaulis* L. (1771).
Jordan, *Kürschner* 5130 (E00026332); Azraq, Wadi Bottom (South west the Castle Amra), *El-Oqlah* (NHJ 2013.0571).

Launaea procumbens (Roxb.) Rajagopal & Ramayya, Kew Bull. 23. 465. 1969.
Syn. *Prenanthes procumbens* Roxb. (1832).
Ref. RSCN database; Al-Eisawi 52 (2013).

Launaea spinosa (Forssk.) Kuntze, Revis. Gen. Pl. 1: 350. 1898.
Aqaba, Wadi Yutum, upper stream, *Jallad, Lahham & Abu Hmaidan* 7498 (E00484133); Aqaba, Wadi Yatum (MSB-K 558109).

Leontodon laciniatus (Bertol.) Widder ex Bornm., Phyton (Horn) 12: 207. 1967.
Wadi Themed to Ziga (Moab), *Davis* 9651 (E00491533); Northeastern Badia, *AlBeshriya, Al Shudiefat & Khalidi* (NHJ 2013.0009).

Leontodon tuberosus L., Sp. Pl. 2: 799. 1753.
Jarash, main highway from Amman to Jarash, beside Jarash nurseries (left side) beyond Bardran farm (MSB-K 227764).

Leuzea pusilla Spreng., Syst. Veg. (ed. 16) [Sprengel] 3: 382. 1826.
Zarqa, Alouk, before Alouk village, 1 km after Mabis (MSB-K 176415); Tal Al-Rumman RBG, Sector No. 15 *Al Mfadi & Al Khawaldah* (NHJ 2013.0032).

Leysera leyseroides (Desf.) Maire, Bull. Soc. Hist. Nat. Afrique N., 20. 18. 1929.
Syn. *Asteropterus leyseroides* (Desf.) Rothm. (1944).
Desertum Syriacum, inter Amman et Rutba, *Rechinger* 12920 (E00542942).

Limbarda crithmoides (L.) Dumort, Fl. Belg. (Dumortier) 68. 1827.
Syn. *Inula crithmoides* L. (1753).
Amman, Azraq, *Frey & Kürschner* VO 5688 (E00589166).

Logfia davisii Holub ex Grierson, Notes Roy. Bot. Gard. Edinburgh 33(3): 425. 1975.
Above Ain Fit, *Meyers & Dinsmore* 1784 (E00538551).

Logfia gallica (L.) Coss. & Germ., Ann. Sci. Nat., Bot. sér. 2, 20: 291, t. 13, f. A. 1843.
Ref. RSCN database.

Matricaria aurea (Loefl.) Sch.Bip., Bonplandia 8: 369. 1860.
Petra, *Davis* 8965 (E00580846); Tal Al-Rumman, *Taifour & Aldamen* (NHJ 2009.0180).

Micropus supinus L., Sp. Pl. 2: 927. 1753.
Amman, RSCN database; Al-Eisawi 53 (2013).

Notobasis syriaca (L.) Cass., Dict. Sci. Nat., ed. 2. [F. Cuvier] 25: 225 1822; 35: 170. 1825.
Syn. *Carduus syriacus* L. (1753).
Madaba, Mount Nebo to Suweimeh (MSB-K 114079); Tal Al-Rumman, *Taifour et al.* (NHJ 2013.0465).

Onopordum alexandrinum Boiss., Diagn. Pl. Orient. ser. 1, 10: 93. 1849.
6 km E of Um Qeis, 13 May 1976, *Tackholm et al.* (K).

Onopordum ambiguum Fresen., Mus. Senckenberg. 1. 85. 1833.
Madaba, Mae'n (MSB-K 152013).

Onopordum carduiforme Boiss., Diagn. Pl. Orient. ser. 1, 10: 92. 1849.
Ref. Ajloun (RSCN database); Al-Eisawi 53 (2013).

Onopordum cynarocephalum Boiss. & Blanche, Diagn. Pl. Orient. ser. 2, 3: 48. 1856.
Balqa, Kufer Houda, Al Anda road near Alsafeh cross (MSB-K 176725).

Onopordum heteracanthum C.A.Mey., Verz. Pfl. Casp. Meer.: 68. 1831.
Syn. *Onopordum canum* Eig (1942); *Onopordum anisacanthum* Boiss. (1849).
Azraq, Wadi Bottom (South west the Castle Amra), *El-Oqlah* (NHJ 2013.0560).

Onopordum jordanicola Eig, Palestine J. Bot., Jerusalem Ser. 2: 196. 1942.
Along Zarqa River, 2004, *Al-Dwiri* s.n. (AMM); Maan, Rum-Jabal Umm Dami, 1998 (HUJ).

Onopordum macrocephalum Eig, Palestine J. Bot., Jerusalem Ser. 2: 194. 1942.
Syn. *Onopordum transjordanicum* Eig (1942).
Tafila, outside Qadisiya on Tafila-Sharbah road (MSB-K 151887.); Al Mafraq, Balama, 23/05/1983 (YAR-H).

Onopordum palaestinum Eig, Palestine J. Bot., Jerusalem Ser. 2: 191. 1942.
Wadi Mahib, *Dinsmore* 11541 (E00462003).

Pallenis hierochuntica (Michon) Greuter, Fl. Medit. 7: 47. 1997.
Syn. *Saulcya hierochuntica* Michon (1854); *Asteriscus pygmaeus* (DC.) Coss. & Durieu (1857); *Asteriscus hierochunticus* (Michon) Wiklund (1985).
Amman, from Nao'ur toward Zara, opposite Dead Sea first wadi (MSB-K 413705); Rum, *Taifour* (NHJ 2013.0394).

Pallenis spinosa (L.) Cass., Dict. Sci. Nat., ed. 2. [F. Cuvier] 37: 276. 1825.
Syn. *Buphthalmum spinosum* L. (1753).
Ajloun, South of Ajloun, *Hunting Aero Survey*106B (E00589573); Tal Al-Rumman, *Taifour & Aldamen* (NHJ 2009.0006).

Parthenium hysterophorus L., Sp. Pl. 2: 988. 1753.
(YAR-H). Introduced.

Phagnalon barbeyanum Asch. & Schweinf., Mém. Inst. Égypte 2: 87. 1887.
Jarash, Souf village, Al Taiat farm (MSB-K 200062).

Phagnalon nitidum Fresen., Mus. Senckenberg. 1. 81. t. 4. f. 2. 1833.
Schellah, *Barsotti* (AMM); Jebel Umm Adami, 55 km ESE of Aqaba, *Danin* (HUJ).

Phagnalon rupestre (L.) DC., Prodr. [A. P. de Candolle] 5: 396. 1836.
Syn. *Conyza rupestris* L. (1767).
Madaba (MSB-K 114013); Tal Al-Rumman *Al-Eisawi* (NHJ 2006.0017).

Picnomon acarna (L.) Cass., Dict. Sci. Nat., ed. 2. [F. Cuvier] 40: 188. 1826.
Syn. *Carduus acarna* L. (1753).
Jarash, Wadi kufour khal, 26/07/1983 (YAR-H); Tal Al-Rumman *Al- Eisawi, Abu-Sondos, Abu Anza & Shishani* (NHJ 2006.0107).

Picris altissima Delile, Fl. Egypte 260. t. 141. 1814.
Ref. RSCN database.

Picris amalecitana (Boiss.) Eig, Palestine J. Bot., Jerusalem Ser. 1: 70. 1938.
Irbid, Ishtafeina to Deir Abu Said (MSB-K 114312).

Picris asplenioides L., Sp. Pl. 2: 793. 1753.
Wadi Rum, *Boulos & Lahham* 6098 (AMM, BM, E).

Picris babylonica Hand.-Mazz., Ann. Nat. Hofmus. Wien 27: 453. 1913.
Wadi Rum, *Boulos & Lahham* 7779 (AMM, BM, E).

Picris cyanocarpa Boiss., Diagn. Pl. Orient. ser. 1, 11: 37. 1849.
Tal Al-Rumman, *Al-Eisawi* (NHJ 2006.0323).

Picris galilaea (Boiss.) Eig, Pal. J. Bot. 1(1): 69. 1938.
Syn. *Hagioseris galilaea* Boiss. (1849).
Ajloun, Umelyanabia, 26/03/1982 YAR-H); Tal Al-Rumman, *Al-Eisawi* (NHJ 2006.0200).

Picris longirostris Sch.Bip., Mus. Senckenberg. 3. 60. 1839.
Syn. *Picris damascena* Boiss. & Gaill. ex Boiss. (1875); *Picris blancheana* Boiss. (1875); *Picris desertorum* Nábělek (1925); *Picris damascena* Boiss. & Gaill. subsp. *blancheana* (Boiss.) Eig (1939); *Picris damascena* Boiss. & Gaill. subsp. *desertorum* (Nábělek) Agnew (1961).
Tal Al-Rumman, *Al-Eisawi* (NHJ 2006.0435); Aqaba, Wadi Rum, *Davis* 9086 (E00494561).

Picris strigosa M.Bieb., Fl. Taur.-Caucas. 2: 250. 1808.
Syn. *Picris glaucescens* DC. (1838); *Picris glomerata* C.Koch (1851); *Picris persica* Gaud. (1918); *Picris subinflata* (Bornm. & Gauba) Rech.f. (1950).
Ref. RSCN database; Al-Eisawi 54 (2013).

Pluchea dioscoridis (L.) DC., Prodr. 5: 450. 1836.
Salt, Wadi Shuaib, Jordan valley entrance (MSB-K 276795); North Ghour Shouneh-Irbid (cross), *Taifour, Al Khawaldah & Al Shwaer* (NHJ 2013.0675).

Podospermum canum C.A.Mey., Verz. Pfl. Casp. Meer. (C.A. von Meyer). 62. 183. 1831.
Syn. *Scorzonera cana* (C.A.Mey.) Grossh. (1927).
Amman, opposite Jordan University toward Tlaa', *Ali & Khalda* (MSB-K 228015).

Pulicaria arabica (L.) Cass., Dict. Sci. Nat., ed. 2. [F. Cuvier] 44: 94. 1826.
Syn. *Inula arabica* L. (1767).
Zarqa, Azraq (YAR-H); Wadi Zarqa, river bank, *Simpson* (K).

Pulicaria dysenterica (L.) Bernh., Syst. Verz. Pfl. 153. 1800.
Syn. *Inula dysenterica* L. (1753).
Irbid, Tal Al-Nafaq (IBBU-H); Upper and Lower Jordan Valley (RSCN database).

Pulicaria incisa (Lam.) DC., Prodr. [A. P. de Candolle] 5: 479. 1836.
Syn. *Inula incisa* Lam. (1789).
Aqaba, 6 km north Aqaba, *Boulos, Al-Eisawi & Jallad* 6101 (E00587210).

Pulicaria inuloides (Poir.) DC., Prodr. [A. P. de Candolle] 5: 480. 1836.
Syn. *Erigeron inuloides* Poir. (1817).
Ma'an, Wadi Musa, Petra, *Frey & Kürschner* (E00587215).

Pulicaria undulata (L.) C.A.Mey. subsp. **undulata**, Verz. Pfl. Casp. Meer. (C.A. von Meyer) 79. 1831.
Syn. *Pulicaria crispa* (Forssk.) Oliv. (1873); *Inula undulata* L. (1767); *Aster crispus* Forssk. (1775); *Francoeuria crispa* (Forssk.) Cass. (1825); *Francoeuria undulata* (L.) Lack (1980).
Aqaba, Wadi Rum, 5 km after main junction of desert highway (MSB-K 497837); Azraq, Azraq oasis, 17/05/1983 (YAR-H).

Reichardia intermedia (Sch.Bip.) Samp., Bol. Soc. Brot. 24. 1909.
Syn. *Picridium intermedium* Sch.Bip. (1850).
Dana (RSCN database).

Reichardia tingitana (L.) Roth, Bot. Abh. 35. 1787.
Syn. *Scorzonera orientalis* L. (1759).
Najdeh, 1 km west the village, *El Oqlah, Taifour & Bani Said* (NHJ 2013.0122).

Rhagadiolus edulis Gaertn., Fruct. Sem. Pl. ii. 354. 1791.
Ajloun, Al Sakheneh, 15/05/1983 (YAR-H).

Rhagadiolus stellatus (L.) Gaertn., Fruct. Sem. Pl. ii. 354. 1791.
Syn. *Lapsana stellata* L. (1753).
Ajloun, 2 km north of Zageb towards Ainjara on Jerash-Ajloun road (MSB-K 151681); Tal Al-Rumman, *Taifour* (NHJ 2013.0198).

Scolymus hispanicus L., Sp. Pl. 2: 813. 1753.
Between Wadi Musa village & the Siq, *Davis* 9487 (E00487804); Amman, from Amman toward Dead Sea (MSB-K 244307).

Scolymus maculatus L., Sp. Pl. 2: 813. 1753.
Irbid, Jordan Univ. of Science & Technology, 08/05/2002 (YAR-H).

Scorzonera lanata Schrank., Baier, Fl. 2: 306 (1789).
Ras Naqab, 1985 (K).

Scorzonera mollis M.Bieb., Fl. Taur.-Caucas. 3: 522. 1819.
Wadi Rum, *Boulos & Lahham* 7683 (AMM, BM, E).

Scorzonera papposa DC., Prodr. [A. P. de Candolle] 7(1): 119. 1838.
Syn. *Scorzonera multiscapa* Boiss. (1849); *Scorzonera kurdica* Boiss. & Noë (1856).
Amman, *Dinsmore* 12618 (K000286333); Tal Al-Rumman, *Taifour* (NHJ 2013.0252).

Scorzonera psychrophila Boiss. & Hausskn. ex Boiss. & Hausskn., Fl. Orient. 3: 777. 1875.
Syn. *Scorzonera judaica* Eig (1931) nomen.; *Scorzonera pseudolanata* Grossh. (1934).
Ma'an, Rajif-Ras Al-Naqab Road, 0.5 km before the junction to the Desert Highway (MSB-K 413509); Al Mafraq, Umaljamal, 19/04/1983 (YAR-H).

Scorzonera pusilla Pall., Reise Russ. Reich. ii.: 744, 329. 1773.
Shobek to Gebel Aneizi (Aqaba), *Davis* 9541 (E00476799).

Scorzonera schweinfurthii Boiss., Fl. Orient. Suppl.: 320. 1888.
Syn. *Scorzonera rawii* Rech.f. et Guest (1957).
Above Wadi Musa on the road to Moan, *Davis* 8665 (E00224237).

Scorzonera semicana DC., Prodr. [A. P. de Candolle] 7(1): 119. 1838.
Syn. *Scorzonera syriaca* Boiss. & C.I.Blanche (1856); ?*Scorzonera abdulazizii* Eig.
Al Mafraq, Bwadhah, (IBBU-H).

Scorzonera subintegra (Boiss.) J.Thiébaut, Bull. Soc. Bot. France 95: 18. 1948.
Amman, Site 22, University station at Muwaggar (MSB-K 554710).

Scorzonera tortuosissima Boiss., Fl. Orient. 3: 775. 1875.
Ma'an, 20 km NE Quweirah, *Frey & Kürschner* 80–657 (E00476992).

Senecio aegyptius L., Sp. Pl. 2: 867. 1753.
Ref. RSCN database; Al-Eisawi 56 (2013).

Senecio flavus (Decne.) Sch.Bip., Webb & Berth. Phyt. Canar. 2. 319. t. 107. 1845.
Gillett 15569 (K000852046); Irbid, Deir Abu Said, 09/08/1981 (YAR-H).

Senecio glaucus L. subsp. **coronopifolius** (Maire) C.Alexander, 412. 1979.
Wadi Rum, *Davis* 9095 (E00516181).

Senecio leucanthemifolius Poir. var. **vernalis** (Waldst. & Kit.) Alexander, Notes Roy. Bot. Gard. Edinburgh 37(3): 403. 1979.
Syn. *Senecio vernalis* Waldst. & Kit. (1800).
Tafila, Dana Reserve, Rumman camp site, *Taifour et al.* (NHJ 2013.0548).

Senecio vulgaris L., Sp. Pl. 2: 867. 1753.
Almafraq, Umaljamal, 19/04/1983 (YAR-H).

Siebera nana (DC.) Bornm., Beih. Bot. Centralbl. lx. E. 201. 1939.
Syn. *Siebera pungens* (Lam.) DC. var. *nana* DC. (1838).
Aqaba, Bir ed-Dabbaghat, 13 km SSW of Shoubak, *Baierle* (E00460917).

Silybum marianum (L.) Gaertn., Fruct. Sem. Pl. 2(3): 378, t. 168. 1791.
Syn. *Carduus marianus* L. (1753).
Irbid, outskirt of Irbid (MSB-K 114264).

Sonchus asper(L.) Hill. subsp. **glausescens** (Jordan) Ball ex Ball, J. Linn. Soc. Bot. 16: 548. 1878.
Tal Al-Rumman RBG, Sector No. 16, *El-Oqlah & Al Mfadi* (NHJ 2013.0376).

Sonchus maritimus L., Syst. Nat., ed. 10, 2: 1192. 1759.
Ma'an, Wadi Musa, Ain Musa (MSB-K 418423).

Sonchus oleraceus L., Sp. Pl. 2: 794. 1753.
Syn. *Sonchus laevis* (L.) Gars. (1789).
Ma'an, Ma'an Rest House, *Boulos, Jallad & El-Gohary* 8056 (E00483705); Tal Al-Rumman, *Al Eisawi, Abu Sondos, Abu Anza & Shishani* (NHJ 2006.0073).

Sonchus suberosus Zohary & P.H.Davis, Kew Bull. 1947: 87. 1947.
Wadi Heidan (Moab), *Davis* 9450 (E00483732).

Sonchus tenerrimus L., Sp. Pl. 2: 794. 1753.
Hadi-Nadid, *Dinsmore* 12535 (E00483743); Tal Al-Rumman, *Al-Eisawi* (NHJ 2006.0382).

Steptorhamphus tuberosus (Jacq.) Grossh., Fl. Kavkaza [Grossheim] 4: 258. 1934.
Syn. *Lactuca tuberosa* Jacq. (1770).
Jarash, Dibbeen, 1 km after tourist site, right turn to an oak forest (MSB-K 276751).

Stizolophus coronopifolius (Lam.) Cass., Dict. Sci. Nat., ed.2, 51: 51. 1827.
Ras Naqab, 1978 (K).

Symphyotrichum subulatum (Michx.) G.L.Nesom, Phytologia 77(3): 293. 1995.
Syn. *Aster subulatus* Michx. (1803).
Karak, at entrance, 2 km from Site 45 (MSB-K 395601).

Tagetes minuta L., Sp. Pl. 887. 1753.
Ma'an, Petra, up the cliff at the court site, Site no 8 (MSB-K 497963).

Tanacetum sinaicum Delile ex Bremer & Humphries, Bull. Nat. Hist. Mus. Lond. (Bot.) 23 (2): 103. 1993.
Syn. *Tanacetum santolinoides* (DC.) Feinbrun & Fertig (1976).
Ref. Al-Eisawi 57 (2013) as *Tanacetum santolinoides.*

Taraxacum cyprium H.Lindb., Acta Soc. Sci. Fenn., Ser. B, Opera Biol. 2(7): 36. 1946.
Al Mafraq (YAR-H).

Tolpis virgata (Desf.) Bertol., Mem. Soc. Emul. Genova 2: 135. 1803.
Syn. *Crepis virgata* Desf. (1792).
Amman, *Davis* (E00503270); Irbid, Jordan Univ. Science & Technology, 08/05/2002 (YAR-H).

Tragopogon buphthalmoides (DC.) Boiss., Fl. Orient. 3: 750. 1875.
Syn. *Scorzonera buphthalmoides* DC. (1838).
Amman, Site 34, (MSB-K 557995); Tal Al-Rumman, *RBG-Ire & RBG-Jo* (NHJ 2008.0094).

Tragopogon coelesyriacus Boiss., Diagn. Pl. Orient. ser. 1, 11: 47. 1849.
Tal Al-Rumman, *Taifour & Aldamen* (NHJ 2009.0195).

Tragopogon collinus DC., Prodr. [A. P. de Candolle] 7(1): 115. 1838.
Tafila, Dana, jst before the station, gate on right (MSB-K 416223); Jarash, Thagret Asfoor, 21/04/1999 (YAR-H).

Tragopogon porrifolius L. subsp. **longirostris** (Sch.Bip.) Greuter, Willdenowia 33(2): 237. 2003.
Syn. *Tragopogon coelesyriacus* Boiss. (1849); *Tragopogon longirostris* Bisch. ex Sch.Bip. (1850).
Jarash, 5 km after site 9 on the way to Khanasry, crossroad to Medwar (MSB-K 227834).

Tripleurospermum auriculatum (Boiss.) Rech.f., Fl. Lowland Iraq: 629. 1964.
Shobek to Gebel Aneizi (Aqaba), *Davis* 9551 (E00584768); Dead sea area, 22/05/1952 (YAR-H).

Tripleurospermum oreades (Boiss.) Rech.f., Denkschr. Akad. Wiss. Wien, Math.-Nat. cv. 1. 633. 1943.
Syn. *Chamaemelum auriculatum* Boiss. (1849).
Jordan, 26.III.1895, *Fox* (AUB).

Tripteris vaillantii Decne., Ann. Sci. Nat., Bot. sér. 2, 2: 260. 1834.
Syn. *Osteospermum vaillantii* (Decne.) Norl. (1943.).
Petra in the Sik, *Davis* 8770 (E00503874); Dana Nature Reserve, rocky area behind Rumana campsite, BOT ERA project (NHJ 2013.0550).

Urospermum picroides (L.) Scop. ex F.W.Schmidt, Samml. Phys.-Oekon. Aufsatze i.: 275. 1795.
Syn. *Tragopogon picroides* L. (1753).
Shuna, Al Shoneh al Shamalieh, Wadi Al Arab, Wadi Zeglab (MSB-K 176404); Tal Al-Rumman, *Taifour* (NHJ 2011.0049).

Varthemia iphionoides Boiss. & C.I.Blanche, Diagn. Pl. Orient. ser. 2, 3: 9. 1856.

Amman, Khalda, from the Al Waha Circle toward Sweleh (MSB-K 211819); Tal Al-Rumman, *RBG-Ire & RBG-Jo* (NHJ 2008.0100).

Verbesina encelioides (Cav.) Benth. & Hook.f. ex A.Gray, Geological Survey of California. Botany 1. 1876.
Syn. *Ximenesia encelioides* Cav. (1793).
Aqaba, Wadi Rum, 5 km after main junction of desert highway towards Rum, site no 34 (MSB-K 497789).

Veronica biloba Schreb. ex L., Mant. Pl. 2: 172. 1771.
Petra, Falahat (1999).

Volutaria crupinoides (Desf.) Maire, Cat. Pl. Maroc 3: 818. 1934.
Syn. *Centaurea crupinoides* Desf. (1799); *Amberboa crupinoides* DC. (1838).
Al Mafraq, 19/04/1983 (YAR-H).

Volutaria lippii (L.) Cass. ex Maire, Cat. Pl. Maroc 3: 817. 1934.
Syn. *Centaurea lippii* L. (1753); *Amberboa lippii* DC. (1838).
Ref. RSCN database.

Xanthium italicum Moretti, Giorn. Fis. Chim. Storia Nat. Med. Arti Dec. 2, 5: 326. 1822.
Lower Jordan Valley (YAR-H).

Xanthium spinosum L., Sp. Pl. 2: 987. 1753.
Amman, Shafa Badran, Tab Kraq village, main road from Tab Kraq toward Um Rumman (MSB-K 200028).

Xanthium strumarium L., Sp. Pl. 2: 987. 1753.
Along Zarqa River, 2004, *Al-Dwiri* s.n. (AMM).

Xeranthemum annuum
S Jordan (pers. comm. Fragman-Sapir, 2016).

Xeranthemum cylindricum Spreng., Syst. Veg. (ed. 16) [Sprengel] 3: 485. 1826.
Ref. Al-Eisawi 58 (2013).

Zoegea purpurea Fresen., Mus. Senckenberg. 1. 86. t. 5, 1833.
Syn. *Zoegea aristata* DC. (1838).
7.5 km N of the Safawi-Ruwaishid rd, 1.5 km from Safawi, 24 Apr. 1992, *Cope & Eisawi* (AMM, K).

Balanophoraceae

Cynomorium coccineum L., Sp. Pl. 2: 970. 1753.
Syn. *Cynomorium songaricum* Rupr. (1869).
Karak, Wadi Araba, 5 km SW of Ain Al-Fidan, *Frey & Kürschner* (BSB); Lower Jordan valley, RSCN database.

Berberidaceae

Bongardia chrysogonum (L.) Spach, Hist. Nat. Vég. 8: 65. 1839.
Zarqa, 10–15 km NW Zarqa, *Boulos & Al-Eisawi* 6425 (E00440223); Tal Al-Rumman, *Taifour & Aldamen* (NHJ 2010.0018)

Leontice leontopetalum L., Sp. Pl.: 312. 1753.
Jordan, *Lamond* 43 (E00153849); Najdeh 1 km west the village, *El-Oqlah et al.* (NHJ 2013.0120).

Biebersteiniaceae

Biebersteinia multifida DC., Prodr. 1: 708. 1824.
1–4 km S of Ras Naqb, *Boulos & Jallad* 7377 (E00100286).

Boraginaceae

Alkanna galilaea Boiss., Diagn. Pl. Orient. ser. 1, 11: 118. 1849.
Tal Al-Rumman RBG, Sector No. 39, *Taifour & Al Khawaldah* (NHJ 2013.0291).

Alkanna orientalis (L.) Boiss., Diagn. Pl. Orient. ser. 1, 4: 46. 1844.
1–4 km South Ras Naqb, *Boulos & Jallad* 7362 (E00669141).

Alkanna strigosa Boiss. & Hohen., Diagn. Pl. Orient. ser. 1, 4: 46. 1844.
Amman, *Hepper* 10013 (K000286093); Tal Al-Rumman, *Taifour* (NHJ 2009.0113).

Alkanna tinctoria Tausch, Flora 7: 234. 1824.
Petra, *Davies* 8647 (K000286097); Tal Al-Rumman, *Shishani* (NHJ 2011.0125).

Anchusa aegyptiaca (L.) A.DC., Prodr. 10: 48. 1846.
Petra, *Gillett* 15981 (K000286102); Tal Al-Rumman, *El-Oqla* (NHJ 2012.0198).

Anchusa azurea Mill., Gard. Dict. ed. 8, 9. 1768.
Syn. *Anchusa italica* Retz (1779).
Tal Al-Rumman RBG, Sector No. 23, *Al Mfadi &Al Khawaldah* (NHJ 2013.0071).

Anchusa humilis I.M.Johnst., Contr. Gray Herb. 73: 55. 1924.
Syn. *Hormuzakia aggregata* Guşul. (1923).
Ref. Al-Eisawi 61 (2013), as *Hormuzakia aggregata*.

Anchusa milleri Willd., Bot. Gart. Halle Nachtr. 1: 11. 1801.
Zarqa, *Gillett* (15621 K000286107); Tal Al-Rumman, *El-Oqlah* (NHJ 2012.0433).

Anchusa strigosa Banks & Sol., Nat. Hist. Aleppo ed. 2, 2: 246. 1794.
Jubeika, *Gillett* 16124 (K000286119); Tafila district, Dana Reserve, Ruman Camp, *Borosova, Ghazanfar & Osborne* (NHJ 2013.0511).

Anchusa undulata L., Sp. Pl.: 133. 1753.
Ain Hesban [Hisban], *Meyers & Dinsmore* (82 K000286127); Tal Al-Rumman, *Taifour & Aldamen* (NHJ 2010.0041).

Anchusa azurea Mill., Gard. Dict. ed. 8, 9. 1768.
Syn. Anchuisa italica Retz. (1779).
Along Zarqa River, 2004, *Al-Dwiri* s.n. (AMM).

Arnebia decumbens (Vent.) Coss. & Kiral., Bull. Soc. Bot. France 4: 398. 1857.
(Incl. *Arnebia decumbens* var. *macrocalyx*)
Jordan, Wadi Sir, *Poore* 1113 (K000286136).

Arnebia hispidissima (Lehm.) A.DC., Prodr. 10: 94. 1846.
Tal Al-Rumman RBG, Sector No. 95, *El-Oqlah, Taifour & Al Khawaldah* (NHJ 2013.0263).

Arnebia linearifolia A.DC., Prodr. 10: 95. 1846.
subsp. **linearifolia**
Jordan, Ain Hesban [Hisban], *Townsend* 152 (K000286144).

Arnebia tinctoria Forssk., Fl. Aegypt.-Arab.: 63. 1775.
Syn. *Arnebia tetrastigma* Forssk. (1775).
Borqu castle (IBU-H); Karak, Wadi Araba, *Frey & Kürschner* (BSB).

Asperugo procumbens L., Sp. Pl.: 138. 1753.
Wadi Aseikhim, *Townsend* 414 (K000286146); Jaresh, Najdeh, 1 km west of the village. *Taifour* (NHJ 2013.0133).

Brunnera orientalis (Schenk) I.M.Johnst., Contr. Gray Herb. 73: 55. 1924.
Syn. *Myosotis orientalis* Schenk (1840).
Tafela, Lahda, 04/05/1984 (YAR-H); Tafila, Dana Reserve (MSB-K 416186).

Buglossoides arvensis (L.) I.M.Johnst., J. Arnold Arbor. 35: 42. 1954.
Syn. *Lithospermum arvensis* L. (1753).
Ajluon, Rasoun, 04/12/1982; Irbid, Bani Kanana, *El-Oqlah & Bani Said* (NHJ 2013.0136).

Buglossoides incrassata (Guss.) I.M.Johnst., J. Arnold Arbor. 35: 43. 1954.
Syn. *Lithospermum incrassatum* Guss. (1826).
Dibeen Forest Reserve, 2004, *Mahklouf* (AMM).

Buglossoides tenuiflora (L.f.) I.M.Johnst., J. Arnold Arbor. 35: 42. 1954.
Syn. *Lithospermum tenuiflorum* L.f. (1782).
Zarqa, Site 4. Alouk, (MSB-K 552624); Azraq Wadi Bottom (South west the Castle Amra), *El-Oqlah* (NHJ 2013.0578).

Cerinthe palaestina Eig. & Sam., Bot. Not. 1943.: 353. 1943.
Amman, *Al-Eisawi* 7825 (K000286155).

Cordia sinensis Lam., Tabl. Encycl. 1: 423 1792.
Balqa, Dead Sea, 1941 (HUJ); Irbid (RSCN database).

Cynoglossum creticum Mill., Gard. Diet. ed. 8, no. 3, 1768.
Irbid, Malaka, 13/5/1983 (YAR-H).

Cynoglossum foliosum (Paine) Greuter & Burdet, Willdenowia 11: 34. 1981.
Syn. *Cerinthopsis foliosa* Paine (1875); *Trachelanthus foliosus* (Paine) Tristram (1884).
Ajloun (RSCN database).

Cynoglossum montanum L., Demonstr. Pl. 51. 753.
Ref. RSCN database.

Echiochilon fruticosum Desf., Fl. Atlant. 1: 166. 1798.
Jordan, *Frey & Kürschner* VO 5102 (E00115558).

Echiochilon kotchyi (Boiss. & Hohen.) I.M.Johnst., J. Arnold Arbor. 38: 283. 1957.
Kotschy (15 K001067005).

Echiochilon persicum (Burm.f.) I.M.Johnst., J. Arnold Arbor. 38: 288. 1957.
Kotschy (15 K001067002).

Echium angustifolium Mill., Gard. Dict. ed. 8, 6. 1768.
Jarash, Ain Jamlah, 04/10/1983 (YAR-H).

Echium glomeratum Poir., Encycl. 8: 670. 1808.
Ajloun, 15 km north of Ajloun to Irbid (MSB-K 165602); Jarash, King Talal dam, 04/10/1983 (YAR-H).

Echium italicum L., Sp. Pl.: 139. 1753.
Jarash, Tadros, 30.viii.1972 (AUB).

Echium judaeum Lacaita, Linn. Soc., Bot. 44: 366 1919.
Wadi Zarqa, *Simpson* (53308 K000286167); Tal Al-Rumman, *Taifour & Aldamen* (NHJ 2009.0012).

Echium longifolium Delile, Descr. Egypte, Hist. Nat. 184. 1813.
N of Aqaba, *Boulos & Lahham* 6095 (AMM, BM, E).

Echium plantagineum L., Mant. Pl. 2: 202. 1771.
Irbid, Abu-logos, 05/04/1983 (YAR-H).
Ref. Al-Eisawi 60 (2013) as *Echium lycopsis* L.

Echium rauwolfii Delile, Descr. Egypte, Hist. Nat. 195. 1813.
Tal Al-Rumman, *Al-Eisawi* (NHJ 2006.0330).

Gastrocotyle hispida (Forssk.) C.B.Clarke, Ann. Sci. Nat., Bot. 3, 12: 363. 1849.
Syn. *Anchusa hispida* Forssk. (1775).
Ma'an, Site 16. Fujaij (MSB-K 554422); N of Aqaba, *Boulos et al.* 6158 (AMM, BM, E).

Heliotropium arbainense Fresen., Mus. Senckenberg. 1: 168. 1833.
Aqaba, 8 km south of Aqaba, *Boulos & Jallad* 7419 (E00432854).

Heliotropium bacciferum Forssk., Fl. Aegypt.-Arab.: 38. 1775.
Ma'an: 15 km S of Aqaba, *Frey & Kürschner* 80–570 (E00558224, TUB).

Heliotropium bovei Boiss., Diagn. Pl. Orient. 11: 87. 1849.
Tal Al-Rumman, *Taifour & Aldamen* (NHJ 2009.0111).

Heliotropium digynum (Forssk.) Asch. ex C.Chr., Dansk Bot. Ark. 4(3): 14. 1922.
Camp on the way to Kilwa, *Hunting Aero Survey* 88A (E00432972).

Heliotropium europaeum L., Sp. Pl.: 130. 1753.
Amman, Aug–45, *Davis* (E00558038); Tal Al-Rumman, *Taifour* (NHJ 2013.0292).

Heliotropium hirsutissimum Grauer, Pl. Min. Cogn. Dec. 1, 1784.
Tal Al-Rumman, *Shishani* (NHJ 2011.0118).

Heliotropium lasiocarpum Fisch. & C.A.Mey., Ind. Sem. Horti Petrop. 4: 38. 1838.
Syn. *Heliotropium europaeum* L. var. *tenuifolium* Boiss. (1875); *Heliotropium eichwaldii* Steud. var. *lasiocarpum* (Fisch. & Mey.) C.B.Clarke (1840); *Heliotropium ellipticum* Ledeb. var. *lasiocarpum* (Fisch. & Mey.) M.Popov (1931).
Amman, 25 km S of Amman, *Danin* (HUJ).

Heliotropium maris-mortui Zohary, Palestine J. Bot. Ser. 2: 177. 1941.
Wadi Reseifeh, Aug. 1945, *Davis* s.n. (E00558165).

Heliotropium rotundifolium Sieber ex Lehm., Icon. Descr. Nov. Stirp. 2: t. 19, 1822.
Amman, 5 km W Zarqa, *Frey & Kürschner* 5602 (E00558184, TUB); Tal Al-Rumman, Royal Botanic Garden Team, (NHJ 2008.0009).

Heliotropium suaveolens M.Bieb., Fl. Taur.-Caucas. 3: 116. 1819.
Aqaba, Wadi Araba, on way to Aqaba, just before the Finan Junction (MSB-K 496830).

Heliotropium supinum L., Sp. Pl.: 130. 1753.
Burqu, 30 km North West of H4, *Woods* 9 (E00558272).

Heterocaryum subsessile Vatke, Zeitschr. Naturwiss. 11: 129. 1875.
East of Azraq (RSCN database).

Heterocaryum szovitsianum (Fisch. & C.A.Mey.) A.DC., Prodr. 10: 145. 1846.
Tafila Province, 2014, *Oran* (AMM).

Lappula barbata (M.Bieb.) Gurke, Nat. Pflanzenfam. 4(3a): 107 1893.
Tafila Province, 2014, *Oran* (AMM).

Lappula sessiliflora (Boiss.) Gurke, Nat. Pflanzenfam. 4(a): 107. 1894.
Tafila, Fujaij Station outside fence on the main road on right after the second entrance, Site no 21 (MSB-K 497402).

Lappula sinaica (A.DC.) Asch. & Schweinf., Mém. Inst. Égypt. 2: 111. 1887.
Ref. RSCN database; Al-Eisawi 61 (2013).

Lappula spinocarpos (Forssk.) Asch. & Kuntze, Trudy Imp. S.-Peterburgsk. Bot. Sada 10: 215. 1887.
Ma'an, Site 16. Fujaij, just outside the station offices beside the interance gate (MSB-K 554411).

Lobostemon strigosus (Sw.) H.Buek, Linnaea 11: 136. 1837.
Syn. *Echium strigosum* Sw. (1817).
Najdeh, 1 km west the village, *El-Oqlah Taifour & Bani Said* (NHJ 2013.0132).

Microparacaryum intermedium (Fresen.) Hilger & Podl., Pl. Syst. Evol. 148(3–4): 302. 1985.
Syn. *Cynoglossum intermedium* Fresen. (1834).
Ma'an, 5 km NW of Rum railway station, *Frey & Kürschner* (BSB).

Moltkiopsis ciliata (Forssk.) I.M.Johnst., J. Arnold Arbor. 34: 3. 1953.
Tafila, Dhana (YAR-H); Jebel Rahebi, 15 Apr. 1987, *Garbari & Giordani* (AMM).

Myosotis discolor Pers., Syst. Veg. ed. 15, 190. 1798.
Ajloun (RSCN database).

Myosotis ramosissima Rochel, Oestr. Fl. ed. 2, 1: 366. 1814.
Dibeen Forest Reserve, May 2006, *Taifour* (FOR-H).

Myosotis refracta Boiss., Voy. Bot. Espagne 2: 433. 1841.
4 km S of Wadi Sabat, 2 km N of Jabal Umm Adami, 55 km ESE of Aqaba, *Danin* (HUJ).

Myosotis stricta Link ex Roem. & Schult., Syst. Veg. 4: 104. 1819.
Ajloun (RSCN database).

Myosotis uncata Boiss., Diagn. Pl. Orient. ser. 2, 3: 139. 1856.
Dibeen National Park, 2005, *Mahklouf* s.n. (AMM); Eastern slopes of Jordan Valley (NCARE).

Nonea melanocarpa Boiss., Diagn. Pl. Orient., ser. 1, 2(11): 96. 1849.
Jordan, *Davis* 9237 (E00153272).

Nonea obtusifolia (Willd.) DC., Fl. Franç. ed. 3, 3: 626. 1805.
Syn. *Melanortocarya obtusifolia* (Willd.) Selvi, Bigazzi, Hilger & Papini (2006).
Nr the village Makhraba (ca 20 km W of Irbid), *Böcker* RB 994 (B100520883); Madaba, Jebel Siaghah, 12.IV.1877 (AUB).

Nonea philistaea Boiss., Diagn. Pl. Orient. 11: 96. 1849.
Tal Al-Rumman, *Al-Eisawi, Abu Sondos, Abu Anza & Shishani* (NHJ 2006.0343).

Nonea ventricosa Griseb., Spic. Fl. Rumel. 2(4): 98. 1844.
Syn. *Nonea echioides* Roem. & Schult. (1819).
Jarash, Ain al Deek, 04/10/1983 (YAR-H).

Ogastemma pusillum (Coss. & Durieu ex Bonnet & Barratte) Brummitt, Kew Bull. 36: 680 1982.
Syn. *Megastoma pusillum* Coss. & Durieu ex Bonnet & Barratte (1895).
Wadi Araba (RSCN database).

Onosma aaronsohnii Feinbrun, Israel J. Bot. 26: 79. 1976.
Jordan, *Hunting Aero Survey* 7 (E00276838).
Ref. Al-Eisawi 63 (2013) as *Onosma roussaei* DC.

Onosma aleppica Boiss., Diagn. Pl. Orient. 11: 10.7 1849.
Jordan, Tadros, 5.iv.1972 (AUB).

Onosma echinata Desf., Fl. Atlant. 1: 161. 1798.
Tafila, 20 km west of Jorf el Dorawish to Tafila (MSB-K 151809); Jordan valley, Salelzarqa, 26/2/1954 (YAR-H).

Onosma frutescens Lam., Tabl. Encycl. 1: 407. 1792.
Ajloun, Eshtafeana, 04/07/1999 (YAR-H).

Onosma giganteum Lam., Tabl. Encycl. 1. 407. 1792.
Tal Al-Rumman, *Al-Eisawi* (NHJ 2006.0239).

Onosma orientalis L., Sp. Pl. (ed. 2) 1: 196. 1762.
Syn. *Podonosma orientalis* (L.) Feinbrun (1976).
Petra, *Hepper* 3076 (K000286188); Tafila, Afra, *Taifour* (NHJ 2012.0506).

Onosma stellulata Waldst. et Kit., Descr. Icon. Pl. Hung. 2: 189. 1804.
Tafila, 27 Mar. 1955, *Hunting Aero Survey* (E); RSCN database.

Paracaryum intermedium (Fresen.) Lipsky, Trudy Imp. S.-Peterburgsk. Bot. Sada 26(2): 487. 1910.
10–12 knm N of Aqaba, *Boulos & Lahham* 6126 (AMM, BM, E).

Paracaryum rugulosum (DC.) Boiss., Diagn. Pl. Orient. 11: 129. 1849.
Syn. *Cynoglossum rugulosum* (DC.) Greuter & Burdet (1981).
[As] Safawi, *Cope* 273 (K000286191); 10–12 knm N of Aqaba, *Boulos & Lahham* 7618 (AMM, BM, E); Tafila, Wadi Araba, Tafila Road (MSB-K 412742).

Podonosma galalens Schweinf. ex Boiss., Fl. Orient. 4: 1199. 1879.
Ref. RSCN database; Al-Eisawi 63 (2013).

Rochelia disperma (L.f.) C.Koch, Linnaea 22: 649. 1849.
Syn. *Lithospermum dispermum* L.f. (1762).
Tafila, Fujaij Station outside fence, Site no 21 (MSB-K 497398).

Symphytum brachycalyx Boiss., Diagn. Pl. Orient. 4: 43. 1844.
Syn. *Symphytum palestinum* Boiss. (1849).
Amman, on the way to Bahhath, left lateral road opposite to private farm of King Abdallah (MSB-K 275097); Ajloun, *Bocker* RB890 (B100520991).

Trichodesma africanum(L.) Sm., Cycl. 36: no. 2, 1817.
Syn. *Trichodsma africanum* var. *heterotrichum* Bornm. & Kn.
8 km S of Aqaba, *Hemsley* s.n. (AMM, BM, E); Wadi Yutum, 21 Mar. 1975, *Boulos* (AMM).
Dana (RSCN database).

Trichodesma boissieri Post, J. Linn. Soc., Bot. 24: 436. 1888.
Wadi Shuaib, *Simpson* 53344 (K000286197); Tal Al-Rumman, *Taifour & Aldamen* (NHJ 2010.0033)

Trichodesma ehrenbergii Schweinf. ex Boiss., Fl. Orient. 4: 281. 1879.
Aqaba, after main entrance to Aqaba toward Wadi Al Yutum (MSB-K 228358).

Brassicaceae

Aethionema carneum (Banks & Sol.) B.Fedtsch., Fl. Zap. Tian Shan: 176. 1905.
Syn. *Aethionema cristatum* DC. (1821); *Thlapsi carneum* Banks & Sol. (1794).
Frey & Kürschner VO 5116 (E00040443).

Aethionema gileadense Post, J. Linn. Soc., Bot. 24: 421. 1888.
Ref. RSCN database; Al-Eisawi 64 (2013).

Aethionema heterocarpum J.Gay, Index Seminum (LE) 4: 1, 1837.
Zarqa, Site 4, Alouk (MSB-K 553919).

Alyssum aureum (Fenzl) Boiss., Fl. Orient. 1: 286. 1867.
Syn. *Meniocus grandiflorus* Jaub. et Spach (1843).
Ref. RSCN database; Al-Eisawi 64 (2013).

Alyssum damascenum Boiss. & Gaill., Diagn. Pl. Orient. 2, 6: 18. 1859.
Ma'an, in Shaubah at grounds of NCARTT station (MSB-K 151924); Dana Reserve, *RBG-Ire & RBG-Jo* (NHJ 2008.0088).

Alyssum dasycarpum Stephan ex Willd., Sp. Pl. 3: 469. 1800.
Ma'an, 25 km W Ma'an, *Frey & Kürschner* 80–706 (E00379029).

Alyssum desertorum Stapf., Denkschr. Kaiserl. Akad. Wiss., Wien. Math.-Naturwiss. Kl. 51: 33. 1886.
Syn. *Alyssum calycinum* L. (1763).
Ref. RSCN database; Al-Eisawi 64 (2013).

Alyssum homalocarpum (Fisch. & C.A.Mey.) Boiss., Fl. Orient. 1: 285. 1867.
Syn. *Psilonema homalocarpum* Fisch. & C.A.Mey. (1840).
Balqa, Naqb Ishtar, *Hunting Aero Survey* 1B (E00379068).

Alyssum iranicum Hausskn.ex Baumg., Jahresber. Landes-Gymn. Baden bei Wien 48: 9. 1911.
Karak, Jebel Ataita, *Frey & Kürschner* 80–246 (E00328197).

Alyssum linifolium Steph. ex Willd., Sp. Pl. 3: 467. 1800.
Syn. *Meniocus linifolius* (Steph. ex Willd.) DC. (1821); *Alyssum cupreum* Freyn & Sint. (1903).
Above Wadi Musa on the road to Ma'an, *Davis* 8691 (E00374877).

Alyssum meniocoides Boiss., Ann. Sci. Nat., Bot. 2, 17: 158. 1842.
Syn. *Alyssum linifolium* (non Bornm.) Dudley var. *teheranicum*; *Alyssum tetraspermum* Bertol. (1844); *Alyssum kermanshahense* Cowan ex Parsa (1951).
Salt, El Ghundhur, 12.iv.1890 (AUB); Qal'at-Ziza, near the pool, (Moab), 27.03.1936, *Feinbrun & Zohary* 327 (B).

Alyssum simplex Rudolphi, J. Bot. (Schrader) [1799] (2): 290. 1800.
Syn. *Alyssum minus* (L.) Rothm. (1941).
Ma'an 5 km NW Basta, along the road to Petra, *Boulos, Jallad* & *El-Gohary* 8036 (E00379330).

Alyssum strigosum Banks & Sol., Nat. Hist. Aleppo ed. 2, 2: 257. 1794.
Syn. *Alyssum micropetalum* Fisch. ex DC. (1819); *Alyssum campestre* (M.B.) W.D.J.Koch var. *hirsutum*; *Alyssum parviflorum* (non M.B.) Blakelock; *Alyssum minus* (Banks & Sol.) Zohary var. *strigosum; Alyssum subspinosum* Dudl. (1962).
Ma'an, Rajif Ras Al-Naqab road, 5. km after Rajif junction, Site no 37 (MSB-K 497929).

Alyssum szovitsianum Fisch. & C.A.Mey., Index Seminum (LE) 4: 31. 1837.
Syn. *Alyssum marginatum* Steud. ex Boiss. (1842); *Alyssum pyramidatulum* Bornm. (1921); *Alyssum campestre* (C.A.Mey.) Boiss. var. *micranthum.*
Near village Makhraba (ca 20 km W of Irbid), *Böcker* RB 993 (B).

Anastatica hierochuntica L., Sp. Pl., ed. 2. 2: 895. 1763.
Ma'an, Shoubak (YAR-H).

Arabis aucheri Boiss., Ann. Sci. Nat., Bot. 2. 17: 52. 1842.
Bir Dabaghat, *Davis* 9360 (E00380913).

Arabis alpina L., Sp. Pl.: 664. 1753.
Ref. Tafila and Ma'an, Danin 2005; RSCN database.

Arabis nova Vill., Prosp. Hist. Pl. Dauphiné: 39. 1779.
Syn. *Arabis auriculata* Lam. (1783); *Arabis montbretiana* Boiss. (1842).
Al-Jubaiha, Univ. Campus, near Amman, *Jallad* (E00380925).

Arabis recta Vill., Hist. Pl. Dauphiné 3: 319. 1788.
Ref. Tafila and Ma'an, Danin 2005; RSCN database.

Arabis turrita L., Sp. Pl. 2 : 665. 1753.
Irbid, Yarmouk village (YAR-H).

Arabis verna W.T.Aiton, Hort. Kew. ed. 2. 4: 105. 1812.
Dibeen National Park, 2004, *Mahklouf* s.n. (AMM); ibid, *Taifour & Othman* (DFR-H).

Arabidopsis thaliana (L.) Heynh., Fl. Sachsen 538. 1842.
Eastern desert, 1983 (YAR-H).

Biscutella didyma L., Sp. Pl. 2: 653. 1753.
Syn. *Biscutella apula* L. (1771); *Biscutella ciliata* DC. (1811); *Biscutella columnae* Ten. (1811).
Jarash, 15 km SE Jerash, along Alouk road, *Jallad* (E00374251); Tal Al-Rumman, *Taifour & Aldamen* (NHJ 2010.0065)

Brassica nigra (L.) W.D.J.Koch, Fl. ed. 3, 4: 713. 1833.
Mafraq, Al Mazzah (IBBU-H).

Brassica oleracea L., Sp. Pl.: 667. 1753.
Mafraq and Jerash (NCARE).

Brassica tournefortii Gouan, Ill. Observ. Bot. 44. 1773.
Jallad 184 (E00111530); Tal Al-Rumman, *Al-Eisawi* (NHJ 2006.0390).

Cakile arabica Velen. & Bornm., Repert. Spec. Nov. Regni Veg. 9: 114 1911.
Jebel Gulzan, *Borzatti & Barsotti* (AMM).

Cakile maritima Scop., Fl. Carniol., ed. 2. 2: 35. 1772.
Syn. *Cakile euxina* Pobed. ex Grossh. (1953).
Irbid (YAR-H).

Calepina irregularis (Asso) Thell., Fl. Schweiz ed. 2: 218. 1905.
Syn. *Crambe corvini* All. (1785); *Calepina corvini* (All.) Desv. (1814).
Az Zarqam 10–15 km west Zarqa, *Jallad* (E00449773).

Camelina hispida Boiss., Ann. Sci. Nat., Bot. II, 17: 176. 1842.
Syn. *Myagrum irregulare* Asso (1779).
Jarash, 15 km southeast Jerash, along Alouk road, *Jallad* (E00377498).

Capsella bursa-pastoris (L.) Medik., Pfl.-Gatt. 85. 1792.
Syn. *Thlaspi bursa-patoris* L. (1753).
Jallad (E00018959); Tal Al-Rumman, *Taifour* (NHJ 2010.0001).

Capsella rubella Reut., Compt.-Rend. Trav. Soc. Hallér 1853–1854: 18. 1853.
Ref. RSCN database; Al-Eisawi 65 (2013).

Cardamine hirsuta L., Sp. Pl.: 655. 1753.
Ref. Al-Eisawi 66 (2013).

Carrichtera annua (L.) DC., Mém. Mus. Hist. Nat. 7: 250. 1821.
Syn. *Carrichtera vellae* DC. (1821); *Vella annua* L. (1753).
12 km north Wadi Wala, *Jallad* 134 (E00372177); Tal Al-Rumman, *Taifour* (NHJ 2013.0173).

Chorispora purpurascens (Banks & Sol.) Eig, J. Bot. 75: 189. 1937.
Al-Jubaiha, Univ. Campus, near Amman, *Jallad* 207 (E00328170).

Clypeola aspera (Grauer) Turrill, J. Bot. 60: 269. 1922.
Syn. *Bergeretia echinata* Desv. (1813); *Clypeola echinata* (Desv.) DC. (1821); *Clypeola chaetocarpa* Jaub. et Spach (1847); *Peltaria aspera* Grauer (1784).
Al Balqa, SE of Salt, *Meyers & Dinsmore* (E00376756).

Clypeola jonthlaspi L., Sp. Pl.: 652. 1753.
Syn. *Alyssum minimum* Willd. (1800) non L.; *Ionthlaspi clypeolatum* var. *macrocarpum* Caruel (1893); *Clypeola microcarpa* Moris (1843).
Madaba, *Meyers & Dinsmore* M397 (E00376708).

Conringia orientalis (L.) Andrz., Fl. Belg. 123. 1827.
Syn. *Brassica orientalis* L. (1753).
Karak, Wadi Dana, between Dana and Lahta, *Frey & Kürschner* (B).

Conringia clavata Boiss. Ann. Sci. Nat. Biot. 2, 17: 84. 1842.
Syn. *Conringia perfoliata* (C.A.Mey.) N.Busch (1939).
Ref. Al-Eisawi 66 (2013) as *Coringia perfoliata*.

Crambe hispanica L., Sp. Pl. 671. 1753.
Balqa, Site 7 (MSB-K 552738); Salt, iv.1886 (AUB).

Crambe orientalis L., Sp. Pl.: 671. 1753.
Tafila (MSB-K 277873); Tal Al-Rumman RBG, Sector No. 57, *Taifour & Al Khawaldah* (NHJ 2013.0315).

Descurainia sophia (L.) Webb & Berth., Nat. Pflanzenfam. 3(2): 192. 1892.
Syn. *Sisymbrium sophia* L. (1753).
Ma'an, Wadi Musa, near Petra, *Boulos & Jallad* 8119 (E00383302).

Diplotaxis acris (Forssk.) Boiss., Fl. Orient. 1: 389. 1867.
Syn. *Hesperis acris* Forssk. (1775); *Hesperis angustifolia* Lam. (1789).
Zarqa, Azraq Oasis, *Jallad* 391 (E00449026).

Diplotaxis erucoides (L.) DC., Syst. Nat. 2: 631. 1821.
Syn. *Sinapis erucoides* L. (1756).
Deir 'alla Road, North Ghor, *Jallad* 43 (E00449076); Tal Al-Rumman RBG, Sector No. 16, *El-Oqlah & Al Mfadi* (NHJ 2013.0385).

Diplotaxis harra (Forssk.) Boiss., Fl. Orient. 1: 388. 1867.
Syn. *Sinapis harra* Forssk. (1775).
Zarqa, 10 km east of Azraq Oasis, *Jallad* 410 (E00449366); nr. Aqaba, *Boulos & Lahham* 7427 (AMM, BM, E).

Diplotaxis villosa Boulos, Bot. Not. 128: 367. 1975 publ. 1976.
Wadi Shaumari, c. 40 km NE El-Jafr., *Boulos* 7994 (E00386049).

Diplotaxis viminea (L.) DC., Syst. Nat. 2: 635. 1821.
Syn. *Sisymbrium vimineum* L. (1763).
Ref. Al-Eisawi 67 (2013).

Eigia longistyla (Eig) Soják, ?as. Nár. Mus., Odd. P?ír. 148: 193. 1979 publ. 1980.
Syn. *Stigmatella longistyla* Eig (1938).
Wadi Rum, *Jarrar* (8036 K000283163); ibid., *Boulos, Jallad, Lahham & Abu Hmaidan* 7625 (E00083374).

Enarthrocarpus arcuatus Labill., Icon. Pl. Syr. 5: 4. 1812.
Al Mafraq, Um Al Surab, 19 Apr. 1983 (YAR-H).

Enarthrocarpus strangulatus Boiss., Diagn. Pl. Orient. 8: 44. 1849.
Madaba 3 km north Wadi Wala, *Jallad* 126 (E00449815); Tal Al-Rumman RBG, Sector No. 23, *El-Oqlah & Taifour* (NHJ 2013.0162).

Eremobium aegyptiacum (Spreng.) Asch. ex Boiss., Fl. Orient. Suppl.: 30. 1888.
Syn. *Malcolmia aegyptiaca* Spreng. (1825).
Ma'an,5 km NW Rum station, *Frey & Kürschner* VO 5040 (E00384248).

Eremobium aegyptiacum var. **lineare** (Delile) Zohary, Fl. Palaestina 1: 267. 1966.
Syn. *Eremobium lineare* (Delile) Asch. & Schweinf. ex Boiss. (1867); *Eremobium diffusum* (Decne.) Botsch. (1964).
Salt, 24 Nov. 1893 (AUB); sin. loc., *Jallad* (AMM).

Erophila minima C.A.Mey., Verz. Pfl. Casp. Meer. 184. 1831.
Dibeen National Park, 2005, *Mahklouf* s.n. (AMM).

Erophila verna (L.) DC., Mém. Mus. Hist. Nat. 7: 251. 1821.
Syn. *Erophila spathulata* F.Lang (1820); *Erophila vulgaris* DC. (1821); *Erophila praecox* (Stev.) DC. (1821); *Erophila setulosa* Boiss. & Blanche (1856); *Erophila virescens* Jord. (1864).
Ref. Al-Eisawi 67 (2013).

Eruca vesicaria (L.) Cav. Descr. Pl. 426. 1802.
Syn. *Eruca sativa* Mill. (1768); *Eruca cappadocica* Reut. (1857); *Eruca cappadocica* Reut. var. *eriocarpa* Boiss. (1867); *Eruca lativalvis* Boiss. (1867); *Eruca eruca* (L.) Asch. &Graebn. (1898); *Eruca sativa* Mill. var. *eriocarpa* (Boiss.) Post (1883); *Eruca vesicaria* (L.) Cav. subsp. *sativa* (Mill.) Thell. (1918).
Balqa, Salt Aira Agricultural reserve (MSB-K 176390); North of Quoraymeh, North Ghor, *Jallad* 64 (E00449444); Tal Al-Rumman, *Al-Eisawi* (NHJ 2006.0205).

Erucaria crassifolia (Forssk.) Delile, Descr. Egypte, Hist. Nat. 244. 1813.
Syn. *Brassica crassifolia* Forssk. (1775).
Wadi Thalaja, 27 Mar 1999, *Borzatti & Barsotti* (AMM).

Erucaria hispanica (L.) Druce, Rep. Bot. Exch. Club Soc. Brit. Isles 3: 418. 1913.
Syn. *Erucaria aleppica* Gaertn. (1791); *Brassica purpurascens* Banks & Sol. (1794); *Erucaria lineariloba* Boiss. (1842); *Erucaria grandiflora* Boiss. (1846); *Erucaria myagroides* (L.) Halácsy (1900).
Desertum Syriacum, inter Amman et Rutba, *Rechinger* 12914 (E00449987); Safi, *Hepper* 10040 (K000318574); Tal Al-Rumman, *RBG-Ire & RBG-Jo* (NHJ 2008.0065).

Erucaria microcarpa Boiss., Diagn. Pl. Orient. 8: 47. 1849.
Jarash, 30.iii.1964, *Mrs. Kurani* B–155(AUB).

Erucaria pinnata (Viv.) Täckh. & Boulos, Publ. Cairo Univ. Herb. 5: 37. 1972.
Syn. *Raphanus pinnatus* Viv. (1824); *Erucaria uncata* (Boiss.) Asch. & Schweinf. (1889); *Reboudia pinnata* (Viv.) O.E.Schulz (1916).
Amman, Site 11. Dabala site on the desert highway (MSB-K 558039); Tal Al-Rumman, *Taifour* (NHJ 2012.0330)

Erucaria rostrata (Boiss.) A.W.Hill ex Greuter & Burdet, Willdenowia 15: 419. 1986.
Syn. *Erucaria boveana* Coss. (1884).
30 km W of Azraq, 3 Apr. 1980 (HUJ); Amman Mashare'a, North Ghor, *Jallad* 74 (E00449912); Tal Al-Rumman, *Al-Eisawi* (NHJ 2006.0002).

Erysimum crassipes Fisch. & C.A.Mey., Ind. Sem. Hort. Petrop. 1: 27. 1835.
Ma'an, Musa, *Davis* 9234 (E00378645).

Erysimum oleifolium J.Gay, Erysim. Nov. 6. 1842.
Ma'an, above Wadi Musa, road to Maan, *Davis* 8692 (E00384657).

Erysimum purpureum J. Gay, Erysim. Nov. 6. 1842.
Ma'an, Bir Dabaghat, *Davis* 9342 (E00384690).

Erysimum repandum L., Demonstr. Pl.: 17. 1753.
Ajloun, *Meyers, Fred & Dinsmore* G1650 (E00381699); Amman Citadel (Herkules temple), *Böcker* RB 405 (B).

Erysimum scabrum DC., Syst. Nat. 2: 505. 1821.
Syn. *Erysimum goniocaulon* Boiss. (1849).
Castle of Ajloun, *Meyers, Fred & Dinsmore* G1749 (E00378697).

Euclidium syriacum (L.) R.Br., Hortus Kew. 4: 74. 1812.
Syn. *Anastatica syriaca* L. (1763).
10 km NE of the Safawi-Mafraq rd. (BRDC-H).

Farsetia aegyptiaca Turra, Farsetia: 5. 1765.
Wadi Petra, *Jallad* 195 (E00377037); nr. Aqaba, *Boulos & Lahham* 7389 (AMM, BM, E).

Fibigia clypeata (L.) Medik., Pfl.-Gatt.: 90. 1792.
Syn. *Alyssum clypeatum* L. (1753); *Farsetia rostrata* Schenk (1840); *Fibigia rostrata* (Schenk) Boiss.(1867).
Dana Reserve, Rumana trail behind Rumana camp, *Al Awaji et al.* (NHJ 2013.0455).

Fibigia eriocarpa (DC.) Boiss., Fl. Orient. 1: 258. 1867.
Ajloun castle, *Davis* 9669 (E00374678).

Glastaria glastifolia (DC.) Kuntze, Revis. Gen. Pl. 1: 30. 1891.
Syn. *Peltaria glastifolia* DC. (1821); *Texiera glastifolia* (DC.) Jaub. & Spach (1842).
As Safawi, *Cope* 370 (K000283174).

Hesperis biscuspidata (Willd.) Poir., Encycl. Suppl. 3: 195. 1813.
Syn. *Cheiranthus bicuspidatus* Willd. (1800).
Ref. RSCN database; Al-Eisawi 68 (2013).

Hesperis laciniata All., Fl. Pedem. 1: 271. 178.
Ref. RSCN database; Al-Eisawi 68 (2013).

Hesperis pendula DC., Rech. Anat. Taxon. Fam. Crucifer. 2: 457. 1821.
Tafila, Dana Reserve, (MSB-K 419235); Salt, 30.iv.1886 (AUB).

Hirschfeldia incana (L.) Lag.-Foss., Fl. Tarn Garonne: 19. 1847.
Syn. *Hirschfeldia adpressa* Moench (1794); *Brassica adpressa* (Moench) Boiss. (1839).
Al Balqa' Deir 'Alla Road, north Ghor, *Jallad* 42 (E00439706); Tal Al-Rumman, *Taifour & Aldamen* (NHJ 2010 .0014)

Hornungia procumbens (L.) Hayek, Repert. Spec. Nov. Regni Veg. Beih. 30(1): 480. 1925.
Syn. *Hymenolobus procumbens* (L.) Nutt. (1838); *Hutchinsia procumbens* (L.) Desv. (1814); *Capsella procumbens* (L.) Fr. (1832).
Recorded by (YAR-H). Ref. Al-Eisawi 68 (2013) as *Hymenolobus procumbens*.

Horwoodia dicksoniae Turrill,J. Bot. 77: 117. 1939.
Kilwa, *Hunting Aero Survey* 90B (E00374399).

Iberis odorata L., Sp. Pl.: 649. 1753.
Syn. *Iberis acutiloba* Bertol. (1843).
Ajloun Forest Reserve (RSCN database).

Isatis glauca Aucher, Ann. Sci. Nat., Bot., sér. 2, 17: 201. 1842.
Ref. Al-Eisawi 69 (2013).

Isatis lusitanica L., Sp. Pl.: 670. 1753.
Syn. *Isatis aleppica* Scop. (1787).
Boulos, Jallad & El-Gohary 8041 (E00050752); Azraq castle (Qasr Azraq), *R. Böcker* RB 412 (B).

Isatis microcarpa J.Gay ex Boiss., Ann. Sci. Nat., Bot. II, 17: 201. 1842.
Ref. Al-Eisawi 69 (2013).

Lepidium aucheri Boiss., Ann. Sci. Nat., Bot. II, 17: 195. 1842.
Ma'an El-Jafr, in the vicinity of the Police Station, *Boulos & Jallad* 7929 (E00372254); Northeastern Badia Biqyaweea, *Al Shudiefat & Khalidi* (NHJ 2013.0601).

Lepidium cartilagineum (J.Mayer) Thell., Ill. Fl. Mitt.-Eur. 4: 88. 1913.
Syn. *Thlaspi cartilagineum* J.Mayer (1786); *Lepidium crassifolium* Waldst. & Kit. (1799); *Lepidium caespitosum* Desv. (1815); *Lepidium cartilagineum* (J.Mayer) Thell. subsp. *crassifolium* (Waldst. & Kit.) Thell.; *Lepidium crenatifolium* Boiss. & Balansa (1859); *Lepidium cartilagineum* (J.Mayer) Thell. subsp. *caespitosum*, Mitt. (1906).
Azraq, 1963 *Mountfort et al.* RSCN database (Azraq Wetland Reserve Herbarium, 2008).

Lepidium coronopus (L.) Al-Shehbaz, Novon 14: 152. 2004.
Syn. *Coronopus squamatus* (Forssk.) Asch. (1860).
Dibeen National Park, 2005, *Mahklouf* s.n. (AMM).

Lepidium didymum L., Mant. Pl.: 92. 1767.
Syn. *Coronopus didymus* (L.) Sm. (1800).
Madaba, *Danin* (http://flora.org.il/en/plants/CORDID/).

Lepidium draba L., Sp. Pl. 2: 645. 1753.
Syn. *Cardaria draba* (L.) Desv. (1813).
Ma'an, Al Shoubak, Al Hisheh, Royal palace reserve (MSB-K 158819); Mafraq, Beqaweya (NHJ 2013.0601)

Lepidium latifolium L., Sp. Pl. 2: 644. 1753.
Amman, Jabal Amman, *Jallad* 475 (E00372263).

Lepidium sativum L., Sp. Pl.: 644. 1753.
Syn. *Lepidium sativum* L. subsp. *spinescens* (DC.) Thell. (1906); *Lepidium spinescens* DC. (1821).
Amman, Wadi Sir, *Meyers & Dinsmore* M1684 (E00372289).

Lepidium spinosum Ard., Animadv. Bot. Spec. Alt. 34. 1764.
Ref. Karak, Danin (2005); RSCN database; http://flora.org.il/en/plants/LEPSPN.

Leptaleum filifolium (Willd.) DC., Syst. Nat. 2: 511. 1821.
Syn. *Sisymbrium filifolium* Willd. (1800).
Zarqa, Site 8 (MSB-K 554145).

Lobularia arabica (Boiss.) Muschl., Man. Fl. Egypt 1: 421. 1912.
Syn. *Koniga arabica* Boiss. (1849).
Ref. Wadi Araba *Danin* (2005); RSCN database.

Lobularia libyca (Viv.) Webb & Berth., Hist. Nat. Iles Canaries 1: 90. 1837.
Syn. *Lunaria libyca* Viv. (1824).
Karak, Wadi Araba, 5 km E of Ain el-Fidan, *Frey & Kürschner* (BSB).

Malcolmia africana (L.) R.Br., Hortus Kew. 4: 121. 1812.
Syn. *Hesperis africana* L. (1753).
Ma'an Wadi Petra, *Jallad* 201 (E00378392); Tafila, Dana Reserve, Wadi Ahmer, *Al Awaji et al.* (NHJ 2013.0430).

Malcolmia chia (L.) DC., Syst. Nat. 2: 440. 1821.
Syn. *Cheiranthus chius* L. (1753).
Petra, 17 Apr. 1929 (HUJ).

Malcolmia crenulata (DC.) Boiss., Fl. Orient. 1: 229. 1867.
Syn. *Hesperis crenulata* DC. (1821).
25 km northeast H–4, *Jallad* 350 (E00384034).

Malcolmia maritima (L.) R.Br., Hortus Kew. 4: 121. 1812.
Ajloun (YAR-H).

Malcolmia pygmaea (DC.) Boiss., Fl. Orient. 1: 222. 1867.
Syn. *Maresia pygmaea* (DC.) O.E.Schulz (1924).
Ma'an, *Dinsmore* 7333 (E00384958).

Maresia pulchella(DC.) O.E.Schulz, Opred. Viss. Rast. Baskirsk. ASSR IV–105(86): 211. 1924.
Syn. *Malcolmia pulchella* (DC.) Boiss. (1842).
El-Hesma, 22 km N of Quira, 30 Mar. 1936(HUJ).

Matthiola alyssifolia (DC.) Bornm., Repert. Spec. Nov. Regni Veg. 39: 80 1935.
Syn. *Hesperis alyssifolia* DC. (1821).
Ref. Al-Eisawi 70 (2013) as *Matthiola albicaulis* Boiss.

Matthiola arabica Boiss., Ann. Sci. Nat., Bot. II, 17: 49. 1842.
Ref. Al-Eisawi 70 (2013).

Matthiola incana (L.) R.Br., Hortus Kew. 4: 119. 1812.
Irbid, Abu Logus (YAR-H).

Matthiola longipetala (Vent.) DC., Syst. Nat. 2: 174. 1821.
Syn. *Cheiranthus bicornis* Sm. (1813); *Matthiola bicornis* (Sm.) DC. (1821); *Matthiola oxyceras* DC. (1821); *Matthiola oxyceras* DC. var. *forcipifera* Boiss.; *Matthiola oxyceras* DC. var. *pendula* Nab.; *Matthiola bicornis* (Sm.) DC. var. *oxyceras* (DC.) Bornm.; *Matthiola humilis* Anth. (1821); *Matthiola longipetala* (Vent.) DC. var. *bicornis* (Sm.) Zohary; *Matthiola longipetala* (Vent.) DC. var. *oxyceras* (DC.) Zohary; *Matthiola aspera* Boiss., (1849); *Matthiola longipetala* (Vent.) DC. subsp. *pendula* (Nab.) Hadac et Chrtek; *Cheiranthus longipetalus* (1803).
9 km East Arda-Ghor Road, *Jallad* 47 (E00377407); Azraq, *Gillett* (15638 K000318253); Dana Reserve, plateau near top of Rumana mountain, *Al Awaji et al.* (NHJ 2013.0451).

Matthiola longipetala (Vent.) DC. subsp. **livida** (Delile) Maire, Fl. Afr. Nord 14: 23. 1977.
Syn. *Matthiola livida* (Delile) DC. (1821).
Ghour Safi, Apr. 1895 (AUB).

Matthiola parviflora (Schousb.) R.Br., Hortus Kew. 4: 121. 1812.
Syn. *Cheiranthus parviflorus* Schousb. (1800).
Tafila Province, 2014, *Oran* (AMM); Zarqa, 20 Mar. 1983 (YAR-H).

Matthiola tricuspidata (L.) R.Br., Hortus Kew. 4: 120. 1812.
Syn. *Cheiranthus tricuspidatus* L. (1753).
Al Mafraq, Borqu (IBBU-H).

Morettia canescens Boiss., Diagn. Pl. Orient. 8: 17. 1849.
Wadi Rum, *Boulos & Lahham* 7733 (AMM, BM, E).

Morettia parviflora Boiss., Ann. Sci. Nat., Bot. II, 17: 60. 1842.
Nr. Aqaba, *Boulos & Lahham* 7400, 7420 (AMM, BM, E).

Morettia philaeana DC., Syst. Nat. 2: 426. 1821.
Wadi Araba (RSCN database); Jebel Marsad, 24 Apr. 1999, *Barsotti & Borzatti* (AMM).

Moricandia arvensis (L.) DC., Syst. Nat. 2: 626. 1821.
Syn. *Brassica arvensis* L. (1767).
Salt, Um Ennaj (FOR-H).

Moricandia nitens (Viv.) Durand & Barratte, Fl. Libyc. Prodr. 15. 1910.
Syn. *Hesperis nitens* Viv. (1824); *Moricandia dumosa* Boiss. (1849).
Petra area, Jebel Harun, *Frey & Kürschner* (BSB).

Moricandia sinaica (Boiss.) Boiss., Fl. Orient. 1: 386. 1867.
Syn. *Brassica sinaica* Boiss. (1842).
Rum, 1940, *Dinsmore* 13871 (AUB).

Myagrum perfoliatum L., Sp. Pl.: 640. 1753.
Irbid, Maru village, along the road just before Maru agricultural station (MSB-K 200187).

Nasturtiopsis coronopifolia (Desf.) Boiss. subsp. **arabica** (Boiss.) Greuter & Burdet, Willdenowia 13(1): 94. 1983.
Syn. *Nasturtiopsis coronopifolia* (Desf.) Boiss. (1867); *Nasturtiopsis arabica* Boiss. (1867).
Al-Mafraq, Borqu (IBBU-H).

Nasturtium officinale R.Br., Hortus Kew. 4: 110. 1812.
Jarash, opposite olive extraction factory (MSB-K 2003.61); Tal Al-Rumman, *RBG-Ire & RBG-Jo* (NHJ 2008.0104).

Neotorularia torulosa (Desf.) Hedge & Léonard, Bull. Jard. Bot. Natl. Belg. 56: 395. 1986.
Syn. *Sisymbrium torulosa* Desf. (1798); *Malcolmia torulosa* (Desf.) Boiss. (1867); *Malcolmia torulosa* (Desf.) Boiss. var. *leiocarpa* Boiss.; *Malcolmia torulosa* (Desf.) Boiss. var. *contortuplicata* Boiss.; *Malcolmia ledebourii* Boiss. (1867); *Torularia torulosa* (Desf.) O.E.Schulz (1924).
Jordan, Site 52. Jordan University Research Station (MSB-K 395771); Al Mafraq, Al-Mansoorah (IBBU-H); Musherfa Site, Zarqa River, *Al-Dwiri* (AMM).

Neslia paniculata (L.) Desv. subsp. **thracica** (Vel.) Bornm., Oesterr. Bot. Z. 44: 125. 1894.
Syn. *Myagrum paniculatum* L. (1753); *Vogelia paniculata* (L.) Hornem. (1815); *Neslia apiculate* Fisch., C.A.Mey. & Avé-Lall. (1842); *Neslia thracica* Vel. (1891); *Vogelia apiculate* (Fisch., C.A.Mey. & Avé-Lall.) Vierh (1921); *Vogelia paniculata* (L.) Hornem. var. *thracica* (Vel.) Bornm.
Irbid, Melka village (MSB-K 114183); Tal Al-Rumman, *Al-Eisawi* (NHJ 2006.0418).

Notoceras bicorne (Aiton) Amo, Fl. Fan. Penins. Iberica 6: 536. 1873.
Syn. *Erysimum bicorne* Aiton (1789); *Notoceras canariense* R.Br. (1812); *Notoceras hispanicum* DC. (1821).
Azraq ash Shishan, *Gillett* 15675 (K000283170).

Ochthodium aegyptiacum (L.) DC., Syst. Nat. 2: 423. 1821.
Syn. *Bunias aegyptiaca* L. (1767).
Ajloun, Rajeb village, 2 km along the side of the river (MSB-K 200006).

Olimarabidopisis pumila (Celak.) Al-Shehbaz, Novon 9: 303. 1999.
Syn. *Alyssum spinosum* L. (1753).
Ref. Al-Shehbaz et al., Novon 9: 303. 1999.

Peltaria angustifolia DC., Syst. Nat. [Candolle] 2: 329. 1821.
Tafila, outside Dana village (MSB-K 151832).

Pseuderucaria clavata (Boiss. & Reut.) O.E.Schulz., Bot. Jahrb. Syst. 54(119): 54. 1916.
Lower Jordan Valley (RSCN database).

Raphanus raphanistrum L., Sp. Pl. 2: 669. 1753.
Jordan, coll ignot. 209 (K000568166).

Rapistrum rugosum (L.) All., Fl. Pedem. 1: 257. 1785.
Syn. *Myagrum rugosum* L. (1753).
Yarmuk river, *Meyers & Dinsmore* (E); c. 9 km NW Irbid, along the road to Al-Hemma, 27 Apr. 1975, *Boulos, Jallad & Lahham* (AMM, E, YAR-H).

Ricotia lunaria (L.) DC., Syst. Nat. 2: 284. 1821.
Syn. *Cardamine lunaria* L. (1753).
Irbid, Aqraba (YAR-H); Upper Jordan Valley (RSCN database).

Robeschia schimperi O.E.Schulz, Pflanzenr. IV, 105(86): 361. 1924.
Syn. *Sisymbrium schimperi* Boiss. (1842).
RSCN database; Ref. Al-Eisawi 72 (2013).

Sameraria armena (L.) Desv., J. Bot. Agric. 3: 161. 1815.
Syn. *Isatis arnema* L. (1753).
10 km NE of the Safawi-Mafraq rd., 33 km from Safawi (BRDC-H).

Savignya parviflora (Delile) Webb, Fragm. Fl. Aethiop.-Aegypt.: 14. 1854.
Syn. *Lunaria parviflora* Delile (1813); *Savignya aegyptiaca* DC. (1821).
Amman, Wadi Mudeisisat S Azraq, *Frey & Kürschner* (BSB, E); Wadi Turayfiya, 30 Mar. 1997, *Borzatti & Barsotti* (AMM); Karak, Wadi Araba, 5 km E of Ain el-Fidan, *Frey & Kürschner* (E).

Schimpera arabica Hochst. & Steud., Nomencl. Bot. ed. 2, 2: 530. 1841.
Syn. *Schimpera persica* Boiss. (1846); *Schimpera arabica* Hochst. et Steud. var. *persica* (Boiss.) Bornm.; *Schimpera arabica* Hochst. et Steud. subsp. *persica* (Boiss.) Hadac.
Rum, 26.iv.1942, *Dinsmore* 13744 (AUB); Wadi Rum, *Boulos & Lahham* 7580 (AMM, BM, E).

Schouwia thebaica Webb, Parl. Giorn. Bot. Ital. ii. 219. 1847.
Syn. *Schouwia schimperi* Jaub. & Spach (1850).
Ref. RSCN database; Al-Eisawi 72 (2013).

Sinapis alba L., Sp. Pl.: 668. 1753.
Syn. *Brassica hirta* Moench (1802).
Amman, Na'or - Al Haj Al Garbi, opposite water spring (MSB-K 200095); Tal Al-Rumman, *Al- Eisawi et al.* (NHJ 2006.0095).

Sinapis arvensis L., Sp. Pl.: 668. 1753.
Syn. *Sinapis mesopotamica* Spreng. (1801).
Tafila, nr. Karak (MSB-K 114493).

Sisymbrium damascenum Boiss. et Gaill., Diagn. Pl. Orient., ser. 2 3(6): 11. 1859.
Between Nadrah and Jerash (IBBU-H); Amman (RSCN database).

Sisymbrium erysimoides Desf., Fl. Atlant. 2: 84. 1798.
Tal Al-Rumman, *Shishani & Alkoran* (NHJ 2011.0009)

Sisymbrium irio L., Sp. Pl.: 659. 1753.
Tafila, behind Dana Hotel rooms (MSB-K 416175); Tal Al-Rumman, *Al-Eisawi, Abu Sondos, Abu Anza & Shishani* (NHJ 2006.0069).

Sisymbrium officinale (L.) Scop., Fl. Carniol. ed. 2, 2: 26. 1772.
Syn. *Erysimum officinale* L. (1753).
Along Zarqa River, 2004, *Al-Dwiri* s.n. (AMM); Al Mafraq, Sabha (IBBU-H) Dead Sea Area (RSCN database).

Sisymbrium orientale L., Cent. Pl. II, 24. 1756.
Syn. *Sisymbrium columnae* Jacq. (1776).
Ma'an, Al Hisheh, Al Shoubak, Royal palace reserve (MSB-K 158808).

Sisymbrium runcinatum Lag. ex DC., Syst. Nat. 2: 478. 1821.
Syn. *Sisymbrium runcinatum* Lag. ex DC. var. *hirsutum* (Lag. ex DC.) Coss.
Azraq, Site 29, Shomari Reserve (MSB-K 557308); Tal Al-Rumman, *Al-Eisawi* (NHJ 2006.0160).

Sisymbrium septulatum DC., Syst. Nat. 2: 471. 1821.
Syn. *Sisymbrium rigidulum* Tchich. (1866); *Sisymbrium grandiflorum* Post (1890); *Sisymbrium erucastroides* Bornm. (1904); *Sisymbrium bilobum* (C.Koch) Grossh.(1927).
Dibeen National Park, 2005, *Mahklouf* s.n. (AMM); Wadi Rum, *Boulos & Lahham* 7455 (AMM, BM, E).

Sterigmostemum sulphureum subsp. **sulphureum** Bornm., Auctuar. ad Flor. Pedemont. 28(2): 110. 1911.
Syn. *Cheiranthus sulphureus* Banks & Soland (1794); *Sterigma sulphureum* (Banks & Soland.) DC. (1821).
Ref. Warwick, Francis & Al-Shehbaz, Pl. Syst. Evol. 259: 249–258. 2006; Ref. RSCN database; Al-Eisawi 73 (2013).

Thlaspi arvense L., Sp. Pl.: 641. 1753.
Sbout 20 km S of Irbid, *Böcker* RB 411 (B).

Thlaspi perfoliatum L., Sp. Pl.: 641. 1753.
Syn. *Microthlaspi perfoliatum* (L.) F.K.Mey. (1973).
Madaba, nr. Naur (MSB-K 113876); Ajloun, Ishtafena (IBBU-H).

Torularia contortuplicata O.E.Schulz., Pflanzenr. (Engler) Crucif.-Sisymbr.: 219. 1924.
Ref. Al-Eisawi 73 (2013).

Turritis laxa (Sm.) Hayek, Repert. Spec. Nov. Regni Veg. Beih. 30(1): 402. 1925.
Ref. Amman, Balqa, Danin (2004).

Zilla spinosa (Turra) Prantl, Nat. Pflanzenfam. 3(2): 175. 1891.
Syn. *Zilla myagroides* Forssk. (1775).
Azraq, Wadi Bottom, *Taifour et al.* (NHJ 2013.0727).

Zuvanda exacoides Askerova, Bot. Stud. 70: 524. 1985.
Syn. *Malcolmia conringioides* Boiss. (1867); *Malcolmia exacoides* (DC.) Spreng. (1825).
Wadi Musa, Bir Dabaghat, *Davis* 8949 (E00384092); Tal Rimah (BRDC-H).

Butomaceae

Butomus umbellatus L., Sp. Pl.: 372. 1753.
Jordan, Um Qais, Karak, Marej 1886 (AUB).

Campanulaceae

Campanula camptoclada Boiss., Diagn. Pl. Orient. 11: 63. 1849.
Balqa, Tell Ar Rumman, 500 m after the left side junction, right side of road (MSB-K 339625).

Campanula dulcis Decne., Ann. Sci. Nat., Bot., sér. 2, 2: 258. 1834.
Aqaba, RSCN database.

Campanula erinus L., Sp. Pl.: 169. 1753.
Tal Al-Rumman, *Al-Eisawi* (NHJ 2006.0413).

Campanula heirosolymitana Boiss., Sp. Pl.: 169. 1753.
Zerka, Ma'an, *Davis* 9368 (K000318455).

Campanula lyrata Lam., Encycl. 1: 588. 1785.
Ref. RSCN database; Al-Eisawi 74 (2013).

Campanula phrygia Jaub. & Spach, Ill. Pl. Orient. 3: 42. 1848.
Upper Jordan Valley, RSCN database.

Campanula ramosissima Sm. in Sibthorp & Smith, Fl. Graec. Prodr. 1: 137. 1806.
Ref. RSCN database; Al-Eisawi 74 (2013).

Campanula rapunculus L., Sp. Pl.: 164. 1753.
Jarash, Souf, from Souf Camp junction up to Souf village, right side to ward Ajloun (MSB-K); Ajloun district, outside Ajloun reserve, BOT-ERA project (NHJ 2013.0533).

Campanula retrorsa Labill., Icon. Pl. Syr. 5: 5. 1812.
Tabaget Fahel, 24.v.1967, *Edgecombe* 216 (AUB); Salt, Remameen, 24/04/1983 (YAR-H).

Campanula stellaris Boiss., Diagn. Pl. Orient. 11: 63. 1849.
Tal Al-Rumman, *Al-Eisawi* (NHJ 2006.0214); Tal Al-Rumman, *Taifour & Aldamen* (NHJ 2009.0134).

Campanula strigosa Banks & Sol. in A.Russell, Aleppo, ed. 2, 2: 246. 1794.
Irbid, Bani Kenana, Al Rafeed village (MSB-K 176482); Tal Al-Rumman, RBG, Sector No. 71, *Taifour* (NHJ 2013.0196).

Campanula sulphurea Boiss., Diagn. Pl. Orient. 11: 64. 1849.
Irbid, Yarmouk University Campus, 02/05/1981 (YAR-H).

Legousia falcata (Ten.) Fritsch, Mitt. Naturwiss. Vereins Univ. Wien n.s., 5: 100. 1907.
Syn. *Prismatocarpus falcatus* Ten. (1811); *Campanula falcata* Roem. & Schult. (1819); *Specularia falcata* (Ten.) A.DC. (1830).
Irbid, Barkash (MSB-K 277770).

Legousia hybrida (L.) Delarbre, Fl. Auvergne ed. 2: 4.7 1800.
Ajloun, Um-Elyanabia, 26/03/1982 (YAR-H).

Legousia pentagonia (L.) Thell., Vierteljahrsschr. Naturf. Ges. Zürich 46: 465. 1908.
Syn. *Campanula pentagonia* L. (1753); *Specularia pentagonia* (L.) A.DC. (1830).

Amman, Wadi-Essir, on the way to Bahath (MSB-K 416337).

Legousia speculum-veneris (L.) Durand ex Vill., Hist. Pl. Dauphiné 2: 338. 1786.
Syn. *Campanula speculum-veneris* L. (1753).
Jordan, *Meyers & Dinsmore* 711 (K000286209).

Michauxia campanuloides L'Her., Michauxia I, 1788.
Irbid, Deir Elseneh, 13/05/1983 (YAR-H).

Capparaceae

Capparis aegyptia Lam., Encycl. 1: 605.1785.
Syn. *Capparis sinaica* Veill. (1801).
Wadi Rum, *Jallad & Lahham & Hmaidan* 7636 (E00065905).

Capparis cartilaginea Decne., Ann. Sci. Nat., Bot. sér. 2, 3: 273. 1835.
Syn. *Capparis galeata* Fresen. (1837); *Capparis spinosa* (Fresen.) var. *galeata* Hook.f. & Thomson (1872); *Capparis spinosa* L. subsp. *cartilaginea* (Decne.) Maire & Weiller (1965).
Aqaba, Wadi Rum, 1 km after the main entrance (MSB-K 243506); Wadi Yutum, *Taifour, Al Khawaldah & Al Mogherbi* (NHJ 2013.0699).

Capparis decidua (Forssk.) Edgew., J. Linn. Soc., Bot. 6: 184. 1862.
Amman, Dead Sea area, (MSB-K 417747).

Capparis parviflora Boiss., Diagn. Pl. Orient. 1: 4. 1843.
Tafila; Al Azraq; Borqu; Wadi Yutura (FOR-H).

Capparis sicula Duh., Traité Arbr. Arbust. 1: 159. 1755.
Wadi Yutura (FOR-H).

Capparis spinosa L., Sp. Pl.: 503. 1753.
Syn. *Capparis sicula* Duhamel (1755); *Capparis ovata* Desf. (1798); *Capparis leucophylla* DC. (1824); *Capparis spinosa* L. var. *canescens* Coss. (1849); *Capparis spinosa* (DC.) Boiss. var. *leucophylla*; *Capparis spinosa* (Boiss.) Boiss. var. *parviflora*; *Capparis spinosa* L. var. *deserti* Zohary (1960); *Capparis spinosa* L. var. *arvensis* Zohary (1960); *Capparis cartilaginea* (non Decne.) auctt. incl. Blakelock; *Capparis parviflora* Boiss. var. *glaberrima* Hand.-Mazz.; *Capparis ovata* Desf. var. *kurdica* Zohary; *Capparis leucophylla* DC. var. *parviflora* (Boiss.) Zohary.
Jordan, *Davies* 9398 (K000315654); El-Jafr, *Al-Eisawi* 2104 (K000315572); Northeastern Badia Biqyaweea *Al Shudiefat & Khalidi* (NHJ 2013.0005); Azraq, *Townsend* 415 (K000315571); Tal Al-Rumman, *Taifour & Aldamen* (NHJ 2009.0091); Azraq, Wadi Bottom (South west the Castle Amra), *El-Oqlah* (NHJ 2013.0595); Wadi Rum, *Frey & Kürschner* 5856 (E00443617).

Maerua crassifolia Forssk., Fl. Aegypt.-Arab.: 104. 1775.
Karak, Jordan Valley, Ghor Asal (MSB-K 341446); Ghour, Feefa, 04/04/1984 (YAR-H).

Caprifoliaceae

Cephalaria joppensis (Rchb.) Coult. ex DC., Prodr. 4: 648. 1830.
Syn. *Scabiosa joppensis* Rchb. (1824).
Karak, Wadi Al Arab, Wadi Zeglab (MSB-K 185659).

Cephalaria setosa Boiss. & Hohen., Diagn. Pl. Orient. 2: 107. 1843.
Irbid (YAR-H).

Cephalaria syriaca (L.) Schrad. ex Roem & Schult., Syst. Veg. 3: 45. 1818.
Syn. *Scabiosa syriaca* L. (1753).
Amman, Amman-Madaba road, 3 km after Um El-Basatein junction (MSB-K 277806).

Cephalaria tenella Payne ex Boiss., Fl. Orient. Suppl.: 285. 1888.
Jebel Awsha, nr. Mezar al Nabi, 8.iv.1973. (28961-HUJ); Salt (YAR-H).

Knautia bidens Boiss., Diagn. Pl. Orient. 2: 108. 1843.
Irbid (YAR-H).

Knautia integrifolia (Honck. ex L.) Bertol., Fl. Ital. 2: 32. 1835.
Syn. *Scabiosa integrifolia* Honck. ex L. (1753).
Ajloun district, outside Ajloun reserve near Elynabi village, BOT-ERA (NHJ 2013.0535).

Lomelosia argentea (L.) Greuter & Burdet, Willdenowia 15(1): 73. 1985.
Syn. *Scabiosa argentea* L. (1753).
Amman, Khalda, (MSB-K 200268); Wadi Musa, Wadi Geloakh, *Ananbeh et al.* (NHJ 2013.0645).

Lomelosia aucheri (Boiss.) Greuter & Burdet, Willdenowia 15: 73. 1985.
Syn. *Scabiosa aucheri* Boiss. (1843).
Ref. Al-Eisawi 94 (2013).

Lomelosia olivieri (Coult.) Greuter & Burdet, Willdenowia 15: 75. 1985.
Syn. *Scabiosa olivieri* Coult. (1823).
Um Dami, Rum, 2016, *Ori Fragman-Sapir* (NHJ). Ref. Al-Eisawi 94 (2013) as *Scabiosa olivieri*.

Lomelosia palaestina (L.) Raf., Fl. Tellur. 4: 95. 1838.
Syn. *Scabiosa palaestina* L. (1767).
Tal Al-Rūmman, *RBG-Ire & RBG-Jo* (NHJ 2008.0080).

Lomelosia porphyroneura (Blakelock) Greuter & Burdet, Willdenowia 15: 75. 1985.
Syn. *Scabiosa porphyroneura* Blakelock (1849).
Shoubak, Al Fujaig agricultural reserve, (MSB-K 199924); Tafila, Dana Reserve, Wadi Ahmer, BOT ERA Project (NHJ 2013.0433).

Lomelosia prolifera (L.) Greuter & Burdet, Willdenowia 15: 75. 1985.
Syn. *Scabiosa prolifera* L. (1759).
Amman, Wadi-Essir, on the way to Bahath (MSB-K 416359); Tal Al-Rumman, *Al- Eisawi, Abu Sondos, Abu Anza & Shishani* (NHJ 2006.0067).

Lonicera etrusca Santi, Viagg. Montamiata 113. 1795.
Jarash, Jarash - Al Mansora, site no 16, (MSB-K 497778); Ajloun, Rasoon, 13/05/1983, Wadi Orjan Royal Botanic Garden Team (NHJ 2011.0018).

Pterocephalus arabicus Boiss., Diagn. Pl. Orient. 10: 78. 1849.
Irbid, Surra (FOR-H).

Pterocephalus brevis Coult., Mém. Dipsac. 32. 1823.
Syn. *Pterocephalus involucratus* (Sibth. & Sm.) Spreng. (1824, dated 1825; publ. in late 1824).
Ajloun, *Jarrar* 3769 (K000286292); Jordan (E00681145).

Pterocephalus papposus (L.) Coult., Mém. Dipsac. 32. 1823.
Syn. *Scabiosa papposa* L. (1753); *Knautia plumosa* L. (1771); *Pterocephalus plumosus* (L.) Coult. (1823);
Tal Al-Rumman, *Al-Eisawi* (NHJ 2006.0444).

Pterocephalus pulverulentus Boiss. & Balansa ex Boiss., Fl. Orient. 3: 152. 1875.
Amman, Daba'a (inside Daba'a Agricultural Reserve), Site no 12 (MSB-K 497480).

Pterocephalus sanctus Decne., Ann. Sci. Nat., Bot. sér. 2, 2: 266. 1834.
Jordan (E00681228); Tafila Province, 2014, *Oran* (AMM).

Sixalix arenaria (Forssk.) Greuter & Burdet, Willdenowia 15: 76. 1985.
Syn. *Scabiosa arenaria Forssk.* (1775).
(YAR-H).

Sixalix eremophila(Boiss.) Greuter & Burdet, Willdenowia 15: 76. 1985.
Syn. *Scabiosa eremophila* Boiss. (1849).
Ref. RSCN database; Al-Eisawi 94 (2013) as *Scabiosa eremophila.*

Valeriana dioscorides Sm., Fl. Graec. Prodr. 1: 21. 1806.
(YAR-H).

Valeriana italica Lam., Tabl. Encycl. 1. 92. 1791.
Dibeen Forest Reserve, May 2006, *Taifour & Othman* (FOR-H).

Valeriana sisymbriifolia Vahl, Enum. Pl. 2: 7. 1805.
Ref. RSCN database; Al-Eisawi 171 (2013).

Valerianella coronata (L.) DC., Fl. Franç. ed. 3, 4: 241. 1805.
Syn. *Valeriana locusta* L. var. *coronate (*1753).
Salt, iv.1886 (AUB); Dibeen National Park, 2005, *Mahklouf* s.n. (AMM).

Valerianella cymbocarpa C.A.Mey., Verz. Pfl. Casp. Meer. 49. 1831.
Azraq (RSCN database).

Valerianella dactylophylla Boiss. & Hohen., Diagn. Pl. Orient. 10: 75. 1849.
Ref. RSCN database; Al-Eisawi 171 (2013).

Valerianella dentata (L.) Pollich, Hist. Pl. Palat. 1: 30. 1776.
Ref. RSCN database; Al-Eisawi 171 (2013).

Valerianella dufresnia Bunge ex Boiss., Fl. Orient. 3: 109. 1875.
Tafila, Dana Reserve, Wadi Ahmer, BOT ERA project (NHJ 2013.0432).

Valerianella echinata (L.) DC., Fl. Franç. ed. 3, 4: 242. 1805.
Syn. *Valeriana echinata* L., Syst. Nat., ed. 10. 2: 861. 1759.
Jebel Siaghah, Madaba, 12.iv.1877 (607AUB).

Valerianella kotschyi Boiss., Diagn. Pl. Orient. ser. 1, 3: 60. 1843.
Dibeen National Park, 2005, *Mahklouf* s.n. (AMM).

Valerianella muricata (Stev.) W.Baxt., Suppl. Hort. Brit. [Loudon] 654. 1850.
Syn. *Valerianella truncata* (Rchb.) Betcke (1826.).
Jarash, Site 6, Dibbein Reserve, 1 km Najda after Najoba Junction (MSB-K 552716); Ajloun district, outside Ajloun reserve near Elynabi village, BOT ERA project (NHJ 2013.0529).

Valerianella orientalis Boiss. & Balansa, Diagn. Pl. Orient. II, 2: 120. 1856.
Jarash, 5.v.1886 (AUB).

Valerianella oxyrhyncha Fisch. & C.A.Mey., Index Seminum (LE) 4: 51. 1838.
Syn. *Valerianella diodon* Boiss. (1843).
Ajloun & Amman (RSCN database).

Valerianella pumila (L.) DC., Fl. Franç. ed. 3, 5: 494. 1815.
Ajloun Forest Reserve (RSCN database).

Valerianella sclerocarpa Fisch. & C.A.Mey., Index Seminum (LE) 2: 53. 1836.
Irbid (YAR-H).

Valerianella szovitsiana Fisch. & C.A.Mey., Index Sem. (St. Petersburg) 3: 48. 1836.
Burqu (FOR-H).

Valerianella tuberculata Boiss., Diagn. Pl. Orient. ser. 1, 3: 59. 1843.
Aqaba (RSCN database).

Valerianella vesicaria (L.) Moench, Methodus: 493. 1794.
Zarqa, Site 4. Alouk. (MSB-K 552680); Tal Al-Rumman, *Al-Eisawi, Abu Sondos, Abu Anza & Shishani* (NHJ 2006.0075).

Caryophyllaceae

Agrostemma githago L., Sp. Pl., ed. 2. 1: 435. 1762.
Upper & lower Jordan valley, RSCN database.

Agrostemma gracile Boiss., Diagn. Pl. Orient. ser. 2, 1: 80. 1854.
Ref. Ma'an, Danin (2004); RSCN database.

Ankyropetalum gypsophiloides Fenzl, Bot. Zeitung (Berlin) 1: 393. 1843.
Ref. Al-Eisawi 76 (2013); sin. loc. (YAR-H).

Arenaria deflexa Decne., Ann. Sci. Nat., Bot. 2, 3: 277. 1834.
Along Zarqa River, 2004, *Al-Dwiri* s.n. (AMM).

Arenaria graveolens Schreb., Nova Acta Phys.-Med. Acad. Caes. Leop.-Carol. Nat. Cur. 3: 478. 1767.
Ajloun, 5 May 1886 (AUB).

Arenaria serpylliflora subsp. **leptocladus** (Rchb.) Nyman, Consp. Fl. Eur. 115. 1878.
Syn. *Arenaria leptocldus* (Rchb.) Guss. (1844).
Ma'an, Petra, *Davis* 8974 (E00554898); Tal Al-Rumman, *Taifour & Aldamen* (NHJ 2010.0007).

Arenaria serpylliflora subsp. **tremula** (Boiss.) Govaerts (1994).
Syn. *Arenaria tremula* Boiss. (1849).
Irbid, Deir Abu Said, Apr. 1981, *El-Oqlah* (YAR-H).

Bolanthus filicaulis (Boiss.) Barkoudah, Wentia 9: 167 1962.
Syn. *Gypsophila filicaulis* (Boiss.) Bornm. (1914).
Amman, Daba'a, (inside Daba'a Agricultural Reserve), Site no 12, (MSB-K 497479); Tal Al-Rumman, *Taifour & Aldamen* (NHJ 2009.0045).

Bufonia ephedrina Sam. ex Rech.f., Ark. Bot. a.s., 1: 304. 1950.
Ref. RSCN database; Al-Eisawi 76 (2013); Ma'an & Ajloun, *Danin (2004)*.

Bufonia virgata Boiss., Fl. Orient. 1: 665. 1867.
Karak, 16.vii.1906 (104-AUB).

Cerastium dichotomum L. subsp. **inflatum** Cullen, Notes Roy. Bot. Gard. Edinb. 27: 211. 1967.
Balqa, Abu Hamed village, west of Um Rananeh village near Mabes (MSB-K 200039); Tal Al-Rumman, *Al-Eisawi* (NHJ 2006.0341).

Cerastium dubium (Bastard) Guépin, Fl. Maine et Loire, ed. 2, 1: 267. 1838.
Upper Jordan Valley, Danin 2005; RSCN database.

Cerastium glomeratum Thuill., Fl. Env. Paris, ed. 2. 226. 1799.
Dibeen National Park, 2005, *Mahklouf* s.n. (AMM).

Cometes abyssinica R.Br., Pl. Asiat. Rar. 1: 18. 1829.
Ref. Al-Eisawi 77 (2013).

Dianthus cyri Fisch. & C.A.Mey., Index Seminum (LE) 4: 34. 1838.
Upper Jordan Valley; Ajloun (RSCN database).

Dianthus libanotis Labill., Icon. Pl. Syr. 1: 14. 1791.
Ref. Karak (RSCN database).

Dianthus monadelphus Vent. subsp. **judaicus** (Boiss.) Greuter & Burdet, Willdenowia 12(2): 186.1982.
Syn. *Dianthus judaicus* Boiss. (1849); *D. auraniticus* Post (1888).
Jarash, Site 22. Muwaggar, University station at Muwaggar, (MSB-K 556862); Northeastern Badia, Biqyaweea, *Alshudiefat & Khalidi* (NHJ 2013.0003).

Dianthus pallens Sibth. & Sm., Fl. Graec. Prodr. 1(2): 286. 1809.
Azraq, 23 km NW Azraq Mar. 1955, *Long* (YAR-H).

Dianthus pendulus Boiss. & Blanche, Diagn. Pl. Orient. 2, 6: 28. 1859.
Upper Jordan valley, Ajloun (RSCN database).

Dianthus sinaicus Boiss., Diagn. Pl. Orient. 1: 23. 1843.
Tafila Province, 2014, *Oran* (AMM); Wadi Rum, 1996, *Al-Eisawi et al.* (AMM).

Dianthus strictus Banks & Sol., Nat. Hist. Aleppo ed. 2, 2: 252. 1794.
Syn. *Dianthus polycladus* Boiss. (1949).
Nr. Zarqa, *Frey, Hilger & Kürschner* (E00475315).

Dianthus strictus subsp. **multipunctatus** (Ser.) Mouterde ex Greuter & Burdet, Willdenowia 12: 187. 1982.
Syn. *Dianthus multipunctatus* Ser. (1824).
Wood nr. Maqita, *Hunting Aero Survey* (E).

Dianthus tripunctatus Sibth. & Sm., Fl. Graec. Prodr. 1: 286. 1809.
Azraq, SW of Azraq midway between Azraq and Qasr Amra, Mar. 1979 (YAR-H); Dana (RSCN database).

Dianthus zonatus Fenzl, Pug. Pl. Nov. Syr.: 11. 1842.
Ref. RSCN database; Al-Eisawi 77 (2013).

Gymnocarpos decandrus Forssk., Fl. Aegypt.-Arab.: 65. 1775.
Wadi Mujib, *Karim & Al-Qaran* 2329 (K000315567); Wadi Rum, *Boulos & Lahham* 7704 (AMM, BM, E).

Gymnocarpos sclerocephalus (Decne.) Dahlgren & Thulin, Edinburgh J. Bot. 59: 233. 2002.
Syn. *Paronychia sclerocephala* Decne.(1835); *Sclerocephalus arabicus* Boiss. (1843).
Tafila, Wadi Araba, *Al Oqlah* (NHJ 2012.0360).

Gypsophila arabica Barkoudah, Wentia 9: 139. (1962).
Along Zarqa River, 2004, *Al-Dwiri* s.n. (AMM); Wadi Yutum, *Boulos & Lahham* 56 (AMM).

Gypsophila capillaris (Forssk.) C.Chr., Dansk Bot. Ark. 4(3): 19. 1922.
Syn. *Rokejeka capillaris* Forssk. (1775); *Gypsophilla rokejeka* Delile (1842); *Gypsophila antari* Post & Beauv. (1932).

Two subspecies are recognised:

Perianth ± persistent in fruit; seeds 4–8 in per capsule; coat smooth . . . . .subsp. *capillaris*
Perianth caducous in fruit; seeds 1–4 in per capsule; coat tuberculate . . . . subsp. *confusa*

subsp. **capillaris** is found in Egypt, Iraq, Kuwait & NE Saudi Arabia.
subsp. **confusa** Zmarzty, Kew Bull. 48(4): 694. 1993.
Syn. *Gypsophila arabica* Bark. (1962).
Jordan, *Frey & Kürschner* 5126 (E00219287); Tal Al-Rumman, *Al-Eisawi* (NHJ 2006.0412).
Amman, Bei Zarqa Ma'in, *Frey & Kürschner* 5762 (E00502654).

Gypsophila damascena Boiss., Diagn. Pl. Orient. 8: 57. 1849.
Mafraq, Umm Jemal, 1875 (AUB).

Gypsophila pilosa Hudson, Philos. Trans. 56: 252 .1767.
Tafila, Ain Lahda to Doma Tower at Nature Reserve (MSB-K 151876).

Gypsophila viscosa Murray, Philos. Trans. 56: 252. 1767.
40 km SW of Ma'an, *Dinsmore* 11745 (E00631843).

Habrosia spinuliflora Fenzl, Bot. Zeitung (Berlin) 1: 323. 1843.
Tafila, Dana, 12.IV.1900 (AUB).

Herniaria hemistemon J.Gay, Rev. Bot. Recueil Mens. 2: 371. 1847.
Madaba, 2.iv.1897 (AUB).

Herniaria hirsuta L., Rev. Bot. Recueil Mens. 2: 371. 1847.
Jerash, *Hepper* 3168 (K000286211); Tal Al-Rumman, *Taifour & Aldamen* (NHJ 2009.0088).

Holosteum umbellatum L., Sp. Pl.: 88. 1753.
Tafila, Dana, opposite Dana Hotel, between orchards (MSB-K 412672).

Holosteum umbellatum L. subsp. **glutinosum** (M.Bieb.) Nyman, Consp. Fl. Europ. 112. 1878.
Syn. *Holosteum glutinosum* (M.Bieb.) Fisch. & C.A.Mey. (1840).
Amman, Jubiaha, Mar. 1973., *Boulos* (AMM); Eastern slopes of Jordan valley (NCARE).

Loeflingia hispanica L., Sp. Pl.: 35. 1753.
Lower Jordan Valley (RSCN database); Ma'an, *Danin* (2004).

Minuartia decipiens (Fenzl) Bornm. subsp. **decipiens**
Jarash, Site 32. Souf (MSB-K 557973); Tal Al-Rumman, *Taifour & Aldamen* (NHJ 2009.0086).

Minuartia formosa (Fenzl) Mattf., Bot. Jahrb. Syst. 57(126): 33. 1921.
Syn. *Alsine formosa* Fenzl (1843).
Tal Al-Rumman, *Al-Eisawi* (NHJ 2006.0342).

Minuartia globulosa (Labill.) Schinz & Thell., Bull. Herb. Boissier 2, 7: 403. 1907.
Ref. Ajloun (RSCN database); Al-Eisawi 78 (2013).

Minuartia hamata (Hausskn. & Bornm.) Mattf., in Bot. Jahrb. 57 Beibl. 126: 29. (1921).
(YAR-H).

Minuartia hybrida (Vill.) Schischk., Fl. URSS 6: 488. 1936.
Syn. *Arenaria tenuifolia* L. (1753).*Alsine tenuifolia* (L.) (1766); *Arenaria hybrida* Vill. (1779).
Jebel Siaghah, Madaba,12.IV.1877 239 (AUB); Dibeen National Park, 2005, *Mahklouf* s.n. (AMM).

Minuartia intermedia (Boiss.) Hand.-Mazz., Ann. Nat. Hofmus. Wien 26. 148. 1912.
Syn. *Alsine intermedia* Boiss. (1867); *Arenaria intermedia* (Boiss.) Fernald (1919).
Ref. RSCN database; Al-Eisawi 78 (2013); Ma'an, Danin (2004).

Minuartia mediterranea (Ledeb.) K.Maly, Glasn. Zemaljsk. Muz. Bosni Hercegovini 20: 563. 1908.
Tal Al-Rumman, Royal Botanic Garden, *Al-Eisawi* (NHJ 2006.0281).

Minuartia meyeri (Boiss.) Bornm., Beih. Bot. Centralbl. 27(2): 313. 1910.
Syn. *Alsine brevis* Boiss. (1849); *Alsine Meyeri* Boiss. (1849).
Ref. Moab (RSCN database); Ma'an (Moab), Danin *(2004)*.

Minuartia picta (Sibth. & Sm.) Bornm., Notizbl. Bot. Gart. Berlin-Dahlem 7: 143. 1917.
Syn. *Arenaria picta* Sibth. et Smith (1806); *Alsine picta* (Sibth. & Sm.) Fenzl (1833).
Jordan, *Davis* 9626 (E00201916); Tafila district, Dana Reserve, Ruman Camp, BOT ERA (NHJ 2013.050).

Minuartia picta var. **sinaica** (Boiss.) Bornm., Beih. Bot. Centralbl. 28(2): 148. 1911.
Syn. *Minuartia sinaica* (Boiss.) Danin (1987).
Ref. Aqaba, Danin (2004).

Paronychia arabica (L.) DC., Cat. Pl. Horti Monsp. 130. 1813.
Syn. *Illecebrum arabicum* L. (1767); *Paronychia desertorum* Boiss. (1843); *Paronychia lenticulata* (Forssk.) Asch. & Schweinf. (1889).
800 m W of the Safawi-Azraq rd., 26 km from Safawi (BRDC-H); Karak Governorate, Al Kaser, Ibn Hammad, *Taifour* (NHJ); nr. Rum rest house, 23 Mar. 1975, *Boulos et al.* (AMM).

Paronychia argentea Lam., Fl. Franç. 3: 230. 1779.
Jerash, *Hepper* 3164 (K000286218); Tal Al-Rumman, *Taifour* (NHJ 2009.0083).

Paronychia capitata (L.) Lam., Fl. Franç. 3: 229. 1779.
Syn. *Paronychia kurdica* Boiss. (1843).
Irbid, Um Qais (YAR-H); Azraq (RSCN database).

Paronychia jordanica Chaudhri, Rev. Paronychiinae 280. 1968.
Ref. RSCN database; Al-Eisawi 79 (2013).

Paronychia macrosepala Boiss., Diagn. Pl. Orient. 3: 11. 1843.
Al Mafraq, 14 Mar. 1955 (FOR-H).

Paronychia palaestina Eig, P.Z.E. Inst. Agric. Nat. Hist., Agric. Rec. 2: 205. 1929.
Azraq Wadi Bottom (South west the Castle Amra), *El-Oqlah* (NHJ 2013.0570).

Paronychia sinaica Fresen., Mus. Senckenberg. 1: 180. 1833.
Wadi Bayir [Ba'ir], *Boulos, Addiss & Qumsiyeh* 8698 (K000286237);Tafila district, Dana Reserve, Ruman Camp, BOT ERA project (NHJ 2013.050).

Petrorhagia arabica (Boiss.) P.W.Ball & Heywood, Bull. Brit. Mus. (Nat. Hist.) Bot. 3: 143. 1964.
Syn. *Tunica arabica* Boiss. (1849).
Madaba (HUJ); Lower Jordan valley (RSCN database).

Petrorhagia cretica (L.) Ball & Heywood, Bull. Brit. Mus. (Nat. Hist.) Bot. 3: 142. 1964.
Syn. *Saponaria cetica* L. (1762).
Tafila Province, 2014, *Oran* (AMM).

Petrorhagia dubia (Rafin.) G.Lopez & Romo, Anales Jard. Bot. Madrid 45: 363. 1988.
Syn. *Petrorhagia velutina* (Guss.) P.W. Ball & Heywood (1964).
Salt, iv.1886 Salt (AUB).

Petrorhagia zoharyana Liston, Candollea 41: 179. 1986.
Lower Jordan Valley, *Liston & Sagi* 7–85–317/1 (E00301861).

Polycarpaea repens (Forssk.) Asch. & Schweinf., Oesterr. Bot. Z. 39: 126. 1889.
Syn. *P. fragilis* Del. (1814.) nom. illegit.; *Corrigiola repens* Forssk. (1775).
Aqaba, Aqaba-Saudi border highway (MSB-K 412638).

Polycarpaea robbairea (Kuntze) Greuter & Burdet, Willdenowia 12: 189. 1982.
Syn. *Polycarpon robbairea* Kuntze (1891); *Robbairea delileana* Milne-Redh. (1949).
Ghour Safi, iv.1895, *Fox* (AUB); 6 km N of Aqaba, *Boulos & Lahham* 6085 (AMM, BM, E).

Polycarpon succulentum(Del.) J.Gay in Rev. Bot. Bull. Mens. 2: 373. 1846.
Syn. *Alsine succulenta* Delile (1814); *Polycarpon arabicum* Boiss. (1849).
Karak, Wadi Araba, *Frey & Kürschner* (BSB); Dana (RSCN database).

Polycarpon tetraphyllum (L.) L., Syst. Nat. ed. 10 2: 881. 1759.
Tal Al-Rumman, *Al-Eisawi* (NHJ 2006.0313); Tal Al-Rumman, *Taifour & Aldamen* (NHJ 2009.0087).

Pteranthus dichotomus Forssk., Fl. Aegypt.-Arab.: 62. 1775.
Aqaba, 5 km after Beer Mazkor Resthouse (MSB-K 419143); Wadi Hassa, Afra hot spring, BOT ERA project, (NHJ 2013.0502).

Sagina apetala Ard., Animadv. Bot. Spec. Alt. 2: 22. 1764.
Ref. Ajloun & Amman, Danin (2004).

Saponaria mesogitana Boiss., Diagn. Pl. Orient. 1: 16. 1843.
Ref. Al-Eisawi 80 (2013).

Silene aegyptiaca L.f., Suppl. Pl. 241. 1782.
Jordan, *Frey & Kürschner* 4880 (E 00292051); Tal Al-Rumman, *Al- Eisawi et al.* (NHJ 2006.0053).

Silene apetala Willd., Sp. Pl. 2: 307. 1799.
Irbid, 10m before Burma municipality sign, from Gazeh Camp to Burma (MSB-K 176714); Tal Al-Rumman *Al-Eisawi* (NHJ 2006.0385).

Silene arabica Boiss., Fl. Orient. 1: 593. 1867.
Syn. *Silene affinis* Boiss. (1854).
Jordan, *Frey & Kürschner* 5026 (E00057078); Tal Al-Rumman, *Taifour & Aldamen* (NHJ 2009.0084).

Silene arenosa C.Koch, Linnaea 15: 711. 1841.
Lower Jordan valley (RSCN database).

Silene behen L., Sp. Pl.: 418. 1753.
Syn. *Ebraxis behen* (L.) Tzevlev (2001).
Jebel Siaghah, Madaba, 28.iv.1886(AUB); Tal Al-Rumman, *Taifour & Aldamen* (NHJ 2009.0061).
Some authorities (Tropicos) list *Ebraxis behen* (L.) Tzevlev as an accepted name and *Silene behen* L. as its synonym.

Silene colorata Poir., Voy. Barbarie 2: 163. 1789.
Syn. *Silene decipiens* Barceló (1879).
Tafila, Dana Reserve (MSB-K 416245); Tafila, Dana Reserve, Wadi Ahmer, BOT ERA project (NHJ 2013.0435).

Silene coniflora Nees ex Otth, DC., Prodr. 1: 371. (1824).
Jordan, *Frey & Kürschner* 4998 (E00124410.); Tal Al-Rumman, *Al-Eisawi* (NHJ 2006.0025).

Silene conoidea L., Sp. Pl.: 418. 1753.
Madaba (MSB-K 113979); Tal Al-Rumman, *Taifour & Aldamen* (NHJ 2010.0006).

Silene crassipes Fenzl, Pug. Pl. Nov. Syr. 8. 1842.
Stream area, *Royal Botanic Garden Team* (NHJ 2011.0110).

Silene damascena Boiss. & Gaill., Diagn. Pl. Orient. 2, 6: 34. 1859.
Ajloun district, outside Ajloun reserve near Elynabi village, BOT ERA project (NHJ 2013.0521).

Silene danaensis Danin, Willdenowia 27: 166. 1997.
Endemic. Type: Jordon, Dana Reserve, *Danin* 963502 (holo. HUJ; iso. B, E!, K!).

Silene dichotoma Ehrh., Beitr. Naturk. 7: 144. 1792.
Ajloun district, outside Ajloun reserve near Elynabi village, BOT ERA project (NHJ 2013.0523).

Silene dichotoma subsp. **racemosa** (Otth) Graebn., Syn. Mitteleur. Fl. 5(2): 93. 1920.
Syn. *Silene sibthorpiana* Rchb. (1832).
Madaba, Jebel Siaghah, 5 Apr. 1877 (AUB).

Silene fuscata Link ex Brot., Fl. Lusit. 2: 187. 1805.
Jordan, *Meyers & Dinsmore* 1606 (E00292041).

Silene gallica L., Sp. Pl.: 417. 1753.
Irbid, Abu Logus (YAR-H).

Silene grisea Boiss., Diagn. Pl. Orient. 8: 88. 1849.
Irbid, Abu Logus (YAR-H).

Silene hussonii Boiss., Diagn. Pl. Orient. 8: 76. 1849.
Dana Reserve (RSCN database).

Silene italica (L.) Pers., Syn. Pl. 1: 498. 1805.
Irbid, Agraba (YAR-H).

Silene libanotica Boiss., Diagn. Pl. Orient. 8: 89. 1849.
Ref. Falahat (1999); Petra, access list on the plants of Petra based on Old Petra List (undated); Aqaba, *Danin* (2004).

Silene linearis Decne., Ann. Sci. Nat., Bot. 2, 3: 276. 1835.
Tafila, Tafileh-Ghor road (MSB-K 278272).

Silene longipetala Vent., Descr. Pl. Nouv. t. 83, 1802.
Shoubak, Hisha, Site no 22 (MSB-K 497712); Irbid, Saham, 30/3/1982 (YAR-H).

Silene macrodonta Boiss., Diagn. Pl. Orient. 1: 37. 1843.
Ref. Karak (RSCN database); (Moab), Danin (2004).

Silene moabitica Eig, Repert. Spec. Nov. Regni Veg. Beih. 63(1): 74. 1931.
Wadi Rum, 15 Apr. 1945, *Davis* (K).
This name is unresolved.

Silene modesta Boiss. & Bl., Diagn. Pl. Orient. ser. 2, 6: 33. 1859.
Ref. Al-Eisawi 80 (2013).

Silene muscipula L., Sp. Pl.: 420. 1753.
Irbid, El-Shouneh, 15/3/1983 (YAR-H).

Silene nocturna L., Sp. Pl.: 416. 1753.
Ajloun, Rajeb, Site no 15 (MSB-K 496531); Tal Al-Rumman, *Taifour et al.* (NHJ 2012. 0402).

Silene oxyodonta Barb.-Boiss. & Barbey, Herbor. Levant: 121. 1882.
Balqa, Jordan Valley, 10 km N of Dead Sea (HUJ).

Silene palaestina Boiss., Diagn. Pl. Orient. 8: 80. 1849.
Jordan, *Frey & Kürschner* 4896 (E00057087).

Silene papillosa Boiss., Diagn ser. 1, 1: 39. 1843
Dibeen National Park, 2005, *Mahklouf* s.n. (AMM).

Silene picta Pers., Syn. Pl. 1: 498. 1805.
Salt, Apr. 1895 (AUB).

Silene physalodes Boiss., Diagn. Pl. Orient. 8: 83. 1849.
Ma`an, Al Shoubak, 04/04/1984 (YAR-H).

Silene reinwardtii Roth, Catal. Bot. 3: 42. 1806.
Irbid, Wadi Al Arab, 04/06/1986 (YAR-H).

Silene rubella L., Sp. Pl.: 419. 1753.
Jordan, *Meyers & Dinsmore* B 534 (E00048461); Tal Al-Rumman RBG, Sector No. 71, *Taifour* (NHJ 2013.0195).

Silene schimperiana Boiss., Diagn. Pl. Orient. 1: 31. 1843.
Ref. Petra Falahat (1999).

Silene swertiifolia Boiss., Diagn. Pl. Orient., ser. 1 1: 32. 1842.
Tafila, Dana (MSB-K 41838.6).

Silene tridentata Desf., Fl. Atlant. 1: 349. 1798.
Ref. Lower Jordan Valley (RSCN database); Karak, Aqaba (Moab & Edom), Danin (2004).

Silene trinervis Sol., Nat. Hist. Aleppo ed. 2, 2: 252. 1794.
Tal Al-Rumman, *Al-Eisawi* (NHJ 2006.0241).

Silene villosa Forssk., Fl. Aegypt.-Arab.: 210. 1775.
Ma'an, road to Wadi Rum, 30/11/1955 (YAR-H).

Silene vivianii Steud., Nomencl. Bot. ed. 2, 2: 588. 1841.
Al-Mfraq, Serah, 27 Apr. 1983 (YAR-H).

Silene vulgaris (Moench) Garcke, Fl. N. Mitt.-Deutschland ed. 9: 64. 1869.
Syn. *Silene venosa* Asch. (1860).
Amman, Al Saravo area (MSB-K 228048); Tal Al-Rumman RBG, Sector No. 87, *Taifour* (NHJ 2013.0209).

Spergula arvensis L., Sp. Pl.: 440. 1753.
Lower Jordan Valley (RSCN database).

Spergula fallax (Lowe) E.H.L.Krause, Deutschl. Fl. ed. 2, 5: 19. 1901.
Syn. *Spergularia fallax* Lowe (1856).
Dana, *Taifour* (NHJ 2013.0424).

Spergularia bocconei (Scheele) Asch. & Graebn., Syn. Mitteleur. Fl. 5: 849. 1919.
Azraq, 4 km E of Shehan village, 25 Apr. 1955 (YAR-H).

Spergularia diandra (Guss.) Heldr., Pl. Atticae 1848 s.p. 1851.
Syn. *Spergula diandra* (Guss.) Murb. (1896).
Azraq, Site 9, Shomari Reserve (MSB-K 554260); Tal Al-Rumman, *Taifour & Aldamen* (NHJ 2009.0089).

Spergularia marina(L.) Besser, Enum. Pl. 97. 1822.
Syn. *Spergularia salina* J.Presl & C.Presl (1819).
Amman, Site 2. Dead Sea (MSB-K 552565); Tal Al-Rumman, *Al-Eisawi* (NHJ 2006.0155).

Spergularia media (L.) C.Presl, Enum. Pl. 97. 1822.
Ref. Al-Eisawi 82 (2013).

Spergularia rubra (L.) J.Presl & C.Presl, Fl. Cech. 94. 1819.
Syn. *Arenaria rubra* L.(1753).
Al-Mafraq, Um Al Jimal (IBBU-H); Upper and Lower Jordan valley (RSCN database).

Stellaria media (L.) Vill., Hist. Pl. Dauphiné 3: 615. 1789.
Syn. *Spergularia maritima* (All.) Chiov. (1911).
Ajloun (MSB-K 228370); Azraq Wetland Reserve (IBBU-H); Tal Al-Rumman, *Al-Eisawi* (NHJ 2006.0312).

Stellaria media subsp. **cupaniana** (Jord. & Fourr.) Nyman, Consp. Fl. Eur. 111. 1878.
Syn. *Stellaria cupaniana* (Jordan & Fourr.) Beguinot (1908).
Irbid, Ebdar Mar. 2005, *Swalha* (YAR-H).

Stellaria pallida (Dumort.) Piré, Bull. Soc. Roy. Bot. Belgique 2: 48. 1863.
Syn. *Stellaria media* (L.) Vill subsp. *pallida* (Dumort.) Asch. & Graebn. (1898).
Irbid, Ebdar, Mar. 2005, *Swalha* (YAR-H); Upper Jordan Valley (RSCN database).
Ref. Ma'an, Danin (2004).

Telephium sphaerospermum Steud. ex Boiss., Diagn. Pl. Orient. 4: 12. 1844.
Tal Al-Rumman RBG, *Al Oqlah, Taifour & Al Khawaldah* (NHJ 2013.0270).

Vaccaria hispanica (Mill.) Rauschert, Wiss. Z. Martin-Luther-Univ. Halle-Wittenberg, Math.-Naturwiss. Reihe 14: 496. 1965.
Syn. *Vaccaria pyramidata* Medik. (1789).
Zarqa, Southwest of Alouk (MSB-K 186450); Tal Al-Rumman, *Taifour & Aldamen* (NHJ 2010 .0046).

Velezia rigida L., Sp. Pl.: 332. 1753.
Jarash, Jarash-Anjarah road (MSB-K 275053).

Ceratophyllaceae

Ceratophyllum demersum L., Sp. Pl. 2: 992. 1753.
East of Jordan, 1843, *Paine* (K); Upper Jordan valley, RSCN database.

Cistaceae

Cistus creticus L., Sp. Pl. ed. 2.1: 738. 1762.
Irbid, Burma municipality, after the village, left roadside (MSB-K 176460); Dibeen Forest, *RBG-Ire & RBG-Jo* (NHJ 2008.0105).

Cistus salviifolius L., Sp. Pl. 524. 1753.
Salt to Rumman,*Meyers, Fred & Dinsmore* M500 (E00431702); Dibeen Forest, *RBG-Ire & RBG-Jo* (NHJ 2008.0067).

Fumana arabica (L.) Spach, Ann. Sci. Nat., Bot. 2, 6: 269. 1836.
Syn. *Cistus arabicus* L. (1755).
South of Ajloun, *Hunting Aero Survey* 111A (E00454643);Tal Al-Rumman, *Taifour* (NHJ 2013.0212).

Fumana procumbens (Dunal) Gren. & Godr., Fl. France 1: 173. 1847.
Al Majdal (FOR-H).

Fumana thymifolia (L.) Spach, Ann. Sci. Nat., Bot. 2, 7: 271. 1836.
Syn. *Cistus thymifolius* L. (1753).
Hunting Aero Survey 166A (E00154086).

Helianthemum aegyptiacum (L.) Mill., Gard. Dict. ed. 8, 23. 1768.
Tal Al-Rumman RBG, Sector No. 93, *El-Oqlah, Taifour & Al Khawaldah* (NHJ 2013.0257).

Helianthemum fasciculi Greuter, Boissiera 13: 55. 1967.
Dibeen Forest Reserve, 2004, *Mahklouf* (AMM).

Helianthemum kahiricum Delile, Descr. Egypte, Hist. Nat. 237. 1813.
Near Shobek, *Davis* 9124 (E00431858).

Helianthemum lasiocarpum Desf. ex Willk., Icon. Descr. Pl. Nov. 2: 88, t. 122. 1856.
Syn. *Helianthemum ledifolium* (L.) Mill. var. *lasiocarpum* (Desf. ex Willk.) Stoj. & Stef.
Dibeen National Park, 2005, *Mahklouf* s.n. (AMM).

Helianthemum ledifolium (L.) Mill., Gard. Dict., ed. 8. n. 20. 1768.
Syn. *Cistus ledifolius* L. (1753).
E of Salt, *Meyers, Fred & Dinsmore* M1694 (E00431926);Tal Al-Rumman, *Taifour & Aldamen* (NHJ 2009.0139).

Helianthemum lippii (L.) Dum.Cours., Bot. Cult. 3: 130. 1802.
Syn. *Cistus lippii* L. (1771); *Cistus sessiliflorus* Desf. (1798); *Helianthemum ellipticum* Nábělek (1806); *Helianthemum sessiliflorum* (Desf.) Pers. (1806).
15 km east Ma'an, along the road to El-Jafr, *Boulos & Jallad* 7869 (E00431910); *Taifour & Aldamen* (NHJ 2009.0143).

Helianthemum salicifolium (L.) Mill., Gard. Dict. ed. 8, 21. 1768.
Syn. *Cistus salicifolius* L. (1753).
N of Salt, *Frey & Kürschner* 80–268 (E00454415, TUB); Tal Al-Rumman *Al-Eisawi* (NHJ 2006.0120).

Helianthemum sancti-antonii Boiss., Fl. Orient. Suppl.: 70. 1888.
Petra, *Dinsmore* 11728 (E00431872); Tafila Province, 2014, *Oran* (AMM).

Helianthemum stipulatum (Forssk.) C.Chr., Dansk Bot. Ark. 4(3): 20. 1922.
Syn. *Cistus stipulatus* Forssk. (1775).
Jordan, *Davis* 9058 (K000286330).

Helianthemum syriacum (Jacq.) Dum.Cours., Bot. Cult. 3: 129. 1802.
Syn. *Cistus syriacus* Jacq.(1871).
Amman, Khalda, from the Al Waha Circle toward Sweleh (MSB-K 200279).

Helianthemum ventosum Boiss., Diagn. Pl. Or. 8: 50. 1849.
Wadi Yutum, *Boulos & Lahham* 7596 (AMM, BM, E).

Helianthemum viscarium Boiss. & Reut., Pugill. Pl. Afr. Bor. Hispan. 14. 1852.
Above Wadi Musa on the road to Moan, *Davis* 8670 (E00431849.); Tal Al-Rumman, *Taifour* (NHJ 2009.0140).

Tuberaria guttata (L.) Fourr., Ann. Soc. Linn. Lyon n.s., 16: 340. 1868.
Syn. *Cistus guttatus* L. (1753).
Ref. El-Eisawi 88 (2013).

Cleomaceae

Cleome amblyocarpa Barratte & Murb., Acta Univ. Lund., 2 1(4): 25. 1905.
Syn. *Cleome africana* Botsch. (1964).
Aqaba, *Leonard* 7480 (E00443705); nr. Aqaba, *Boulos & Lahham* 8532 (AMM, BM, E).

Cleome arabica L., Cent. Pl. 1. 20. 1755.
Syn. *Cleome trinervia* Fresen. (1834).
Zerka, *Davis* 9420 (E00443721); Wadi Yutum (FOR-H).

Cleome droserifolia (Forssk.) Delile, Descr. Egypte, Hist. Nat. 250. 1813.
Syn. *Roridula droserifolia* Forssk. (1775).
5 km S of Aqaba, *Frey & Kürschner* 80–553 (E00443798, TUB).

Cleome gynandra L., Sp. Pl. 671. 1753.
Syn. *Gynandropsis gynandra* (L.) & Briq. (1914).
Irbid area (YAR-H).

Colchicaceae

Androcymbium palaestinum Baker, J. Linn. Soc., Bot. 17: 445. 1880.
Deif Alla [Deir Alla] *Simpson* 53011 (K000318360); Tal Al-Rumman, *Al- Eisawi, Abu- Sondos, Abu Anza & Shishani* (NHJ 2006.0055).

Colchicum crocifolium Boiss., Diagn. Pl. Orient. 5: 67. 1844.
North Eastern desert, RSCN database.

Colchicum decaisnei Boiss., Fl. Orient. 5: 157. 1882.
Jordan, 25.I.1963, *Birkinshaw* (AUB).

Colchicum hierosolymitanum Feinbrun, Palestine J. Bot., Jerusalem Ser. 6: 84. 1953.
Ajloun, *Karim &Al-Qaran* 1425 (K000318363).

Colchicum ritchii R.Br., Narr. Travels Africa App.: 241. 1826.
Wadi Rum, *Davis* 10428 (00329004).

Colchicum schimperi Janka ex Stef., Sborn. B'lghar. Akad. Nauk 22: 3.1 1926.
Syn. *Colchicum guessfeldtianum* Asch. & Schweinf. (1889); *Colchicum deserti-syriaci* Feinfrum (1953).
Petra, *Dinsmore* 11716 (E00325699).

Colchicum serpentinum Woronow ex Miscz., Fl. Cauc. Crit. 3(4): 114. 1912.
Syn. *Colchicum tauri* Siehe ex Stefanov (1926).
Tal Al-Rumman, *Taifour & Aldamen* (NHJ 2009.0051).

Colchicum stevenii Kunth, Enum. Pl. 4: 144. 1843.
Near Umm Qeis, NW Irbid, *Boulos, Loutfy & Hauke* 5823 (E00325714).

Colchicum szovitsii Fisch. & C.A.Mey. subsp. **brachyphyllum** (Boiss. & Hausskn.) K.Perss., Bot. Jahrb. Syst. 127: 221 2007.
Syn. *Colchicum brachyphyllum* Boiss. & Hausskn. (1882).
Amman, 2–3 km after Qaser Al-Kharanneh. Site no 11 (MSB-K 496232).

Colchicum triphyllum Kuntze, Flora 29: 755. 1846.
Wadi Rum, 14 Mar. 1974, *Al-Eisawi* (AMM).

Colchicum tunicatum Feinbrun, Palestine J. Bot., Jerusalem Ser. 6: 87. 1953.
Pass between Petra & Mt. Hor, *Davis* 9198 (E00325740).

Colchicum tuviae Feinbrun, Palestine J. Bot., Jerusalem Ser. 6: 79. 1953.
Ref. RSCN database; Al-Eisawi 124 (2013).

Convolvulaceae

Calystegia sepium (L.) R.Br., Prodr. Fl. Nov. Holland.: 483 1810.
Syn. *Convolvulus sepium* L. (1753).
Dibeen Forest Reserve, 2004, *Mahklouf* (AMM).
A polymorphic species where several subspecies are recognized. Widespread in temperate regions.

Convolvulus althaeoides L., Sp. Pl.: 156 1753.
Tafila, Dana Reserve, Rumana trail behind Rumana camp, BOT ERA project (NHJ 2013.0458).

Convolvulus arvensis L., Sp. Pl.: 153. 1753.
Deir Alla farm, 21-Apr–53, *Trought* s.n. (E00465985).

Convolvulus betonicifolius Mill., Gard. Dict., ed. 8. n. 20. 1768.
Syn. *Convolvulus hirsutus* Mill
Irbid, 5.v.1886, *Boshra* 119 (AUB); Dibeen National Park, 2005, *Mahklouf* s.n. (AMM).

Convolvulus cantabrica L., Sp. Pl.: 158. 1753.
Azraq, 17 May 1983 (YAR-H).

Convolvulus coelesyriacus Boiss., Diagn. Pl. Orient. ser. 1, 11: 85. 1849.
Jebel Siaghah, Madaba, 14.IV.1877 (AUB).

Convolvulus dorycnium L.,Syst. Nat., ed. 10 2: 923. 1759.
Wadi Zerqa, *Hunting Aero Survey* 102A (E00465726); Tal Al-Rumman, *Taifour* (NHJ 2009.0107).

Convolvulus fatmensis Kunze, Flora 23(1): t. 172. 1840.
Al-Mafraq- Al-Qnayah, Al-Mdawara and Rehab way (IBBU-H).

Convolvulus glomeratus Choisy, Prodr. 9: 401. 1845.
Karak, on the way from Ghor Safi toward Amman(MSB-K 341435); Dead sea, *Taifour et al.* (NHJ 2013.0629).

Convolvulus humilis Jacq, Collectanea 4: 209, t. 32, 1790.
Amman, Wadi el Seer, after Alsahwa village on the way to Al Bahath (MSB-K 276485).

Convolvulus jordanensis Sa'ad, Convolv. Sp. Canar. Isl., Medit. Reg., etc. 159. 1967.
Jordan, *Robertson* 120 (K000852406).

Convolvulus lanatus Vahl, Symb. Bot. 1. 16. 1790.
(YAR-H); Araba valley, *Danin* (http://flora.org.il/en/plants/CONLAN/).

Convolvulus lineatus L., Syst. Nat., ed. 10 2: 923. 1759.
Karak, Merj, 26 Apr. 1878 (AUB); RSCN database.

Convolvulus palaestinus Boiss., Diagn. Pl. Orient. ser. 1, 11: 84. 1849.
Dibeen Forest Reserve, May 2006 (DFR-H).

Convolvulus pentapetaloides L., Syst. Nat., ed. 12. 2: 229. 1767.
Deir Alla, 21-May–52, *Trought s.n.* (E00499374); Dibeen National Park, 2005, *Mahklouf* s.n. (AMM).

Convolvulus pilosellifolius Desr., Encycl. 3: 551. 1789.
Deir Alla, 21-May–52, *Trought s.n.* (E00499370).

Convolvulus scammonia L., Sp. Pl.: 153. 1753.
Salt, IV.1886, 1274 (AUB); Dibeen National Park, 2005, *Mahklouf* s.n. (AMM).

Convolvulus schimperi Boiss., Diagn. Pl. Orient. ser. 1, 11: 81. 1849.
Wadi Araba, Dana Nature Reserve (RSCN database).

Convolvulus siculus L., Sp. Pl.: 156. 1753.
Tal Al-Rumman, *Taifour H.* (NHJ 2009.0109).

Convolvulus spicatus Peter ex Hallier f., Bot. Jahrb. Syst. 18(1–2): 99. 1893.
NE side of the wadi Sirhan between Azraq and El Hasim, *Hunting Aero Survey* 117A (E00499268).

Convolvulus stachydifolius Choisy, Prodr. 9: 408. 1845.
Dana, *Taifour* (NHJ 2013.0397).

Convolvulus tricolor L., Sp. Pl.: 158. 1753.
Salt, Amman, Karak, Madaba (RSCN database).

Cressa cretica L., Sp. Pl. 223. 1753.
Aqaba, *Dawe* 5 (K000286262).

Cuscuta approximata Bab., Ann. Mag. Hist. 13(84): 253, pl. 4, f. 3. 1844.
Gilead, Danin (2005); RSCN database.

Cuscuta campestris Yunck., Mem. Torrey Bot. Club 18(2): 138. 1932.
Jordan, Site 52. Jordan University Research Station, Mwagar (MSB-K 395759).

Cuscuta epilinum Weihe, Archiv Apoth. 8: 51. 1824.
Ref. RSCN database; Al-Eisawi 90 (2013).

Cuscuta epithymum (L.) L., Syst. Veg. (ed. 14) 140. 1774.
Syn. *Cuscuta europaea* L. var. *epithymum* L. (1753).
Upper Jordan valley, Dana (FOR-H).

Cuscuta europaea L., Sp. Pl.: 124. 1753.
Syn. *Cuscuta brevistyla* A.Braun ex A.Rich. (1850).
Wadi Sirhan, *Hunting Aero Survey* 127A (E00395071)

Cuscuta gennesaretana Sroelov ex Feinbrun & S.Taub, Israel J. Bot. 13: 5. 1965.
Upper Jordan valley, RSCN database.

Cuscuta monogyna Vahl, Symb. Bot. 2. 32. 1791.
Upper Jordan valley, RSCN database.

Cuscuta pedicellata Ledeb., Fl. Altaic. 1: 293, t. 234. 1829.
Upper Jordan valley, Danin (2005); RSCN database.

Cuscuta palaestina Boiss., Diagn. Pl. Orient., ser. 1, 11: 86. 1849.
Dana Reserve, plateau near top of Rumana mountain, SW aspect, BOT ERA project (NHJ 2013.0441).

Cuscuta planiflora Ten., Fl. Napol. 3: 250, t. 220, f. 3. 1824.
Zerka Main to Jebel Attaris, *Davis* 9384 (E00388354).

Ipomoea cairica (L.) Sweet, Hort. Brit. 2: 287. 1826.
Syn. *Convolvulus cairicus* L. (1759).
Upper Jordan valley (RSCN database).

Ipomoea pes-caprae (L.) R.Br., Narr. Exped. Zaire 477. 1818.
Recorded by (YAR-H).

Ipomoea purpurea (L.) Roth, Bot. Abh. Beobacht. 27. 1787.
Ajloun, Rajeb village, Site no 6 (MSB-K 496128).

Crassulaceae

Tillaea alata Viv., Pl. Aegypt. Dec. 16. 1830.
Syn. *Crassula alata* (Viv.) A.Berger (1930).
Upper Jordan valley, RSCN database.

Rosularia sempervivum (Bieb.) Berger in Engler & Prantl., Nat. Pfl. 18a: 466. 1930.
Syn. *Cotyledon sempervivum* Bieb. (1800); *Umbilicus libanoticus* DC. (1828); *Umbilicus sempervivum* (Bieb.) DC. (1929); *Rosularia libanotica* (Labill.) Muirhead (1972); *Rosularia sempervivum* subsp. *libanotica* (Labill.) Eggli (1988).
Possibly subsp. **persica**, also found in Turkey, Lebanon and Syria, and distinguished by its ± glabrous inflorescence as opposed to a glandular-pubescent inforescence found in subsp. *sempervivum*.
Ref. Al-Eisawi 90 (2013) as *Rosularia libanotica* (YAR-H).

Rosularia lineata (Boiss.) A.Berger, Nat. Pflanzenfam. ed. 2 18a: 466. 1930.
Syn. *Umbilicus lineatus* Boiss. (1849).
Amman, Bahath (MSB-K 419187).

Sedum caespitosum (Cav.) DC., Prodr. 3: 405. 1828.
Syn. *Tillaea rubra* L. (1753); *Crassula caespitosa* Cav. (1791); *Sedum rubrum* (L.) Thell. (1912).
Along Zarqa River, 2004, *Al-Dwiri* s.n. (AMM).

Sedum eriocarpum Sibth. & Sm., Fl. Graec. Prodr. 1: 310. 1809.
Between Ajloun and Eidun, *Meyers, Fred & Dinsmore* G696 (E00356427).

Sedum hispanicum L., Cent. Pl. 1. 12. 1755.
Syn. *Sedum glaucum* Waldst. & Kit. nom. illegit.; *S. orientale* Boiss. (1849); *S. armenum* Boiss. (1856).
Petra in the Sek, *Davis* 9309 (E00356485); Jerash, Dibeen, *El- Oqlah et al.* (NHJ 2013.0114).

Sedum laconicum Boiss. & Heldr., Diagn. Pl. Orient. 6: 55. 1846.
Ajloun castle, *Böcker* RB 894 (B100520842) & near the village Makhraba (ca 20 km W of Irbid), *Böcker* RB 991 (B100520735).

Sedum microcarpum (Sibth. & Sm.) Scholand, Nat. Pflanzenfam. 3(2a): 31. 1890.
Syn. *Crassula microcarpa* Sm. (1806).
Dibeen National Park, 2005, *Mahklouf* s.n. (AMM).

Sedum palaestinum Boiss., Diagn. Pl. Orient. 10: 18. 1849.
Tal Al-Rumman RBG, Sector No. 95, *El-Oqlah, Taifour & Al Khawaldah* (NHJ 2013.0264).

Sedum pallidum M.Bieb., Fl. Taur.-Caucas. 1: 353. 1808.
Balqa, main road from Irmaimean, toward Um Jozeh (MSB-K 419202).

Sedum rubens L., Sp. Pl.: 432 .1753.
Tal Al-Rumman, *Taifour & Aldamen* (NHJ 2010 .0051).

Sedum sediforme (Jacq.) Pau, Act. & Mem. Prim. Congr. Nat. Espan. 1908: 246. 1909.
Syn. *Sempervivum sediforme* Jacq. (1771); *Sedum nicaeense* All. (1785).
Eastern upper slopes of the Jordan valley and Shoubak (NCARE).

Telmissa microcarpa (Sm.) Boiss., Fl. Orient. 2: 795. 1872.
Syn. *Crassula microcarpa* Sm. in Sibth. & Sm. (1806); *Telmissa sedoides* Fenzl (1842).
Magdala, 28 Feb. 1912, *Meyers & Dinsmore* (E).

Umbilicus horizontalis DC., Prodr. 3: 400 (1828) var. **intermedius** (Boiss.) Chamberlain, Fl. Turk. 4: 213 (1972).
Syn. *Cotyledon intermedia* (Boiss.) Bornm. (1904); *Umbilicus intermedius* Boiss. (1872); *U. pendulinus* DC. var. *intermedius* (Boiss.) Post (1896).
Between Petra & foot of Mt. Hor, *Davis* 9062 (E00356619); Tal Al-Rumman, *Al- Eisawi, Abu Sondos, Abu Anza & Shishani* (NHJ 2006.034).

Cucurbitaceae

Bryonia cretica L., Sp. Pl. 2: 1013. 1753.
Irbid, 5 km after Hemma towards Irbid (MSB-K 186380); Dair Abusaid Kufur Rakib, *El-Oqlah* (NHJ 2013.0051).

Bryonia multiflora Boiss. & Heldr., Diagn. Pl. Orient. 10: 8. 1849.
Jordan, *Täckholm* 8888 (B 100088638); 1975, *Boulos et al.* (E).

Bryonia syriaca Boiss., Diagn. Pl. Orient. ser. 2, 2: 59. 1856.
Tal Al-Rumman, *Taifour & Aldamen* (NHJ 2009.0166).

Citrullus colocynthis (L.) Schrad., Linnaea 12: 414. 1838.
Syn. *Cucumis colocynthis* L. (1753).
Aqaba, on the way to Wadi Rum (MSB-K 243517); Rum, *Nuaimat, Al Khawaldah & Alshwaer* (NHJ 2013.0656).

Cucumis prophetarum L., Cent. Pl. 1. 32. 1755.
Little Petra, *Taifour, Al Khawaldah & Al Mogherbi* (NHJ 2013.0694).

Ecballium elaterium (L.) A.Rich., Dict. Class. Hist. Nat. 6: 19. 1824.
Syn. *Momordica elaterium* L. (1753).
Mafraq, Um Alo'lo', *Al Shudiefat & Khalidi* (NHJ 2013.0001).

Lagenaria siceraria (Molina) Standl., Publ. Field Mus. Nat. Hist., Bot. Ser. 3: 435. 1930.
Syn. *Lagenaria vulgaris* Ser. (1825).
Along Zarqa River, 2004, *Al-Dwiri* s.n. (AMM). Cultivated.

Momordica balsamina L., Sp. Pl. 2: 1009. 1753.
Karak, Ghor Al-Mazra'a, Site no 26 (MSB-K 497608). Cultivated.

Cymodoceaceae

Halodule uninervis (Forssk.) Boiss., Fl. Orient. 5: 24. 1882.
6 km S of Aqaba Port, in a reef lagoon, 21 Mar. 1975, *Hulings* s.n. (AMM, BM, E).

Cyperaceae

Bolboschoenus glaucus (Lam.) S.G.Sm., Novon 5: 101. 1995.
Syn. *Scirpus tuberosus* Desf. (1798).
Azraq (FOR-H); Karak, Al Safi (FOR-H).

Bolboschoenus maritimus (L.) Palla, Syn. Deut. Schweiz. Fl. ed. 3: 2532. 1905.
Syn. *Scirpus maritimus* L. (1753).
Jordan, *Clarke*, s.n. (K000315884); Burqu 30 km North West of M4, *Woods* 5 (E00449316).

Carex distans L., Syst. Nat., ed. 10. 2: 1263. 1759.
E. of El-Salt, *Meyers & Dinsmore* M958 (E00 356081).

Carex divisa Huds., Fl. Angl. 348.; ed. 2, 405. 1762.
Bin Dabaghat *Davis* 9343 (E00 353812).

Carex divulsa Stokes, Bot. Arr. Brit. Pl. ed. 2 2: 1035. 1787.
Above 'Ain Fit, *Meyers & Dinsmore* 1781 (E00353896).

Carex eremitica Paine, Palestine Explor. Soc. 3: 126. 1875.
Ref. RSCN database.

Carex hispida Willd. ex Schkuhr, Beschr. Riedgräs. 1: 63. 1801.
Amman, Road from Bahath to Adaseyeh, 2 km to the highway, water catchment area (MSB-K276740).

Carex mediterranea C.B.Clarke ex Post, Fl. Syria, Palest. & Sinai 837. 1896.
Jarash, Burma, Jun. 1864 (AUB).

Carex muricata L., Sp. Pl. 2: 974. 1753.
Above Ain Fit, *Meyers & Dinsmore* (AUB).

Carex pachystylis J.Gay, Ann. Sci. Nat., Bot. 2, 10: 301. 1838.
Above Wadi Musa on road to Ma'an, *Davis* 8707 (E0035376).

Carex stenophylla Wahlenb., Kongl. Vetensk. Acad. Nya Handl. 24: 142. 1803.
Aqaba, 99 km north of Aqaba, *Boulos, Al-Eisawi & Jallad* W. 6014 (E00353740).

Cladium mariscus (L.) Pohl., Tent. Fl. Bohem. 1: 32. 1809.
Syn. *Schoenus mariscus* L. (1753).
Azraq, *Dinsmore* 7427 (E00398727).

Cyperus alopecuroides Rottb., Descr. Icon. Rar. Pl. 38. 1773.
(YAR-H).

Cyperus arenarius Retz., Observ. Bot. 4: 9. 1786.
Zarqa, Ma'an, *Davis* 9412 (E00352723).

Cyperus conglomeratus Rottb., Descr. Icon. Rar. Pl. 21. 1773.
Ma'an, Naqab, 23 Apr. 1954 (FOR-H).

Cyperus capitatus Vand., Fasc. Pl. 5. 1771.
Syn. *Cyperus macrorrhizus* Nees ex Boiss. (1882).
Ma'an, 20 km NE of Quweirah, *Frey & Kürschner* (1988).

Cyperus difformis L., Cent. Pl. II, 6. 1756.
Madaba, *Taifour et al.* (NHJ).

Cyperus conglomeratus Rottb., Descr. Icon. Rar. Pl. 21. 1773.
Nr. Aqaba, *Boulos & Lahham* 5189 & 7388 (AMM, BM, E).

Cyperus fuscus L., Sp. Pl.: 46. 1753.
Jordan river, *Meyers & Dinsmore* 8232 (E00352622).

Cyperus glaber L.,ant. Pl. 2: 179. 1771.
Ajloun, Rajib, beside the bridge on the river (MSB-K 200486).

Cyperus jeminicus Rottb., Descr. Icon. Rar. Pl. 25. 1773.
Karak, Dead Sea area, coming from Mujib bridge toward Amman, near the hot water coming from Ma'aeen, (MSB-K 278010).

Cyperus laevigatus L., Mant. Pl. Altera 2: 179. 1771.
Mazr'aa, c. 25 km NW Karak, near the Dead Sea, *Boulos* 5812 (E00352880).

Cyperus laevigatus subsp. **distachyos** (All.) Ball., J. Linn. Soc., Bot. 16: 700. 1878.
Syn. *Cyperus distachyos* All. (1789).
Tabaget Fahel, Jordan; Ghor al Safi, 27.6.1939 (2897-HUJ).

Cyperus longus L., Sp. Pl.: 45. 1753.
Wadi Mahib, *Dinsmore* 10171 (E00356646).

Cyperus michelianus (L.) Delile, Descr. Egypte, Hist. Nat. 3: 50. 1813.
Syn. *Scirpus michelianus* L. (1753).
RSCN database.

Cyperus nutans Vahl var. **eleusinoides** (Kunth) Haines, Bot. Bihar Orissa 5: 898. 1924.
Syn. *Cyperus eleusinoides* Kunth (1837).
Aqaba, *Danin* (http://flora.org.il/en/plants/CYPELE/)

Cyperus odoratus L., Sp. Pl.: 46. 1753.
Al Mafraq, 18 Apr. 1983 (YAR-H).

Cyperus rotundus L., Sp. Pl.: 45. 1753.
Tabaget Fahel, 5.VIII.1898 (AUB); Tal Al-Rumman RBG, Sector No. 35, *El-Oqlah & Al Mfadi* (NHJ 2013.0359).

Eleocharis palustris (L.) Roem. & Schult., Syst. Veg. 2: 151. 1817.
Syn. *Scirpus palustris* L. (1753).
Azraq, *Frey & Kürschner* 80–313 (E00353397).

Fimbristylis bisumbellata (Forssk.) Bubani, Dodecanthea 30. 1850.
Syn. *Scirpus bisumbellatus* Forssk. (1775).
Al Karak: Ghor Safi, southern tip of Dead Sea, *Boulos* 5817 (E00352950).

Fimbristylis dichotoma (L.) Vahl, Enum. Pl. Obs. 2: 287. 1805.
Syn. *Scirpus dichotomus* L. (1753).

Fimbristylis ferruginea (L.) Vahl, Enum. Pl. Obs. 2: 291. 1805.
Syn. *Scirpus ferrugineus* L. (1753).
Karak, Dead Sea area, coming from Mujib bridge toward Amman, near the hot water coming from Ma'aeen (MSB-K 278021).

Pycreus flavescens (L.) P.Beauv. ex Rchb., Fl. Germ. Excurs. 72. 1830.
Syn. *Cyperus flavescens* L. (1753).
(YAR-H).

Pycreus nitidus (Lam.) J.Raynal, Kew Bull. 23: 314. 1969.
Syn. *Cyperus nitidus* Lam. (1791).
RSCN database.

Schoenoplectus lacustris (L.) Palla, Bot. Jahrb. Syst. 10: 299. 1888.
Syn. *Scirpus lacustris* L. (1753).
RSCN database.

Schoenoplectus litoralis (Schrad.) Palla, Bot. Jahrb. Syst. 10: 299. 1888.
Syn. *Scirpus litoralis* Schrad. (1806).
Karak, Hammamat wadi bin Hammad, on the right after the check point (NCARE); Azraq (RSCN database); Azraq Wetland Reserve Herbarium (2008).

Scirpoides holoschoenus (L.) Soják, Cas. Nár. Mus., Odd. Prír. 140 (3–4): 127. 1972.
Syn. *Scirpus romanus* L. (1753); *Scirpus holoschoenus* L. (1753); *Scirpus australis* Murr (1774); *Holoschoenus vulgaris* Link (1827); *Scirpus holoschoenus* (L.) Soják var. *australis* (Murr) W.D.J.Koch; *Holoschoenus australis* (Murr) Rchb. (1830); *Scirpus holoschoenus* L. var. *romanus* (L.) W.D.J.Koch.
6–7 km south Rum Rest House, *Boulos et al.* 7662 (E00355094).

Cytinaceae

Cytinus hypocistis (L.) L., Gen. Pl., ed. 6: 566. 1764.
Syn. *Asarum hypocistis* L. (1753).
Dibeen Forest Reserve, May 2006, *Taifour & Othman* (DFR-H).

Dioscoreaceae

Dioscorea communis (L.) Caddick & Wilkin, Taxon 51(1): 112. 2002.
Syn. *Tamus communis* L. (1753).
Ajloun, Wadi Orjan, *El-Oqlah et al.* (NHJ 2013.0076).

Ericaceae

Arbutus andrachne L., Syst. Nat., ed. 10. 2: 1024. 1759.
Hunting Aero Survey 104 (E00451510); Dibeen, *Royal Botanic Garden Team* (NHJ 2011.0031).

Euphorbiaceae

Andrachne aspera Spreng., Syst. Veg. 16, 3: 884. 1826.
Jordan, *Gillett* 16046 (K000185771); nr. Aqaba, *Boulos & Lahham* 5395 (AMM, BM, E).

Andrachne telephioides L., Sp. Pl. : 1014. 1753.
Jordan, *Field & Lazar*147 (K000185421).

Chrozophora oblongifolia (Delile) A.Juss. ex Spreng., Euphorb. Gen. 28. 1824.
Syn. *Croton oblongifolius* Delile (1812).
Near Aqaba, *Boulos & Lahham* 5409 (AMM, BM, E).

Chrozophora plicata (Vahl) A.Juss.ex Spreng, Syst. Veg. 3: 850. 1826.
Irbid (YAR-H).

Chrozophora tinctoria (L.) Juss., Euphorb. Gen. 110. 1824.
Syn. *Croton tinctorius* L. (1753); *Chrozophora obliqua* (Vahl) A.Juss. ex Spreng. (1824); *Chrozophora verbascifolia* (Willd.) A.Juss. ex Spreng. (1826); *Chrozophora hierosolymitana* Spreng. (1826).
Al Karak, Frey, *Kürschner & Hilger* 80–786 (E00445935).

Euphorbia aleppica L., Sp. Pl.: 458. 1753.
Amman, *Davis* 3772 (E00447540).

Euphorbia arguta Banks & Sol., Nat. Hist. Aleppo ed. 2, 2: 253. 1794.
Tal Al-Rumman, *Taifour & Aldamen* (NHJ 2010 .0012).

Euphorbia aulacosperma Boiss., Diagn. Pl. Orient. ser. 1, 12: 117. 1853.
Hesban, *Meyers & Dinsmore* M 349 (E00447657).

Euphorbia berythea Boiss. & Blanche, Diagn. Pl. Orient. ser. 2, 4: 82. 1859.
Irbid, 1983, *El-Oqlah & Lahham* (YAR-H).

Euphorbia chamaepeplus Boiss. & Gaill., Diagn. Pl. Orient. ser. 2, 4: 88. 1859.
Azraq, *Townsend* 65/25 27827 (K); nr. Aqaba, *Boulos & Lahham* 8524 (AMM, BM, E).

Euphorbia chamaesyce L., Sp. Pl.: 455. 1753.
Irbid, Jordan Univ. Sci. & Tech. (YAR-H); Aqaba, RSCN database.

Euphorbia cheiradenia Boiss. & Hohen., Diagn. Pl. Orient. ser. 1, 12: 112. 1853.
Al Shoubak (YAR-H).

Euphorbia cuspidata Bertol., Misc. Bot. 2: 9. 1843.
Rutba, *Rechinger* 12864 (E00447839).

Euphorbia densa Schrenk, Bull. Phys.-Math. Acad. Petersb. 3: 308. 1845.
Irbid, Jordan Univ. Sci. & Tech. (YAR-H); Aqaba, RSCN database.

Euphorbia exigua L., Sp. Pl.: 456. 1753.
Ajloun, 3 km E of Ishtafena (IBBU-H).

Euphorbia falcata L., Sp. Pl.: 456. 1753.
Jela'at um Binat (nr. Ma'in) to Zarka Ma'in, *Davis* 8629 (E00447634).

Euphorbia falcata var. **galilaea** (Boiss.) Boiss., Prodr. 15(2): 140. 1862.
Syn. *Euphorbia galilaea* Boiss. (1853).
Irbid, Boshra, Sep. 1905 (AUB).

Euphorbia forsskalii J.Gay, Phyt. Canar. 3: 240. 1847.
Dead Sea Area, Wadi Araba, RSCN database.

Euphorbia gaillardotii Boiss. & Blanche, Diagn. Pl. Orient. ser. 2, 4: 84. 1859.
10–15 km north of Sheikh Meskin, 16 Jun 1975, *Boulos* (E).

Euphorbia granulata Forssk., Fl. Aegypt.-Arab.: 94. 1775.
Irbid, Al Ghor, 14 Oct. 1981 (YAR-H).

Euphorbia grossheimii (Prokh.) Prokh., Fl. URSS 14: 391. 1949.
Syn. *Euphorbia isthmia* Täckh. (1932); *Euphorbia mariae* Tamamsch. (1944).
Ref. RSCN database.

Euphorbia helioscopia L., Sp. Pl.: 459. 1753.
10 km N von Suweleh in Richtung Jerash, *Frey & Kürschner* 80–93 (E00447438); Tal Al-Rumman, *Taifour & Aldamen* (NHJ 2009.0007).

Euphorbia heterophylla L., Sp. Pl.: 453. 1753.
Irbid (YAR-H).

Euphorbia hierosolymitana Boiss., Diagn. Pl. Orient. 12: 110. 1853.
Syn. *Euphorbia thamnoides* Boiss. (1860).
Wadi Rum, *Davis* 9103 (E00447367); Tafila, Dana Reserve, *El-Oqlah* (NHJ 2008.0031).

Euphorbia hierosolymitana var. **ramanensis** (Baum) Zohary, Fl. Palaestina 2: 277. 1972.
Syn. *Euphorbia ramanensis* Baum (1963).
(FOR-H).

Euphorbia hirta L., Sp. Pl.: 454. 1753.
Syn. *Euphorbia pilulifera* L. (1862).
Wadi Musa, 2–6 Oct. 2003, *Prausse & Holzapfe* (AMM).

Euphorbia macroclada Boiss., Diagn. Pl. Orient. ser. 1, 5: 54. 1844.
Syn. *Euphorbia tinctoria* Boiss. & Huet. ex Boiss. (1862).
Wadi [Lenka-Jerash], *Davis* 3785 (E00444638).

Euphorbia microsphaera Boiss., Diagn. Pl. Orient. 7: 87. 1846.
Irbid, 22 Apr. 1984, Al Tourah (YAR-H).

Euphorbia nutans Lag., Gen. Sp. Pl. 17. 1816.
Tafila (NCARE).

Euphorbia oxyodonta Boiss., Prodr. 15(2): 1267. 1866.
Jordan, Tabaget Fahel, 20.IV.1967, *Edgecombe* 262 (AUB).

Euphorbia palustris L., Sp. Pl. 462. 1753.
Azraq (FOR-H).

Euphorbia peplus L., Sp. Pl.: 456. 1753.
Jabel Amman, *Boulos* 5820 (E00447693); Tal Al-Rumman, *Al-Eisawi* (NHJ 2006.0391).

Euphorbia petiolata Banks & Sol., Nat. Hist. Aleppo ed. 2, 2: 253. 1794.
Syn. *Euphorbia lanata* Sieber ex Spreng. (1826).
Irbid, *Frey, Hilger & Kürschner* VO 5569 (E00439861).

Euphorbia phymatosperma Boiss. et Gaill., Diagn. Pl. Orient. ser. 2, 4: 83. 1859.
Tal Al-Rumman RBG, Sector No. 90, *El-Oqlah, Taifour & Al Khawaldah* (NHJ 2013.0245).

Euphorbia prostrata Aiton, Hort. Kew. 2: 139. 1789.
Jarash, Al Alouk (YAR-H).

Euphorbia retusa Forssk., Fl. Aegypt.-Arab.: 93. 1775.
Syn. *Euphorbia kahirensis* Raeusch. (1797).
18 km N of the Safawi-Ruwaished rd. (BRDC-H).

Euphorbia reuteriana Boiss., Diagn. Pl. Orient. ser. 1, 12: 115. 1853.
Dibeen National Park, 2005, *Mahklouf* s.n. (AMM).

Euphorbia valerianifolia Lam., Encycl. 2(2): 435. 1788.
Syn. *Euphorbia cybirensis* Boiss. (1846).
Tal Al-Rumman, *Taifour & Aldamen* (NHJ 2009.0103).

Mercurialis annua L., Sp. Pl. 2: 1035. 1753.
Wadi Musa, near Petra, *Boulos & Jallad*, 8090 (E00450114);Tal Al-Rumman, *Taifour & Aldamen* (NHJ 2009.0102).

Phyllanthus rotundifolius Klien ex Willd., Sp. Pl. 4: 584. 1805.
(YAR-H).

Ricinus communis L., Sp. Pl. 2: 1007. 1753.
Tal Al-Rumman, *Al-Eisawi* (NHJ 2006.0331).

Fabaceae

Acacia gerrardii Benth., Trans. Linn. Soc. London 30: 508. 1875.
Vachellia gerrardii (Benth.) P.J.H.Hurter (2008) is also an accepted name.
Aqaba, 6–7 km south of Rum Rest House, *Boulos et al.* 7667 (E00400794); Rum, Al-Masri Farm, *El-Oqlah et al.* (NHJ 2013.0606).

Acacia laeta R.Br. ex Benth., London J. Bot. 1: 508. 1842.
Ref. Wadi Hassa, Danin, http://flora.org.il/en/plants/ACALAE.

Acacia tortilis (Forssk.) Hayne, Getreue Darstell. Gew. 10: I. 31 (1827).
Vachellia tortilis (Forssk.) Galasso & Banfi (2008), is also an accepted name.
Syn. *Mimosa tortilis* Forssk. (1775).
Tal Al-Rumman, *Shishani* (NHJ 2011.0124).

Acacia tortilis (Forssk.) Hayne subsp. **raddiana** (Savi) Brenan, Kew Bull. 12: 87. 1957.
Acacia raddiana Savi (1830), is also an accepted name.
Syn. *Vachellia tortilis* (Forssk.) Galasso & Banfi subsp. *raddiana* (Kyal.) Boatwr. (2013),
Aqaba, *Leonard* 7469 (E00400833); Aqaba Wadi Khshebah, *El-Oqlah et al.* (NHJ 2013.0611).

Alhagi maurorum Medik., Vorles. Churpfälz. Phys.-Öcon. Ges. 2: 397. 1787.
Jordan, Burqu, 30 km North West of M4, *Woods* 3 (E00364468); nr. Aqaba, *Boulos & Lahham* 5383 (AMM, BM, E).

Anagyris foetida L., Sp.Pl. 1: 374. 1753.
Jebel 'Osha, *Meyers & Dinsmore* M616 (E0033569); Tal Al-Rumman, *Taifour & Aldamen* (NHJ 2009.0047).

Anthyllis tetraphylla L., Sp. Pl. 2: 719. 1753.
Tal Al-Rumman, *Taifour & Aldamen* (NHJ 2009.0242).

Anthylis vulneraria L., Sp.Pl. 2: 719. 1753.
Wadi Zerqa, *Hunting Aero Survey* 147B (E00339357).

Argyrolobium crotalarioides Jaub. & Spach, Ill. Pl. Orient. 1: 114, 116. 1843.
Al Aqaba, Wadi Yutum, *Boulos et al.* 7559 (E00334004).

Argyrolobium uniflorum (Decne.) Jaub. & Spach, Ill. Pl. Orient. 1: 115. 1843.
Aqaba, Wadi Yutum, 31 Mar. 1936, *Eig & Zohary* (HUJ).

Astracantha cruentiflora (Boiss.) Podl., Mitt. Bot. Staatssamml. München 19: 8. 1983.
Syn. *Astragalus cruentiflorus* Boiss. (1849).
Ref. RSCN Database.

Astracantha deinacantha (Boiss.) Podlech, Mitt. Bot. Staatssamml. München 19: 9. 1983.
Salt 30.iv.1886, Barbey (AUB); Dana, *Taifour* (NHJ 2013.0414).

Astragalus aaronii (Eig) Zohary, Fl. Palaestina 2: 74. 1972.
Syn. *Astragalus aaronsohnianus* Eig subsp. *aaronii* Eig.
Dinsmore 11707 (K000911375).

Astragalus adpressiusculus Eig, Syst. Stud. Astrag. Near East 126. 1955.
Ma'an, Shoubak, NCARTT regional centre of Shoubak on the hill behind the workshops (MSB-K 339957); Tafila Province, 2014, *Oran* (AMM).

Astragalus aduncus Willd., Sp. Pl., ed. 4. 3(2): 1269. 1802.
Syn. *Astragalus kotschyanus* Boiss. (1843); *Astragalus zemeraniensis* Eig (1955).
Tafila, Dana reserve (MSB-K 341240); nr. Wadi Musa, 29 Mar. 1929 (HUJ).

Astragalus aleppicus Boiss., Syst. Stud. Astrag. Near East 67. 1955.
Syn. *Astragalus feinbrunii* Eig (1940); *Astragalus postii* Eig (1955).
Above Wadi Mujib to Ahiban (Moab), *Davis* 916A (E00340314).

Astragalus amalecitanus Boiss., Diagn. Pl. Orient. 9: 46. 1849.
Karak, nr. Shobek, *Frey & Kürschner* (B); Tafila, Lahda (YAR-H).

Astragalus ancistrocarpus Boiss. & Hausskn.,Fl. Orient. 2: 463. 1872.
Syn. *Astragalus ancistrocarpus* Boiss. & Hausskn. var. *brachycarpus* Eig; *Astragalus ancistrocarpus* Boiss.

& Hausskn. var. *stepposus* Eig; *Astragalus ancistrocarpus* Boiss. & Hausskn. var. *nitidulus* (Hand.-Mazz.) Eig ex Rech.
Aqaba, Ras en Naqb, *Boulos & Jallad* W. 7340 (E00340962).

Astragalus annularis Forssk., Fl. Aegypt.-Arab.: 139. 1775.
Syn. *Astragalus maculatus* Lam. (1783).
Karak, Wadi Araba, *Frey & Kürschner* (B).

Astragalus arpilobus Kar. & Kir. subsp. **hauarensis** (Boiss.) Podlech, Sendtnera 2: 58. 1994.
Syn. *Astragalus hauarensis* Boiss. (1849).
Ma'an: 5 km Rum Eisenbahnstation, *Frey & Kürschner* VO 5020 (E00343180).

Astragalus asterias Steven, Bull. Soc. Imp. Naturalistes Moscou 4: 267. 1832.
Wadi Rum, *Davis* 8980 (E00347350).

Astragalus azraqensis C.C.Towns., Kew Bulletin 21(1): 53. 1967.
Townsend 65/214 (K000911554).

Astragalus berytheus Boiss. & Blanche, Diagn. Pl. Orient. II, 2: 33. 1856.
Ajloun, Halaweah, 24 Feb. 1982 (YAR-H).

Astragalus bethlehemiticus Boiss., Diagn. Pl. Orient. ser. 1, 9: 85. 1849.
Petra Richtung El Haishah, *Frey, Hilger & Kürschner* VO5861 (E00343550).

Astragalus boeticus L., Sp. Pl. 2: 758. 1753.
Al Mafraq, Um Alsurab, 19 Apr. 1983 (YAR-H); Naqab, Aqaba (RSCN database).

Astragalus bombycinus Boiss., Diagn. Pl. Orient. ser. 1, 2: 50. 1843.
Syn. *Astragalus palmyrensis* Post (1890).
Al-Jafr, in the vicinity of the Police Station, *Boulos & Jallad* 7923 (E00347936).

Astragalus brachystachys DC., Prodr. 2: 303. 1825.
Syn. *Astragalus brachystachys* DC. var. *argenteus* Eig.
20 km NE Quweirah *Frey & Kürschner* VO5067 (E00354483); Azraq, Wadi Bottom (South west the Castle Amra) *El-Oqlah* (NHJ 2013.0561).

Astragalus callichrous Boiss., Diagn. Pl. Orient. ser. 1, 2(9): 62. 1849.
Jerash, *Meyers & Dinsmore* G433 (E00347834); Tal Al-Rumman, *Taifour & Aldamen* (NHJ 2009.0227).

Astragalus camelorum Barbey, Herb. Levant 131, pl. 3, 1882.
Ref. RSCN database; Al-Eisawi 98 (2013).

Astragalus campylorrhynchus Fisch. & C.A.Mey., Linnaea 10 (Lit.): 84. 1836.
Jebel Hosha, 16 Apr. 1929 (HUJ).

Astragalus caprinus L., Fl. Sicul. Syn. 2(1): 315. 1844.
Syn. *Astragalus beershabensis* Eig & Samuelson ex Rech.f. (1950); *Astragalus platyraphis* Fisch. (1868); *Astragalus alexandrinus* Boiss. (1849).
Al 'Aqaba, Ras en Naqb, *Boulos & Jallad* 7338 (E00347475).

Astragalus corrugatus Bertol., Rar. Lig. [Ital.] Pl. 3: 33. 1810.
Syn. *Astragalus cruciatus* Link. (1822).
Azraq, *Frey & Kürschner* 80–162 (E00347606).

Astragalus crenatus Schult., Observ. Bot. 86. 1809.
Ref. (YAR-H); Al-Eisawi 98 (2013).

Astragalus cretaceus Boiss. & Kotschy, Diagn. Pl. Orient. ser. 2, 3(5): 84. 1856.
10 km von Suweleh in Richtung Jerash, *Frey & Kürschner* 80–58 (E00347898); Tal Al-Rumman, *Taifour & Aldamen* (NHJ 2009.0040).

Astragalus dactylocarpus Boiss., Diagn. Pl. Orient. ser. 1, 1(2): 60. 1843.
Naqab, approx. 10 km south of En Hazera, *Musselman* 10253 (E00347534).

Astragalus dactylocarpus subsp. **acinaciferus** (Boiss.) Eug. Ott, Phanerogam. Monogr. 9: 62. 1978.
Syn. *Astragalu sacinaciferus* Boiss. (1849).
Nr. Aqaba, *Boulos & Lahham* 74512 (AMM BM, E).

Astragalus echinops Boiss., Diagn. Pl. Orient. ser. 1, 1(2): 57. 1843.
Salt, iv.1888 (AUB).

Astragalus echinus DC., Astragalogia 197, pl. 34, 1802.
Petra, Wadi Musa (YAR-H).

Astragalus emarginatus Labill., Icon. Pl. Syr. 1:19. 1791.
Ref. (YAR-H)

Astragalus epiglottis L., Sp. Pl. 2: 759. 1753.
Dibeen Forest Reserve, *Mahklouf* s.n. (AMM).

Astragalus eremophilus Boiss., Diagn. Pl. Orient. ser. 1, 2: 54. 1843.
6 km N of Aqaba, 13 Mar. 1974, *Boulos, El-Eisawi & Jalad* (AMM).

Astragalus fruticosus Forssk., Fl. Aegypt.-Arab.:139. 1775.
Irbid, 28 Apr. 1999, Jordan. Univ. Sci. Tech. (YAR-H).

Astragalus gileadensis Eig, Syst. Stud. Astragalus Near East 42. 1955.
Azrak, Balama, 23 May 1983 (YAR-H).

Astragalus guttatus Banks & Sol., Nat. Hist. Aleppo 2: 260. 1794.
28 km south Amman, along the road to Aqaba, *Boulos & Jallad* 8169 (E00347876).

Astragalus hamosus L., Sp. Pl. 2: 758. 1753.
Ma'an, Petra, *Davis* 9148 (E00343257); Tal Al-Rumman, *Taifour & Aldamen* (NHJ 2009.0036).

Astragalus harpilobus Boiss., Fl. Orient. [Boissier] 2: 234, in Syn. [Dec 1872 or Jan 1873].
Syn. *Astragalus gyzensis* Delile ex Boiss. (1873).
Ma'an, 5 km NW of railway station, 13 Apr. 1980, *Frey & Kürschner* (E).

Astragalus hispidulus DC., Astragalogia 132, 30, t. 13, 1802.
Ref. Al-Eisawi 98 (2013).

Astragalus intercedens Rech.f., Ark. Bot. ser. 2, 1: 308. 1950.
14 km S of Qal'at-el-Hasa, *Eig, Feinbrun &Zohary* (E00296180).

Astragalus kahiricus DC., Prodr. 2: 292. 1825.
Jordan, *Ferguson-Lees* 254 (K000283233).

Astragalus macrocarpus DC., Astragalogia 179 (ed. quarto), 143 (ed. folio), no. 73, t. 28, 1802.
Dibeen Forest Reserve, 2004, *Mahklouf* s.n. (AMM).

Astragalus macrocephalus Willd., Sp. Pl., ed. 4. 3(2): 1260. 1802.
Ref. Al-Eisawi 98 (2013).

Astragalus mitchellii Post, Pl. Postianae 1: 7. 1890.
Salt, El Ghundhur, 12, Apr. 1890 (AUB).

Astragalus moabiticus Post, Fl. Syria 11. 1896.
Amman, Zizia, Mashitta, 2 Apr. 1895 (AUB).

Astragalus negevensis Zohary & Fertig, Israel J. Bot. 19: 293. 1970.
Ref. Al-Eisawi 96 (2013).

Astragalus oleifolius DC., Astragalogia 192 (ed. quarto), no. 87. 1802.
Syn. *Astragalus deinacanthus* Boiss. (1849); *Astracantha deinacanthus* (Boiss.) Podlech (1983).
Ref. Al-Eisawi 96 (2013).

Astragalus oocephalus Boiss., Diagn. Pl. Orient. ser. 1, 2: 56. 1843.
Salt, fields (YAR-H); Eastern upper slopes of Jordan valley (NCARE).

Astragalus oxyglottis Steven ex M.Bieb., Fl. Taur.-Caucas. 2: 192. 1808.
Karak, Lajjoun, 09 Jun. 1996 (NCARE).

Astragalus oxytropifolius Boiss., Diagn. Pl. Orient. ser. 1, 9: 37. 1849.
Salt, 30.iv.1886, *Barbey* (AUB).

Astragalus palaestinus Eig, J. Bot. 72: 124. 1934.
Jerash, *Meyers & Dinsmore* G409 (E00343264).

Astragalus pelecinus (L.) Barneby, Mem. New York Bot. Gard. 13. Pt. 1, 26. 1964.
Jordan, Tabaget Fahel, 21.iv.1967 (AUB).

Astragalus peregrinus Vahl, Symb. Bot. (Vahl) i. 57. 1790.
Ref. RSCN database; Al-Eisawi 98 (2013).

Astragalus postii Eig., Syst. Stud. Astrag. Near East 44. 1955.
Jebal Qaha, 30 Apr. 1911, *Meyers & Dinsmore* (E).

Astragalus sanctus Boiss., Diagn. Pl. Orient. ser. 1, 9: 47. 1849.
Wadi Musa, Bit Dabaghat, *Davis* 8955 (E00341751); Tal Al-Rumman, *Taifour & Aldamen* (NHJ 2009.0225).

Astragalus schimperi Boiss., Diagn. Pl. Orient. ser. 1, 2: 53. 1843.
30 km N Mudawwa, *Frey & Kürschner* 80–698 (E00347787).

Astragalus scorpioides Pourr. ex Willd., Sp. Pl., ed. 4 [Willdenow] 3(2): 1280. 1802.
Al Mafraq, Borqu (IBU-H).

Astragalus sieberi DC., Prodr. 2: 295. 1825.
Azraq, *Gillett* 15610 (K000286289).

Astragalus sparsus Decne., Ann. Sci. Nat., Bot. sér. 2, 3: 267. 1835.
Wadi Yutum, 21 Mar. 1975, *Boulos, Jallad & Lahham* (AMM).

Astragalus spinosus (Forssk.) Muschl., Verh. Bot. Vereins Prov. Brandenburg 49. 98. 1908.
30 km W von Azraq, *Frey & Kürschner* 80–111 (E00343838).

Astragalus suberosus Banks & Sol. subsp. **suberosus**, Nat. Hist. Aleppo (ed. 2) 2: 260. 1794.
Syn. *Astragalus tuberculosus* DC. (1802).
Methihitha (above Wadi Mojib) to Dheban (Moab) *Davis* 9161 (E00347630).

Astragalus syriacus L., Sp. Pl. 2: 759. 1753.
Syn. *Astragalus fruticosus* Pall. (1800–1803).
Ref. Al-Eisawi 96 (2013).

Astragalus trachoniticus Post, J. Linn. Soc., Bot. 24. 426. 1888.
30 km W of Azraq, *Frey & Kürschner* (B); 0.5 km N of Safawi-Mafraq rd. (BRDC-H).

Astragalus transjordanicus Sam. ex Rech.f., Ark. Bot. ser. 2, 1: 307. 1950.
Ref. Al-Gharaibeh (2005) (AMM).

Astragalus tribuloides Delile**,** Descr. Egypte, Hist. Nat. 70. 1813.
Karak, N von Tafila, *Frey & Kürschner* VO 4967 (E00347719); Azraq Wadi Bottom (South west the Castle Amra) *El-Oqlah* (NHJ 2013.0574).

Astragalus trigonus DC., Astragalogia 186 (ed. quarto), no. 81. 1802.
Ref. RSCN database; Al-Eisawi 98 (2013).

Astragalus trimestris L., Sp. Pl. 2: 761. 1753.
Al-Mafraq, Al-Mazzah (IBBU-H); Irbid, Saham, 30 Mar. 1982 (YAR-H).

Astragalus turbinatus Bunge, Arb. Naturf. Ver. Riga i. 215. 1847.
Balqa, in mountains, just before sign at turn to Kufr Huda (MSB-K 185833).

Astragalus vogelii (Webb) Bornm., Beih. Bot. Centralbl., Abt. 2. 33(2): 233, in adnot. 1915.
Ref. RSCN database.

Astragalus xylobasis Freyn & Bornm., Bull. Herb. Boissier vi. 984.
Shoubak, *Hunting Aero Survey* (E); Tafila, Dana Reserve 13 Jun. 2006 (NCARE).

Astragalus zemeraniensis Eig, Syst. Stud. Astragalus Near East 117. 1955.
Ma'an, 10 km NW Quweirah, *Frey & Kürschner* (B).

Bituminaria bituminosa (L.) C.H.Stirt., Bothalia 13(3–4): 318 1981.
Salt, from Salt city toward Wadi Shuaib at the beginning of Aira (MSB-K 200084); Wadi Hassa, Afra hot spring BOT ERA project (NHJ 2013.0504).

Bituminaria flaccida (Nab.) Greuter, Willdenowia 16(1): 108. 1986.
Syn. *Psorlalea flaccida* Nabelek (1923).
Dana Reserve, Rumana trail behind Rumana camp, BOT ERA project (NHJ 2013.0457).

Calicotome villosa (Poir.) Link, Neues J. Bot. 2(2): 51. 1808.
Tal Al-Rumman, *Taifour & Aldamen* (NHJ 2009.0246).

Ceratonia siliqua L., Sp. Pl. 2: 1026 1753.
Petra, *Dinsmore* 10231 (E00400603); Tal Al-Rumman, *Al-Eisawi* (NHJ 2006.0333).

Cercis siliquastrum L., Sp.Pl. 1: 374. 1753.
Amman, on the way to Salt, 5 km from Sweeteh main circle toward Salt (MSB-K 227731).

Cicer arietinum L., Sp. Pl. 2: 738. 1753.
Ref. RSCN database.

Cicer judaicum Boiss., Diagn. Pl. Orient. ser. 1, 9: 130. 1849.
Irbid (YAR-H).

Cicer pinnatifidum Jaub. & Spach, Ann. Sci. Nat., Bot. sér. 2, 18: 227. 1842.
Upper Jordan valley, RSCN database.

Colutea abyssinica Kunth & C.D.Bouché, Ann. Sci. Nat., Bot. sér. 3, 7: 188. 1847.
Ma'an, in Shaubah (MSB-K 151898).

Colutea cilicica Boiss. & Balansa, Diagn. Pl. Orient. ser. 2, 5: 83. 1856.
Shoubak, *Ananbeh &Alshwaer* (NHJ 2013.0632).

Colutea istria Mill., Gard. Dict., ed. 8. n. 2. 1768.
Petra, *Dinsmore* 7053 (E00354654); Ma'an-Petra, *El-Oqlah et al.* (NHJ 2013.0624).

Coronilla rostrata Boiss. & Spruner, Diagn. Pl. Orient. ser. 1, 2: 100. 1843.
Ref. RSCN database; Al-Eisawi 100 (2013).

Coronilla scorpioides (L.) W.D.J.Koch, Syn. Fl. Germ. 188. 1835.
Medaba, *Meyers & Dinsmore* M609 (E00359814); Tal Al-Rumman, *Taifour* (NHJ 2009.0003).

Crotalaria aegyptiaca Benth., London J. Bot. 2: 473. 1843.
Wadi Yutum, 12 km north of Aqaba, *Boulos et al.* 7560 (E00333885).

Dalbergia sissoo DC., Hort. Bengal. 53; Fl. Ind. iii. 223.
Ayfra, *Taifour, Nuaimat, Al Khawaldah & Alshwaer* (NHJ 2013.0662).

Dorycnium rectum (L.) Ser., Prodr. 2: 208. 1825.
Balqa, Rummaimin area, just beside the waterfall (MSB-K 278227).

Faidherbia albida (Delile) A.Chev., Rev. Bot. Appl. Agric. Trop. 14: 876. 1934.
Syn. *Acacia albida* Delile (1813).
North Ghour, Tell Arabin, *Taifour, Al Khawaldah & Alshwaer* (NHJ 2013.0676).

Genista fasselata Decne., Ann. Sci. Nat., Bot. sér. 2, 4: 360. 1835.
Ref. Al-Eisawi 100 (2013).

Glycyrrhiza echinata L., Sp. Pl. 2: 741. 1753.
Ref. Al-Eisawi 100 (2013).

Glycyrrhiza glabra L., Sp. Pl. 2: 742. 1753.
Jordan bridge, *Meyers & Dinsmore* 1630 (E00359377).

Hedysarum pannosum Boiss., Fl. Orient. 2: 525. 1872.
Ref. RSCN database; Al-Eisawi 100 (2013).

Hippocrepis areolata Desv., Mém. Soc. Linn. Paris 4: 329. 1827.
Syn. *Hippocrepis bicontorta* Loisel. (1827); *Hippocrepis cornigera* Boiss. (1843).
On the road between Quweira and Naqb Ishtar, *Hunting Aero Survey* 50B (E00359883).

Hippocrepis constricta Kuntze, Linnaea 16: 320. 1842.
Aqaba (YAR-H).

Hippocrepis multisiliquosa L., Sp. Pl. 2: 744. 1753.
Valley between Jebel Brera and Umm Sabibe, *Borzatti & Borsatti* (AMM).

Hippocrepis unisiliquosa L., Sp. Pl. 2: 744. 1753.
Medaba, *Meyers & Dinsmore* M446 (E00359989); Tal Al-Rumman, *Taifour* (NHJ 2009.0035).

Hymenocarpos circinnatus (L.) Savi, Fl. Pis. 2. 205. 1798.
Syn. *Medicago circinnata* L. (1753); *Hymenocarpos nummularius* Willd.; *Hymenocarpos circinnatus* (L.) Savi var. *nummularius* (DC.) Bornm.
Kefrein, *Meyers & Dinsmore* 2612 (E00342784).

Indigofera argentea Burm.f., Fl. Ind. (N. L. Burman) 171. 1768.
Tabaget Fahel, Dec. 1889 (AUB).

Indigofera articulata Gouan, Ill. Observ. Bot. 49. 1773.
Ghor es Safiyeh, *Meyers & Dinsmore* 3851 (E00354882); Tal Al-Rumman, *Al-Eisawi* (NHJ 2006.0037).

Indigofera oblongifolia Forssk., Fl. Aegypt.-Arab.: 137. 1775.
Ref. Al-Eisawi 101 (2013).

Lathyrus annuus L., Demonstr. Pl. 20. 1753.
Syn. *Lathyrus hierosolymitanus* Boiss. (1849).
Irbid, Enbah (YAR-H).

Lathyrus aphaca L., Sp. Pl. 2: 729. 1753.
Jerash, *Meyers & Dinsmore* 9548 (E00370641); Tal Al-Rumman, *Taifour* (NHJ 2011.0070).

Lathyrus blepharicarpos Boiss., Diagn. Pl. Orient. ser. 1, 9: 126. 1849.
S of Amiad, *Plitman* 6208 (E00400141); Ajloun district, outside Ajloun reserve near Elynabi village, BOT ERA Project (NHJ 2013.0531).

Lathyrus cicera L., Sp. Pl. 2: 730. 1753.
Ghour Safi Fox, IV.1895 (AUB); Tafila Province, 2014, *Oran* (AMM).

Lathyrus digitatus Fiori & Paol., Fl. Italia 2: 105. 1900.
Ajloun, Estafeana, 15 Apr. 1983 (YAR-H).

Lathyrus gloeosperma Warb. & Eig, Repert. Spec. Nov. Regni Veg. 25: 351. 1928.
Jordan, *Kupicha* (E00199572).

Lathyrus gorgonei Parl., Giorn. Sci. Sicilia 62(184–185): 3 (–4). 1838.
Dier Alla, *Trought* (E00370991).

Lathyrus hierosolymitanus Boiss., Diagn. Pl. Orient. ser. 1, 9: 127. 1849.
Medaba, *Meyers & Dinsmore* M1127 (E00370967).

Lathyrus inconspicuus L., Sp. Pl. 2: 730. 1753.
Syn. *Lathyrus erectus* Lag. (1816); *Lathyrus hispidulus* Boiss. (1846).
Wadi Rum, *Meyers & Dinsmore* M776 (E00370706).

Lathyrus marmoratus Boiss. & Balansa ex Boiss., Fl. Orient. 2: 606. 1872.
Jarash, Deguale forest (YAR-H); Ain Musa, 17 Apr. 1929 (HUJ).

Lathyrus ochrus (L.) DC., Fl. Franc. ed. 3. 4: 578. 1805.
Magdala, *Meyers & Dinsmore* (E).

Lathyrus plitmannii Greuter & Burdet, Willdenowia 19(1): 32. 1989.
Syn. *Lathyrus gorgonei* Parl. var. *lineatus* (Post) C.C.Towns.; *Lathyrus pseudocicera* Pamp. (1924); *Lathyrus amoenus* Fenzl var. *lineatus* Dinsm.
Tal Al-Rumman RBG, Sector No. 14, *Almfadi & Al Khawaldah* (NHJ 2013.0039).

Lathyrus sativus L., Sp. Pl. 2: 730. 1753.
Syn. *Lathyrus sativus* L. var. *stenophyllus* Boiss.
East of Jordan, *Paine* (K).

Lathyrus setifolius L., Sp. Pl. 2: 731. 1753.
Tafila Province, 2014, *Oran* (AMM); Ajloun, 17 Apr. 1982 (YAR-H).

Lathyrus spathulatus Celak., Oesterr. Bot. Z. 38: 6. 1888.
Suf, Ain Jenneh, 4 May 1911 (HUJ).

Lens culinaris Medik., Vorles. Churpfälz. Phys.-Öcon. Ges. 2: 361. 1787.
Syn. *Lens esculenta* Moench (1794); *Ervum lens* L. (1753).
E of Es-Salt, *Meyers & Dinsmore* M1716 (E00370158).

Lens ervoides Grande, Bull. Ort. Bot. Napoli, 5: 58. 1918.
Syn. *Lens lenticula* (Hoppe) Alef. (842) as "*lenticulata*".
Jarash, Thagret Asfoor (YAR-H).

Lens orientalis Popow, Byull. Sredne-Aziatsk. Gosud. Univ. 15. Suppl. 22. 1927.
Near Salt, *Meyers & Dinsmore* M751 (E00370218).

Lotononis platycarpa (Viv.) Pic.Serm., Webbia vii. 331. 1950.
Al 'Aqaba, 6 km north Aqaba, *Boulos, Al-Eisawi & Jallad* 6099 (E00334134).

Lotus hispidus Desf. ex DC., Fl. Franc. [de Candolle & Lamarck], ed. 3. Tome 4(2): 556. 1805.
Syn. *Lotus angustissimus* L. (1753).
Tabaget Fahel (AUB).

Lotus arabicus L., Nat. Hist. Aleppo (ed. 2) t. 14, 1794.
Ref. RSCN database; Al-Eisawi 102 (2013).

Lotus collinus (Boiss.) Heldr., Herb. Graec. Norm. no. 1320. 1897.
Syn. *Lotus judaicus* Boiss. ex Bornm. (1872).
Ajloun, *Davis* 9465 (E00339942); Tal Al-Rumman, *El-Oqlah* (NHJ 2008.0098).

Lotus corniculatus L. var. **tenuifolius** L., Sp. Pl. 2: 776. 1753.
Syn. *Lotus tenuis* Waldst. & Kit. ex Willd. (1809); *Lotus tenuifolius* (L.) Rchb., Fl. Germ. Exc. 22: 506; *Lotus corniculatus* L. subsp. *tenuifolius* (L.) Hartm.; *Lotus noeanus* Boiss. (1856); *Lotus tenuifolius* (L.) Rchb. var. *noeanus* (Boiss.) Boiss.; *Lotus corniculatus* L. subsp. *frondosus* Freyn; *Lotus krylovii* Shishk. et Serg. (1932).
Azraq (RSCN database).

Lotus creticus L., Sp. Pl. 2: 775. 1753.
Ref. Al-Eisawi 102 (2013); (FOR-H).

Lotus edulis L., Sp. Pl. 2: 774. 1753.
Tal Al-Rumman RBG, Sector No. 23, *Al Mfadi & Al Khawaldah* (NHJ 2013.0066).

Lotus gebelia Vent., Descr. Pl. Nouv. t. 57, 1801.
Amman, Wadi Yaduda (FOR-H).

Lotus glinoides Delile, Ann. Sci. Nat., Bot. sér. 2, 7: 286. 1837.
Irbid, Malka (YAR-H).

Lotus halophilus Boiss. & Spruner, Diagn. Pl. Or. Nov. Se, 1, 1 (2): 37. 1843.
Syn. *Lotus villosus* Forssk. (1775); *Lotus pusillus* Viv., Beobacht. (1783); *Lotus aucheri* Boiss. (1843).
Wadi Rum, *Davis* 10450 (E00339850); Tafila, Dana Reserve, Barra, BOT ERA project (NHJ 2013.0476).

Lotus lanuginosus Vent., Jard. Malmaison 92. 1803.
Wadi Botom (NHJ).

Lotus ornithopodioides L., Sp. Pl. 2: 775. 1753.
Azraq, Wadi Bottom (south west of Castle Amra), *El-Oqlah* (NHJ 2013.0598).

Lotus palustris Willd., Sp. Pl., ed. 4. 3(2): 1394. 1802.
Jordan, *Davis* 3778 (E00225267).

Lotus peregrinus L., Sp. Pl. 2: 774. 1753.
Syn. *Lotus carmeli* Boiss. (1849).
5 km NW Rum station, *Frey & Kürschner* VO 5041 (E00339878); Jerash, Dibeen, *El-Oqlah, Taifour & Said* (NHJ 2013.0115).

Lupinus albus L., Sp. Pl. 2: 721. 1753.
Bir Salim (NCARE).

Lupinus angustifolius L., Sp. Pl. 2: 721. 1753.
Ref. RSCN database; Al-Eisawi 102 (2013).

Lupinus luteus L., Sp. Pl. 2: 721. 1753.
Bir Salim (E).

Lupinus palaestinus Boiss., Diagn. Pl. Orient. ser. 1, 9: 9. 1849.
Tal Al-Rumman, *Taifour & Aldamen* (NHJ 2009.0056).

Lupinus pilosus L., Syst. Veg., ed. 13. 545. 1774.
Wadi Shu'ayb, *Simpson* 53345 (K000286284).

Lupinus varius L., Sp. Pl. 2: 721. 1753.
Tal Al-Rumman, *Taifour* (NHJ 2009.0241).

Medicago arabica (L.) Huds., Fl. Angl.: 288. 1762.
Jarash, Kufrkhal 22 Apr. 1984 (YAR-H).

Medicago arborea L., Sp. Pl. 2: 778. 1753.
Ref. RSCN database; Al-Eisawi 103 (2013).

Medicago astroites (Fisch. & C.A.Mey.) Trautv., Bull. Sc. Acad. Petersb. 8. 27.1 1841.
Syn. *Trigonella astroites* Fisch. & C.A.Mey. (1835); *Trigonella sinuata* Boiss. (1849).
Jordan, Petra, *Gillett* 15975 (K000511864); Wadi Rum, *El-Esawi* (AMM).

Medicago blancheana Boiss., Diagn. Pl. Orient. ser. 2, 5: 75. 1856.
E of Es-Salt, *Meyers & Dinsmore* M701 (E00337228).

Medicago constricta Dur., Actes Soc. Linn. Bordeaux 29: 15. 1873.
Tal Al-Rumman RBG, Sector No. 57, *Taifour & Al Khawaldah* (NHJ 2013.0314).

Medicago coronata (L.) Bartal., Cat. Piant. Siena: 61. 1776.
Tal Al-Rumman RBG, Sector No. 95, *El-Oqlah, Taifour & Al Khawaldah* (NHJ 2013.0266).
Amman, env. of Amman, *Eig & Zohary* (E00337276).

Medicago doliata Carmign. in Giorn. Sc. Aoc. Ital. i. 48 1818; ex Bertol. Fl. Ital. viii. 292.
Syn. *Medicago aculeata* Willd. (1802).
About 6 km before Ain Suela 14 Apr. 1929 (HUJ).

Medicago granadensis Willd., Enum. Pl. 2: 803. 1809.
Syn. *Medicago galilaea* Boiss. (1849).
Tal Al-Rumman, *Shishani & Alkoran* (NHJ 2011.0068).

Medicago intertexta (L.) Mill., Gard. Dict., ed. 8. n. 4 1768.
Wadi Kefrein, *Meyers & Dinsmore* 3596 (E00336255)*; Taifour & Aldamen* (NHJ 2009.0256).

Medicago intertexta subsp. **ciliaris** (L.) Ponert, Feddes Repert. 83 (9–10): 640. 1973.
Syn. *Medicago ciliaris* (L.) Krock. (1790).
Ajloun, *Meyers & Dinsmore* (E).

Medicago laciniata (L.) Mill., Gard. Dict., ed. 8. n. 5 1768.
Wadi Rum, *Davis* 10446 (E00337314); Tal Al-Rumman RBG, Sector No. 95, *El-Oqlah, Taifour & Al Khawaldah* (NHJ 2013.0265).

Medicago littoralis Rohde ex Loisel., Not. Pl. Fl. France 118. 1810.
Al Kark, Al Mazrah (YAR-H); Jarash, Jobbeh (YAR-H).

Medicago lupulina L., Sp. Pl. 2: 779. 1753.
Tafila Province, 2014, *Oran* (AMM); Irbid (YAR-H).

Medicago minima (L.) Bartal., Fl. Angl. 21. 1754.
Syn. *Medicago polymorpha* L. var. *minima* L. (1753); *Medicago recta* Desf. (1799).
Zarqa river, 8.5.1927 (19117-HUJ); Ajloun (NCARE); Tell Ar Rumman (NHJ).

Medicago minima (L.) Bartal. var. **minima**, Cat. Piante Siena 61. 1776.
Syn. *Medicago mollissima* Roth (1806).
Tal Al-Rumman, *Taifour* (NHJ 2009.0257).

Medicago monantha (C.A.Mey.) Trautv., Index Seminum 1840.: 6, adnot. 4. 1841.
Syn. *Trigonella monantha* C.A.Mey. (1831); *Trigonella noeana* Boiss. (1856).
Ajloun, Rajib, 15 km NE of Dair-Alla towards Jerash (NCARE); Aqaba (RSCN database).

Medicago monspeliaca Trautv., Bull. Sc. Acad. Petersb. 8. 272. 1841.
Syn. *Trigonella monspeliaca* L. (1753).
Jordan, Tabaget Fahel, 20.iv.1967 (AUB).

Medicago murex Willd., Sp. Pl., ed. 4 [Willdenow] 3(2): 1410. 1802.
Syn. *Medicago sicula* Tod (1868).
Irbid, Malka (YAR-H).

Medicago orbicularis (L.) Bartal., Cat. Piante Siena 60. 1776.
Syn. *Medicago applanata* Willd. (1814).
11 km north of Jarash, along the road to Irbid, *Boulos, Jallad & Lahham* 8286 (E00336303); Tal Al-Rumman, *Taifour* (NHJ 2009.0039).

Medicago polymorpha L., Sp. Pl. 2: 779, 1200. 1753.
Syn. *Medicago lappacea* Desr. (1789); *Medicago hispida* Gaertn. (1791); *Medicago denticulata* Willd. (1802).
Amman, Salt to Amman, 14-Apr–29, *Eig & Zohary* s.n. (E00336005); Tal Al-Rumman, *Al-Eisawi* (NHJ 2006.0139).

Medicago praecox DC., Cat. Pl. Horti Monsp. 123. 1813.
Balqa', Rmaimin, 1.5 km N of town on the road to the water fall (NCARE).

Medicago radiata L., Sp. Pl. 2: 778. 1753.
Syn.*Trigonella radiata* (L.) Boiss. (1872 or 1873).
Petra to Negb. Otoshera (foot of Mt. Hot E), *Davis* 8651 (E00336328); Tal Al-Rumman, *Taifour & Aldamen* (NHJ 2009.0253).

Medicago rigidula (L.) All., Fl. Pedem. 1: 316. 1785.
Amman, ascent to Salt, 29 km from Jordan *Eig, Feinbrun & Zohary* (E00336159).

Medicago rotata Boiss., Diagn. Pl. Orient. ser. 1, 2: 24. 1843.
Zarka main to Jebel attarus, *Davis* 9380 (E00337250); Tal Al-Rumman, *Taifour* (NHJ 2009.0255).

Medicago rugosa Desr., Encycl. 3: 632. 1789.
Syn. *Medicago elegans* Jacq. ex Willd. (1802).
Ajloun, *Meyers & Dinsmore* 103 (E00337216).

Medicago sativa L., Sp. Pl. 2: 778. 1753.
Al Mafraq, Ain Al Beida (FOR-H).

Medicago scutellata (L.) Mill., Gard. Dict., ed. 8. n. 2, 1768.
Suleleh in Richtung Nablus, *Frey & Kürschner* VO 4942 (E00337191); Tal Al-Rumman, *Taifour & Aldamen* (NHJ 2009.0254).

Medicago tornata (L.) Mill., Gard. Dict., ed. 8. n. 3, 1768.
Irbid, Wadi Al Arab (YAR-H).

Medicago truncatula Gaertn., Fruct. Sem. Pl. 2: 150. 1790.
Syn. *Medicago tribuloides* Desr. (1792).
10–15 km NW Zarqa, along the road to Alouk, *Boulos & Al-Eisawi* 6439 (E00336089); Tal Al-Rumman, *Taifour* (NHJ 2009.0249).

Medicago turbinata (L.) All., Fl. Pedem. 1: 315. 1785.
Syn. *Medicago tuberculata* Willd. (1802).
Jerash, *Meyers & Dinsmore* G617 (E00336234); Tal Al-Rumman, *RBG-Ire & RBG-Jo* (NHJ 2008.0046).

Melilotus albus Medik., Vorles. Churpfälz. Phys.-Öcon. Ges. 2: 382. 1787.
Amman, Al Ameer (MSB-K 395748).

Melilotus elegans Salzm. ex Ser., Prodr. 2: 198. 1825.
Upper Jordan valley (RSCN database).

Melilotus indicus (L.) All., Fl. Pedem. 1: 308. 1785.
Ma'an Rest House, in the garden, *Boulos, Jallad & El-Gohary* 8050 (E00337442); Tal Al-Rumman, *Taifour & Aldamen* (NHJ 2009.0238).

Melilotus messanensis (L.) All., Fl. Pedem. 1: 309. 1785.
Syn. *Melilotus sicula* (Turra ex Vitm.) B.D.Jackson.
Upper Jordan valley (RSCN database).

Melilotus sulcatus Desf., Fl. Atlant. 2: 193. 1799.
Kafrein, Ghor Al-Kafrein (MSB-K 113762).

Onobrychis aequidentata (Sibth. & Sm.) Urv., Mém. Soc. Linn. Paris 1: 346. 1822.
Jarash, 4.v.1886 (AUB).

Onobrychis caput-galli (L.) Lam., Fl. Franç. (Lamarck) 2: 651. 1779.
Dibeen Forest Reserve, 2004, Mahklouf (AMM); Amman & Ain Musa (HUJ); Irbid, Aqra (YAR-H).

Onobrychis crista-galli (L.) Lam., Fl. Franç. (Lamarck) 2: 652. 1779.
Syn. *Onobrychis crista-galli* Heyn (1779); *Onobrychis squarrosa* Viv. (1830).
17 km north Jarash, along the road to Irbid, *Boulos, Jallad & Laham* 8258 (E00362602); Tal Al-Rumman, *Al-Eisawi* (NHJ 2006.0109)

Onobrychis kotschyana Fenzl, Pug. Pl. Nov. Syr. 3. 1842.
Ajloun, *Davis* 9437 (E00362712).

Onobrychis montana DC., Fl. Franc. ed. 3. 4: 611. 1805.
Syn. *Onobrychis cadmea* Boiss. (1843).
Above Wadi Musa on the road to Maan, *Davis* 8675 (E00362733).

Onobrychis ptolemaica (Delile) DC., Prodr. 2: 347. 1825.
Wadi Mudeisisat, S Azraq, *Frey & Kürschner* VO 4806 (E00364085).

Onobrychis squarrosa Viv., Pl. Aegypt. Dec. 14. 1830.
40 km E von Amman, *Frey & Kürschner* 80–177 (E00364260).

Onobrychis supina (Chaix) DC., Fl. Franc. ed. 3. 6: 612. 1815.
Jarash (AUB); Ma'an, Eltayibi village (FOR-H); opposite Jordan University (NCARE).

Onobrychis wettsteinii Nábělek, Spisy Prír. Fak. Masarykovy Univ. 35: 91. 1923.
Petra, *Davis* 8971 (E00364206); Tafila, Dana Reserve, BOT ERA project (NHJ 2013.0436).

Ononis alopecuroides L., Sp. Pl. 2: 717. 1753.
Upper Jordan valley (RSCN database).

Ononis biflora Desf., Fl. Atlant. 2: 143. 1798.
Alouk, ca 15 km SE Jerash, *Boulos & Al-Eisawi* 6497 (E00334312); Tal Al-Rumman, *El-Eisawi* (NHJ 2006.0298).

Ononis diffusa Ten., Prod. Fl. Nap. p. xl i. 1805.
Syn. *Ononis serrata* Forssk. (1775).
Ma'an, Wadi Rum, *Frey & Kürschner* VO 5847 (E00336611).

Ononis hirta Desf., Hort. Par. ex Poir. Encyc. Suppl. 1. 741. 1811.
Wadi Kefrein, *Meyers & Dinsmore* 1632 (E00336637).

Ononis mitissima L., Sp. Pl. 2: 717. 1753.
Al Shoubak and Al Khraibah (YAR-H).

Ononis natrix L., Sp. Pl. 2: 717. 1753.
Wadi Zerka, *Davis* 3784 (E00334265); Tal Al-Rumman, *Al-Eisawi* (NHJ 2006.0198).

Ononis ornithopodioides L., Sp. Pl. 2: 718. 1753.
Zerka Main to Jebel Attarus, *Davis* 9390 (E00334311); Tal Al-Rumman, *Al-Eisawi* (NHJ 2006.0133).

Ononis phyllocephala Boiss., Fl. Orient. 2: 63. 1872.
Ref. RSCN database; Al-Eisawi 105 (2013).

Ononis pubescens L., Mant. Pl. Altera 267. 1771.
Balqa, on the way to Wadi Shuaib (MSB-K 185785).

Ononis reclinata L., Sp. Pl., ed. 2. 2: 1011. 1763.
Syn. *Ononis mollis* Savi (1802); *Ononis reclinata* L. var. *minor* Moris; *Ononis reclinata* L. var. *mollis* (Savi) Heldr.
About 9 km NW Irbid, along the road to Al-Hemma, *Boulos, Jallad & Laham* 8305 (E00334319); Tal Al-Rumman, *El-Oqlah* (NHJ 2008.0016).

Ononis sicula Guss., Cat. Pl. Hort. Boccadifalco 10. 1821.
Syn. *Ononis sicula* Bornm. var. *microcarpa*; *Ononis viscosa* (Guss.) Hub.-Mor. subsp. *sicula*
Along Zarqa River, 2004, *Al-Dwiri* s.n. (AMM); 1.4 km from Duhni Qa'a, 45 km from Safawi (IBBU-H).

Ononis spinosa L., Sp. Pl.: 2. 1753.
Irbid, Ramtha Mgier (YAR-H).

Ononis vaginalis Vahl, Symb. Bot. 1. 53. 1790.
Ma'an, Site 58, Al-Rajif to Ras Al-Naqab, 3 km before junction on desert highway (MSB-K 395900).

Ononis viscosa L., Sp. Pl. 2: 718. 1753.
Jerash ruins, *Davis* 9692 (E00334376); Tal Al-Rumman, *Taifour* (NHJ 2009.0228).

Ononis viscosa L. subsp. **breviflora** (DC.) Nyman
Syn. *Ononis breviflora* DC. (1825).
Wadi Lahibi, *Hunting Aero Survey* (E); Balqa Governorate, Tal Al-Rumman, sector No. 30 (NHJ).

Ornithopus pinnatus (Mill.) Druce, J. Bot. 45: 420. 1907.
Ref. RSCN database.

Oxytropis foetida (Vill.) DC., Astragalogia 75, no. 11, 1802.
Syn. *Astragalus foetidus* Vill. (1779).
Ajloun, Halaweah (YAR-H).

Pisum fulvum Sibth. & Sm., Fl. Graec. 7. 79. t. 688, 1806–1840.
Wadi Zarqa, *Simpson* 53275 (K000286297); Tal Al-Rumman, *Taifour* (NHJ 2009.0247).

Pisum sativum L. var. **elatius** (M.Bieb.) Asch. & Graebn.
Syn. *Pisum elatius* M.Bieb. (1808); *Pisum humile* Boiss. & Noe (1856).
10–15 km NW Zarqa, along the road to 'Alouk, *Boulos & Al-Eisawi* 6442 (E00162313).

Pisum syriacum (Berg.) Lehm., Züchter 24(11–12): 318. 1954.
Tal Al-Rumman, *Al- Eisawi, Abu- Sondos, Abu Anza & Shishani* (NHJ 2006.0339).

Prosopis farcta J.F.Macbr., Contr. Gray Herb. 59: 17. 1919.
North Ghour, N Shouneh, *Taifour, Al Khawaldah & Alshwaer* (NHJ 2013.0677).

Retama raetam Webb & Berthel., Hist. Nat. Iles Canaries 2: 56. 1842.
Ras El-Naqb, *Boulos, Al-Eisawi & Jallad* 5996 (E00354757); Ma`an, Wadi Musa, *Taifour* (NHJ 2013.0690).

Scorpiurus muricatus L., Sp. Pl. 2: 745. 1753.
Tal Al-Rumman, *Al-Eisawi* (NHJ 2006.0124).

Securigera securidaca (L.) Degen & Doerfl., Denkschr. Kaiserl. Akad. Wiss., Wien. Math.-Naturwiss. Kl. 64: 718. 1897.
Syn. *Securina lutea* Medic. (1787); *Securigera coronilla* DC. (1805).
Upper Jordan Valley (RSCN database).

Senna italica Mill., Gard. Dict., ed. 8. n. 2, 1768.
Aqaba, On the highway to Aqaba, 8 km before Aqaba (MSB-K 338189); Aqaba Governorate, Wadi Yutum, *Taifour, Alkhawaldah & Almogherbi* (NHJ 2013.0704).

Sesbania sesban (L.) Merr., Philipp. J. Sci., C 7: 235. 1912.
Karak, Ghor Al-Mazraia, the southern shore of the Dead Sea. Site no 26 (MSB-K 497457); Wadi Yutum, *Taifour, Al Khawalda & Al Mogherbi* (NHJ 2013.0704).

Spartium junceum L., Sp. Pl.: 2: 708. 1753.
Salt, 2005 (HUJ).

Tephrosia apollinea (Delile) Link, Enum. Hort. Berol. 5 (2): 252. 1822.
Al 'Aqaba, wadi facing site of Marine Biological Station, c. 8 km south of Aqaba, *Boulos & Jallad* 7418 (E00354254).

Tephrosia nubica (Boiss.) Baker, Fl. Trop. Afr. 2: 125. 1871.
Ref. Al-Eisawi 106 (2013).

Tetragonolobus palaestinus Boiss., Diagn. Pl. Orient., ser. 2, 3 (2): 20. 1856.
Syn. *Tetragonolobus purpureus* Moench subsp. *palaestinus* (Boiss.) Ponert, (1973).
Tal Al-Rumman, *Taifour* (NHJ 2009.0079).

Tetragonolobus requienii (Mauri ex Ten.) Sanguinetti, Atti Accad. Pontif. Sci. Nuovi Lincei 15: 357. 1862.
Tal Al-Rumman, *Al-Eisawi* (NHJ 2006.0043).

Trifolium agrarium L., Sp. Pl. 2: 772. 1753.
Jarash, 17 km north Jarash, along the road to Irbid, *Boulos, Jallad & Lahham* 8245 (E00341715).

Trifolium alexandrinum L., Cent. Pl. 1. 25. 1755.
Ref. RSCN database; Al-Eisawi 106 (2013).

Trifolium angustifolium L., Sp. Pl. 2: 769. 1753.
Dibeen Forest Reserve, May 2006, *Taifour & Othman* (FOR-H).

Trifolium argutum Sol., Nat. Hist. Aleppo (ed. 2) 2: 260. 1794.
Upper Jordan Valley (RSCN database).

Trifolium arvense L., Sp. Pl. 2: 769. 1753.
Irbid, 13 Apr. 1985 (YAR-H).

Trifolium berytheum Boiss. & C.I.Blanche, Diagn. Pl. Orient. ser. 2, 2: 15. 1856.
Mejdel (Sea of Tiberius), *Meyers & Dinsmore* (E).

Trifolium boissieri Guss. ex Soy.-Will. & Godr., Fl. Sicul. Syn. 2(2): 858. 1848.
Jarash, 5 km southwest of Jerash to Dibeen near Min. of Agric. nursery (MSB-K 151739).

Trifolium bullatum Boiss. & Hausskn., Fl. Orient. 2: 138. 1872.
Along Zarqa River, 2004, *Al-Dwiri* s.n. (AMM).

Trifolium bullatum Boiss. & Hausskn. var. **bullatum**, Fl. Orient. 2: 138. 1872.
Syn. *Trifolium tomentosum* L. var. *glabrescens* (Post) Hausskn. & Bornm.; *Trifolium bullatum* Post var. *glabrescens*; *Trifolium tomentosum* L. var. *bullatum* (Boiss. & Hausskn.) Hossain.
Ajloun district, outside Ajloun reserve near Elynabi village BOT ERA project (NHJ 2013.0525).

Trifolium campestre Schreb., Deutschl. Fl. 1, Phanerog. Heft 16. 1804.
Syn. *Trifolium procumbens* L. (1753); *Trifolium glaucescens* Hausskn. (1893).
Madaba, (MSB-K 114057); Tal Al-Rumman, *Taifour* (NHJ 2009.0221).

Trifolium cherleri L., Demonstr. Pl. 21. 1753.
Gamala, *Meyers & Dinsmore* 1609 (E00338795).

Trifolium clusii Godr. & Gren., Fl. France 1: 414. 1848.
Tal Al-Rumman, *Shishani & Alkoran* (NHJ 2011.0037).

Trifolium clypeatum L., Sp. Pl. 2: 769. 1753.
Ruins N of Medaba, *Meyers & Dinsmore* M1401 (E00347007); Tal Al-Rumman, *Taifour* (NHJ 2011.0037).

Trifolium constantinopolitanum Ser., Prodr. [A. P. de Candolle] 2: 193. 1825.
N of Irbid, Kufr-Soum village, 24 Jun. 2003 (NCARE).

Trifolium dasyurum C.Presl, Symb. Bot. (Presl) 1: 53. 1831.
Syn. *Trifolium formosum* d'Urv. (1822).
Tal Al-Rumman, *Taifour & Aldamen* (NHJ 2009.0048).

Trifolium dichroanthum Boiss., Diagn. Pl. Orient. ser. 1, 9: 20. 1849.
Wadi Hanein, *Meyers & Dinsmore* (E).

Trifolium echinatum M.Bieb., Fl. Taur.-Caucas. 2: 216. 1808.
Ref. Al-Eisawi 107 (2013).

Trifolium eriosphaerum Boiss., Diagn. Pl. Orient., ser. 1, 2 (9): 25. 1849.
Jordan, *Simpson* 53367 (K000744624).

Trifolium erubescens Fenzl, Pug. Pl. Nov. Syr. 5. 1842.
Between Ajloun & Mar Elias, *Davis* 9440 (E00339218).

Trifolium fragiferum L. var. **fragiferum**, Sp. Pl. 772. 1753.
Jarash, Site 32. Souf (MSB-K 557940).

Trifolium glanduliferum Boiss., Diagn. Pl. Orient., ser. 1, 1 (2): 30. 1843.
Irbid, outskirts towards Hordan valley, at Jamha village (MSB-K 114323).

Trifolium glanduliferum var. **nervulosum** (Boiss. & Heldr.) Zohary, Fl. Palaestina 2: 175. 1972.
Syn. *Trifolium nervulosum* Boiss. & Heldr. (1849).
Ajloun (RSCN database); Salt, 29 Apr. 1911, *Meyers & Dinsmore* (E).

Trifolium globosum L., Sp. Pl. 2: 767. 1753.
Irbid (FOR-H).

Trifolium grandiflorum Schreb., Nova Acta Phys.-Med. Acad. Caes. Leop.-Carol. Nat. Cur. 3: 477. 1767.
(RSCN database).

Trifolium hirtum All., Auct. Fl. Pedem.: 20. 1789.
Ref. RSCN database; Al-Eisawi 107 (2013).

Trifolium lappaceum L., Sp. Pl. 2: 768. 1753.
Syn. *Trifolium lappaceum* L. var. *brachyodon* Hausskn.
Irbid, Abu Loqos (YAR-H).

Trifolium leucanthum M.Bieb., Fl. Taur.-Caucas. 2: 214. 1808.
Al-Mafraq- Al-Mdawar, (IBBU-H).

Trifolium micranthum Viv., Fl. Libyc. Spec. 45. t. 19. f. 3, 1824.
Jordan valley (RSCN database).

Trifolium nigrescens Viv., Fl. Ital. Fragm.: 12. t. 13, 1808.
Ain Bellata, *Meyers & Dinsmore* 5245 (E00342516).

Trifolium palaestinum Boiss., Diagn. Pl. Orient. ser. 1, 9: 21. 1849.
Tal Al-Rumman, *Al-Eisawi* (NHJ 2006.0015).

Trifolium pauciflorum d'Urv., Mém. Soc. Linn. Paris 1: 350. 1822.
Tal Al-Rumman RBG, Sector No. 20, *El-Oqlah, Taifour & Al Khawaldah* (NHJ 2013.0226).

Trifolium physodes Stev. ex M.Bieb., Fl. Taur.-Caucas. 2: 217. 1808.
Between Suf and Ain Jenneh, *Meyers & Dinsmore* (E); Irbid, Saham (YAR-H).

Trifolium pilulare Boiss., Diagn. Pl. Orient. ser. 1, 2: 29. 1843.
S. of Ajloun, *Hunting Aero Survey* 97D (E00310211); Tal Al-Rumman, *Taifour & Aldamen* (NHJ 2009.0222).

Trifolium plebeium Boiss., Diagn. Pl. Orient. ser. 1, 9: 28. 1849.
Ref. RSCN database; Al-Eisawi 107 (2013).

Trifolium prophetarum Hossain, Notes Roy. Bot. Gard. Edinburgh 23: 421. 1961.
Eastern upper slopes of the Jordan valley, 09 Mar. 2007 (NCARE); Upper Jordan valley (RSCN database).

Trifolium purpureum Lois., Fl. Gall. 2: 484, t. 14, 1807.
South of Ajloun, *Hunting Aero Survey* 108B (E00339040); Tal Al-Rumman, *Taifour* (NHJ 2011.0034).

Trifolium repens L., Sp. Pl. 2: 767. 1753.
Jebel Siaghah, Madaba, (AUB).

Trifolium resupinatum L., Sp. Pl. 2: 771. 1753.
Syn. *Trifolium resupinatum* L. var. *majus* Boiss.; *Trifolium resupinatum* L. var. *microcephalum* Zohary
Deir Alla, Jordan Valley, north of the Dead Sea, *Betts* (E00342345); Tal Al-Rumman, *Al-Eisawi* (NHJ 2006.0108).

Trifolium scabrum L., Sp. Pl. 2: 770. 1753.
Irbid, North Irbid, Soum village (MSB-K 200246);

Trifolium spumosum L., Sp. Pl. 2: 771. 1753.
Salt, Wadi Shuaib (YAR-H); Upper Jordan Valley (RSCN database).

Trifolium stellatum L., Sp. Pl. 2: 769. 1753.
S. of Ajloun, *Hunting Aero Survey* 97E 17-Apr–55 (E00338959); Tal Al-Rumman, *Taifour* (NHJ 2009.0028)

Trifolium subterraneum L., Sp. Pl. 2: 767. 1753.
Upper Jordan valley (RSCN database).

Trifolium tomentosum L., Sp. Pl. 2: 771. 1753.
S of Ajloun, *Hunting Aero Survey* 97C (E00342387); Tal Al-Rumman RBG, Sector No. 23 *El-Oqlah & Taifour* (NHJ 2013.0156).

Trifolium vavilovii Eig, Trudy Prikl. Bot., Ser. 7, Kormov. Kul't. 1: 108. 1934.
Slopes facing Jiser, 10 May 1927 (HUJ).

Trigonella arabica Delile, Voy. Arab. Petr. 84. 1830.
Moab, between Methluthe and Dhiban (Moab), *Davis* 9432 (E00336700).

Trigonella anguina Delile, Descr. Egypte, Hist. Nat. 254, t. 110, f. 2, (1813–1814).
Azraq (RSCN database).

Trigonella astroides Fisch & Mey.
Tafila Province, 2014, *Oran* (AMM).

Trigonella balansae Boiss. & Reut., Diagn. Pl. Orient. ser. 2, 5: 79. 1856.
Ref. Balqa, Danin (2004).

Trigonella berythea Boiss. & C.I.Blanche, Diagn. Pl. Orient. ser. 2, 2: 10. 1856.
Jebel Siaghah, Madaba, iv.1877 (AUB).

Trigonella caelesyriaca Boiss., Diagn. Pl. Orient. ser. 1, 9: 19. 1849.
Syn. *Trigonella gaillardotii* Boiss. (1856); *Trigonella aleppica* Boiss. & Hausskn. (1873).
Ajloun (FOR-H); Irbid, (RSCN database); Jerash, *Meyers & Dinsmore* (E).

Trigonella corniculata L., Syst. Nat., ed. 10. 2: 1180. 1759.
Irbid, Al Ramtha (YAR-H).

Trigonella cylindracea Desv., J. Bot. Agric. 3: 77. 1814.
Irbid, Al Ramtha (YAR-H); Ma'an, 34 km from Ma'an-Tybah junction (NCARE).

Trigonella filipes Boiss., Diagn. Pl. Orient. ser. 1, 9: 16. 1849.
Syn. *Trigonella minima* Paine (1875).
Between Petra & foot of Mt. Hor, *Davis* 9053 (E00336901).

Trigonella foenum-graecum L., Sp. Pl. 2: 777. 1753.
Syn. *Trigonella foenum-graecum* L. var. *haussknechtii* Sirj.
Ajloun Forest Reserve (RSCN database); Ajloun, Kufrangeh (YAR-H).

Trigonella glabra Thunb., Prodr. Pl. Cap. 2: 137. 1800.
Syn. *Trigonella hamosa* L. (1759); *Trigonella hamosa* L. subsp. *hamosa* (1759).
Ref. RSCN database.

Trigonella glabra subsp. **uncata** (Boiss. & Noe) Lassen, Willdenowia 16(2): 447. 1987.
Syn. *Trigonella uncata* Boiss. & Noe (1856).
Azraq (RSCN database).

Trigonella hierosolymitana Boiss., Diagn. Pl. Orient. ser. 1, 9: 15. 1849.
Syn. *Trigonella kotschyi* Fenzl ex Boiss. (1849).
Tal Al-Rumman, *Shishani & Alkoran* (NHJ 2011.0075).

Trigonella lilacina Boiss., Diagn. Pl. Orient. ser. 1, 9: 17 1849.
Between Nadrah and Jerash (IBBU-H); Upper Jordan valley (RSCN database); on sandy slopes near camp on road to Kilwa, 15 Apr. 1955, *Hunting Aero Survey* (E).

Trigonella lunata Boiss., Diagn., Pl. Orient. ser. 1, 2: 19. 1843.
Ref. RSCN database; Al-Eisawi 108 (2013).

Trigonella maritima Delile, Encyc. Suppl. 5: 361. 1817.
Irbid, Al Torrah, 27 Mar. 1983 (YAR-H).

Trigonella moabitica Zohary, Fl. Palaestina 2: 456. 1972.
Azraq, *Frey & Kürschner* VO5029 (E001755.24); Tafila, El Hasa, 1929 (HUJ); Aqaba (RSCN database).

Trigonella monspeliaca L., Sp. Pl. 2: 777. 1753.
Irbid, Ein Aqraba, *El-Oqlah* (NHJ 2013.0153).

Trigonella schlumbergeri Boiss., Fl. Orient. Suppl.: 163. 1888.
Karak, Wadi Mujib, N von Karak, *Frey & Kürschner* VO 5080 (E00336694).

Trigonella spicata Sm., Fl. Graec. Prodr. 2(1): 108. 1813.
Ref. Al-Eisawi 108 (2013).

Trigonella spinosa L., Sp. Pl. 2: 777. 1753.
Along Zarqa River, 2004, *Al-Dwiri* s.n. (AMM).

Trigonella spruneriana Boiss., Diagn. Pl. Orient. ser. 1, 2: 17. 1843.
Syn. *Trigonella torulosa* Griseb. (1843); *Trigonella sibthorpii* Boiss. (1849).
Zerka Main to Jebel Attaris, *Davis* 9379 (E00336871).

Trigonella stellata Forssk., Fl. Aegypt.-Arab.: 140. 1775.
Wadi Rum, *Davis* 9032 (E00336767); Azraq Wadi Bottom (South west the Castle Amra), *El-Oqlah* (NHJ 2013.0564).

Tripodion tetraphyllum (L.) Fourr., Ann. Soc. Linn. Lyon sér. 2, 16: 359. 1868.
Syn. *Physanthyllis tetraphylla* (L.) Boiss. (1840).
Tal Al-Rumman RBG, Sector No. 39, *Taifour & Al Khawaldah* (NHJ 2013.0290).

Vicia cypria Kotschy, Unger & Kotschy, Cypern 384. 1865.
Tal Al-Rumman RBG, Sector No. 23, *Al Mfadi* (NHJ 2013.0088).

Vicia ervilia (L.) Willd., Sp. Pl., ed. 4. 3(2): 1103. 1802.
Jordan, Tabaget Fahel, 20.iv.1967, *Edgecombe* (AUB).

Vicia galeata Boiss., Diagn. Pl. Orient. ser. 1, 2: 103. 1843.
Ajloun, Ishtafena (IBBU-H).

Vicia galilaea Plitmann & Zohary, Israel J. Bot. 14: 91. 1965.
Ajloun Forest Reserve (RSCN database).

Vicia hulensis Plitmann, Israel J. Bot. 14: 96. 1965.
Balqa Governorate, Tal Al-Rumman, Sector No. 13 (NHJ).

Vicia hybrida L., Sp. Pl. 2: 737. 1753.
Above Zarqa river, *Meyers & Dinsmore* G591 (E00364796).

Vicia lathyroides L., Sp. Pl. 2: 736. 1753.
Salt, 30.iv.1886, *Barbey* (AUB).

Vicia lutea L., Sp. Pl. 2: 736. 1753.
Irbid, Saham and Jarash (YAR-H).

Vicia monantha Retz., Observ. Bot. 3: 39. 1783.
Syn. *Vicia gracilis* Banks & Sol. ex Russ (1794); *Vicia cinerea* M.Bieb. (1819 or early 1820); *Vicia calcarata* Boiss. (1799); *Vicia calcarata* Desf. var. *cinerea* (M.Bieb.) Boiss.; *Vicia griffithii* Baker (1876); *Vicia monantha* Tetz subsp. *cinerea* (M.Bieb.) Maire; *Vicia monantha* Retz. var. *cinerea* (M.Bieb.) Dinsm.
Amman, *Hepper* 10015 (K000283232); Tal Al-Rumman, *Taifour & Aldamen* (NHJ 2009.0231).

Vicia narbonensis L., Sp. Pl. 2: 737. 1753.
Medaba, *Meyers & Dinsmore* M845 (E00366313).

Vicia palaestina Boiss., Diagn. Pl. Orient. ser. 1, 9: 116. 1849.
Tal Al-Rumman, *Taifour & Aldamen* (NHJ 2010.0058).

Vicia parviflora Cav., Anales Ci. Nat. iv. 73. 1801.
Jerash, *Meyers & Dinsmore* (E).

Vicia peregrina L., Sp. Pl. 2: 737. 1753.
Medaba, *Meyers & Dinsmore* M1655 (E00366273); Tal Al-Rumman, *Taifour* (NHJ 2009.0230).

Vicia sativa L., Sp. Pl. 2: 736. 1753.
Jerash, *Meyers & Dinsmore* G525 (E00366049); Tal Al-Rumman, *Al-Eisawi* (NHJ 2006.0252).

Vicia sericocarpa Fenzl, Pug. Pl. Nov. Syr. 4. 1842.
Salt, IV.1886, *Barbey* (AUB); Tafila Province, 2014, *Oran* (AMM).

Vicia tenuifolia Roth, Tent. Fl. Germ. 1: 309. 1788.
Jarash, Souf, 5.V.1886 (AUB).

Vicia villosa Roth., Tentamen Florae Germanicae 2(2). 1793.
Irbid, *Al-Gharaibeh* (YAR-H).

Vigna luteola (Jacq) Benth., Fl. Bras. (Martius) 15(1): 194, t. 50, f. 2, 1859.
Upper Jordan Valley (RSCN database).

Vigna unguiculata (L.) Walp., Repert. Bot. Syst. 1: 779. 1842.
Syn. *Dolichos sinensis* L. (1756).
West Dair Abi Said, Bioversity International Database: http://bioversityinternational.org.

Fagaceae

Quercus coccifera L., Sp. Pl. 2: 995. 1753.
Syn. *Quercus calliprinos* Webb (1838).
Ajloun, *Davis* 9698 (E00441883); Dibeen Forest, *RBG-Ire & RBG-Jo* (NHJ 2008.0101).

Quercus infectoria Olivier subsp. **veneris** (A.Kern.) Meikle, Diagn. Pl. Orient. 12: 119. 1853.
Syn. *Quercus boissieri* Reut. (1853); *Quercus veneris* A.Kern. (1904).
Between Ajloun & Mar Elias, *Davis* 9442 (E00404012); Ajloun district, outside Ajloun reserve near Elynabi village, BOT ERA project (NHJ 2013.0538).

Quercus ithaburensis Decne., Ann. Sci. Nat., Bot. 2, 4: 348. 1835.
Irbid: Ajloun-Berge *Frey, Hilger & Kürschner* 80–1012 (E00404301); Tal Al-Rumman, *Taifour & Aldamen* (NHJ 2009.0050).

Frankeniaceae

Frankenia hirsuta L., Sp. Pl.: 331. 1753.
Balqa, Dea Sea area, 16 Apr. 2009 (NCARE); *Hunting Aero Survey* 125A (E00160483).

Frankenia laevis L., Sp. Pl. 331. 1753.
Ref. Azraq (FOR-H).

Frankenia pulverulenta L., Sp. Pl.: 332. 1753.
10 km N von Azraq, *Frey & Kürschner* 80–420 (E00382300).

Gentianaceae

Centaurium pulchellum (Sw.) Druce, Fl. Berks. 342. 1898.
Syn. *Gentiana pulchella* Sw. (1783).
Ref. Al-Eisawi 110 (2013).

Centaurium spicatum (L.) Fritsch, Mitt. Naturwiss. Vereins Univ. Wien 5: 97. 1907.
Syn. *Gentiana spicata* L. (1753).
Zraq, 7/1/1953, *Kaspligil*.

Centaurium tenuiflorum (Hoffmanns. & Link) Fritsch, Mitt. Naturwiss. Vereins Steiermark 5: 97. 1907.
Syn. *Erythraea tenuiflora* Hoffmanns. & Link (1813).
Dead sea, 2 km befor the Hot Water Spring, 12.v.2009, *Abulayla* (NCARE).

Geraniaceae

Erodium acaule (L.) Bech. & Thell., Repert. Spec. Nov. Regni Veg. 25: 215. 1928.
Syn. *Geranium acaule* L. (1759); *Erodium romanum* (Burm.f.) L'Her (1789).
E of Suf, *Meyers, Fred & Dinsmore* G366 (E0033167).

Erodium alnifolium Guss., Fl. Sicul. Prodr. 2: 307. 1828.
Wadi Rum, 1966, *Al-Eisawi, von Lowenstern & Garbari* (AMM).

Erodium arborescens (Desf.) Willd., Sp. Pl., ed. 4, 3(1): 638. 1800.
Syn. *Geranium arborescens* Desf. (1798).
Tafila Province, 2014, *Oran* (AMM).

Erodium botrys (Cav.) Bertol. Amoen. Ital. 35. 1819.
Syn. *Geranium botrys* Cav. (1787).
Tal Al-Rumman, *Al-Eisawi* (NHJ 2006.0392).

Erodium chium (Burm.f.) Willd., Phytographia 10. 1794.
Irbid, Zobia (YAR-H).

Erodium ciconium (L.) L'Hér., Hort. Kew. 2: 415. 1789.
Syn. *Geranium ciconium* L. (1755).
Ma'an, Jebel Harun, *Davis* 9516 (E00331486); Tal Al-Rumman, *Al-Eisawi, Abu Sondos, Abu Anza & Shishani* (NHJ 2006.0049).

Erodium cicutarium (L.) L'Hér. ex Aiton, Hort. Kew. 2: 414. 1789.
Syn. *Geranium cicutarium* L. (1753).
Jordan, *Boulos, Jallad & El-Gohary* M.8063 (E00216466); Tal Al-Rumman, *Taifour & Aldamen* (NHJ 2010.0003).

Erodium crassifolium L'Hér. ex Aiton, Hort. Kew. 2: 414. 1789.
Syn. *Erodium hirtum* (Forssk.) Willd. (1800); *Geranium hirtum* Forssk. (1775).
Jordan, *Boulos, Al-Eisawi & Jallad* 5994 (E00216177).

Erodium gaillardotii Boiss., Diagn. Pl. Orient. II, 6: 41. 1859.
Deir Atiyeh, May 1879. (AUB).

Erodium glaucophyllum (L.) L'Hér., Hort. Kew. 2: 416. 1789.
Syn. *Geranium glaucophyllum* L. (1753).
Al 'Aqaba, 6 km north Aqaba, *Boulos & Al-Eisawi* 6112 (E00331739).

Erodium gruinum (L.) L'Hér. ex Aiton, Hort. Kew. 2. 415. 1789.
Syn. *Geranium gruinum* L. (1753).
Tal Al-Rumman, *Al-Eisawi* (NHJ 2006.0020).

Erodium guttatum (Desf.) Willd., Sp. Pl., ed. 4. 3(1): 636. 1800.
Syn. *Geranium guttatum* Desf. (1798).
Aqaba, RSCN database.

Erodium laciniatum (Cav.) Willd., Sp. Pl., ed. 4. 3(1): 633. 1800.
Syn. *Geranium laciniatum* Cav. (1787).
Wadi Rum, *Davis* 10445 (E00331759);Tal Al-Rumman, *Al-Eisawi* (NHJ 2006.0134).

Erodium malacoides (L.) L'Hér., Hort. Kew. 2: 415. 1789.
Ma'an, Wadi Musa, near Petra, *Boulos & Jallad* 8110(E00331871); Tal Al-Rumman, *Al-Eisawi* (NHJ 2006.0181).

Erodium moschatum (L.) L'Hér., Hort. Kew. 2: 414. 1789.
Syn. *Geranium cicutarium* L. var. *moschatum* L. (1753).
Wadi al Buja, *Dinsmore* 7462 (E00332120);Tal Al-Rumman, *Al-Eisawi* (NHJ 2006.0393).

Erodium neuradifolium Delile ex Godr., Mém. Sect. Med. Acad. Sci. Montpellier 1: 425. 1853.
Aqaba, Wadi Rum, *Davis* 9042 (E00331923).

Erodium oxyrhinchum M.Bieb., Fl. Taur.-Caucas. 2: 133. 1808.
Nr. Aqaba, *Boulos & Lahham* 5415 (AMM, BM, E).

Erodium oxyrhinchum M.Bieb. subsp. **bryoniifolium** (Boiss.) Schönb.-Tem., Fl. Iranica 69: 43. 1970.
Syn. *Erodium bryoniifolium* Boiss. (1843).
Amman, 30 km W von Azraq, *Frey & Kürschner* VO 4820 (E00331899).

Erodium subintegrifolium Eig, Beih. Bot. Centralbl. 50(2): 230. 1932.
Irbid, Um Qais, (IBBU-H).

Erodium touchyanum Delile ex Godr., Mém. Sect. Med. Acad. Sci. Montpellier 1: 423. 1853.
Syn. *Erodium deserti* (Eig) Eig (Beih. Bot. Centralbl. I. II. 237, in obs. 1932).
Jordan, *Frey & Kürschner* VO 4819(E00091235).

Geranium columbinum L., Sp. Pl. 682. 1753.
Wadi Yutum upstream, 15 km N of Aqaba, 21 Mar 1983, *Boulos et al.* (E, AMM).

Geranium dissectum L., Cent. Pl. 1. 21. 1755.
Hesban, *Meyers & Dinsmore* M564 (E00330858); Tal Al-Rumman, *Al-Eisawi* (NHJ 2006.0023).

Geranium lucidum L., Sp. Pl. 2: 682. 1753.
Between Es-Salt and Rumman, *Meyers & Dinsmore* M1360 (E00331036).

Geranium molle L., Sp. Pl. 2: 682. 1753.
Jerash, *Meyers & Dinsmore* G391 (E00330982); Tal Al-Rumman, *Al-Eisawi, Abu Sondos, Abu Anza & Shishani* (NHJ 2006.0089).

Geranium purpureum Vill, Syst. Pl. Eur. 1: 72. 1786.
Ajlum castle, *Davis* 9674 (E00331560); Ajloun district, Wadi Orjan, BOT ERA project (NHJ 2013.0544).

Geranium pusillum L., Syst. Nat., ed. 10. 2: 1144. 1759.
Ref. RSCN database; Al-Eisawi 112 (2013).

Geranium robertianum L., Sp. Pl. 2: 681. 1753.
Ref. RSCN database; Al-Eisawi 112 (2013).

Geranium rotundifolium L., Sp. Pl. 2: 683. 1753.
Ajloun district, Wadi Orjan, BOT ERA project (NHJ 2013.0545).

Geranium trilophum Boiss., Diagn. Pl. Orient. ser. 1, 6: 30. 1846.
Ajloun, Aqaba (YAR-H).

Geranium tuberosum L., Sp. Pl. 2: 680. 1753.
Jordan, *Boulos & Jallad* 7349 (E00129706); Tal Al-Rumman, *Taifour & Aldamen* (NHJ 2010.0023).

Monsonia heliotropoides Boiss., Fl. Orient. 1: 897. 1867.
Wadi Araba, RSCN database.

Monsonia nivea (Decne.) Webb, Fragm. Fl. Aethiop.-Aegypt. 59. 1854.
Syn. *Erodium niveum* Decne. (1835).
Nr. Aqaba, *Boulos & Jallad* 7436 (E00109968).

Gisekiaceae

Gisekia pharnaceoides L., Mant. Pl. Altera 562 1771.
Ref. RSCN database; Al-Eisawi 129 (2013).

Haloragaceae

Myriophyllum spicatum L., Sp. Pl. 2: 992 1753.
(YAR-H); Danin (http://flora.org.il/en/plants/MYRSPI/).

Hydrocharitaceae

Halophila ovalis (R.Br.) Hook.f., Bot. Antarct. Voy. 3. (Fl. Tasman.). 2: 45. 1858.
Syn. *Caulinia ovalis* R.Br. (1810).
6 km S of Aqaba Port, reef lagoon, 21 Mar. 1975, *Hulings* s.n. (AMM, BM, E).

Halophila stipulacea (Forssk.) Asch., Sitzungsber. Ges. Naturf. Freunde Berlin: 3. 1867.
Syn. *Zostera stipulacea* Forssk. (1775).
6 km S of Aqaba Port, reef lagoon, 21 Mar. 1975, *Hulings* s.n. (AMM, BM, E).

Najas marina L., Sp. Pl. 1015. 1753.
Syn. *Najas delilei* Rouy (1912); *Najas marina* var. *intermedia* (Wolfg. ex Gorski) Rendle
Upper Jordan valley (RSCN database).

Najas minor All., Auct. Syn. Stirp. Horti Taur. 3. 1773.
(YAR-H).

Hypericaceae

Hypericum empetrifolium Willd., Sp. Pl. 3: 1452. 1803.
Ajloun Forest Reserve (RSCN database).

Hypericum hyssopifolium Chaix, Hist. Pl. Dauphiné 1: 329. 1786.
Ref. RSCN database.

Hypericum lanuginosum Lam., Encycl. 4(1): 171 .1797.
Ajloun Forest Reserve (YAR-H).

Hypericum libanoticum N.Robson, Nouv. Fl. Liban 2: 524. 1970.
Ajloun, Ishtafanah, *El-Oqlah* (YAR-H).

Hypericum olivieri (Spach) Boiss., Fl. Orient. 1: 803. 1867.
Syn. *Eremosporus olivieri* Spach (1836).
Amman, University of Campus, Al-Jubaiha, *Boulos* 286 (E00371771).

Hypericum sinaicum Hochst. ex Boiss., Fl. Orient. 1: 808. 1867.
Aqaba (HUJ).

Hypericum thymifolium Banks & Sol., Nat. Hist. Aleppo ed. 2, 2: 270. 1794.
Syn. *Hypericum serpyllifolium* Lam. (1797).
Ref. RSCN database; Al-Eisawi 113 (2013).

Hypericum triquetrifolium Turra, Farset. Nov. Gen. 12. 1765.
Syn. *Hypericum crispum* L. (1767).
Tal Al-Rumman, *Royal Botanic Garden Team* (NHJ 2008.0016).

Iridaceae

Crocus aleppicus Baker, Gard. Chron. 1873: 609. 1873.
Wadi Musa, *Birkinshaw* (K000318281).

Crocus cancellatus Herb., Bot. Mag. 67: sub t. 3864. 1841.
Tafila Province, 2014, *Oran* (AMM).

Crocus cancellatus Herb. subsp. **damascenus** (Herb.) B.Mathew, Crocus 68. 1982.
Syn. *Crocus damascenus* Herb. (1845).
8 km E of Wadi Musa, 29.3.1936 (71728-HUJ).

Crocus cartwrightianus Herb., Edwards's Bot. Reg. 29 (Misc.): 82. 1843.
Madaba, RSCN database.

Crocus hermoneus Kotschy ex Maw., Gard. Chron. n.s., 16: 559. 1881.
Jarash, Amman, Salt, Zarqa (RSCN database).

Crocus hyemalis Boiss. & Blanche, Diagn. Pl. Orient. 2, 4: 93. 1859.
Ajloun, *Al-Eisawi* 9770 (K000318285).

Crocus moabiticus Bornm., Repert. Spec. Nov. Regni Veg. 10: 383. 1912.
Madaba, *Kerndorff* 1986 (K000318277).

Crocus naqabensis Ref. Al-Eisawi & Kiswani, Arab Gulf J. Sci. Res. 19: 167. 2001.
Ras Naqab, 2001, *Al-Eisawi* (YAR-H).

Crocus pallasii Goldb., Mém. Soc. Imp. Naturalistes Moscou 5: 157. 1817.
Jordan, sin. loc. 10.12.1911 (71914-HUJ); Al-Mafraq, Al-Bayet University (IBBU-H).

Gladiolus atroviolaceus Boiss., Diagn. Pl. Orient. 13: 14. 1854.
Syn. *Gladiolus aleppicus* Boiss. (1854).
Um Qeis, *Boulos* 8479 (K000283558).

Gladiolus italicus Mill., Gard. Dict. ed. 8, 2. 1768.
Syn. *Gladiolus segetum* Ker-Gawl. (1804).
Al-Ramtha, s.coll. 303 (K000318270).

Iris aqabaensis Sealy, Kew Bull. 4: 561. 1950.
Petra, *Boulos* 8435 (K00031826.2).

Iris atrofusca Baker, Gard. Chron. 1: 384. 1893.
Jordan, *Hayne* (K000464990);Tal Al-Rumman, *Al-Eisawi* (NHJ 2006.0334).

Iris atropurpurea Baker, Gard. Chron. 1: 330. 1889.
Wadi Kufrinja, *Simpson* 53447 (K000318286).

Iris atropurpurea Baker var. **gileadensis** Dinsm., G.E.Post, Fl. Syria, ed. 2, 2: 601. 1933.
Salt, 20 Apr. 1929, *Meyers & Dinsmore* (E).

Iris aucheri (Baker) Sealy, Kew Bull. 4: 562. 1950.
Syn. *Xiphion aucheri* Baker (1871); *Iris fumosa* Boiss. & Hausskn. (1882).
Shobek, *Davis* 9649 (K00031829.4); Petra, Apr. 1945 (AUB).

Iris bismarckiana Damman & Sprenger, Cat. 51: 4. 1890.
Jerash, Dibeen, *El-Oqlah, Taifour & Said* (NHJ 2013.0101).

Iris bostrensis Mouterde, Bull. Soc. Bot. France 101: 420. 1955.
Ref. Al-Eisawi 114 (2013).

Iris grant-duffii Baker, Handb. Irid. 7. 1892.
Recorded by (YAR-H).

Iris haynei Baker, Gard. Chron. n.s., 6: 710. 1876.
Ref. Ajloun (RSCN database).

Iris mariae Barbey,Oesterr. Bot. Z. 41: 207. 1891.
Karak, Al-Walah, 1984, (YAR-H).

Iris nigricans Dinsm., Fl. Syria ed. 2, 2: 600. 1933.
Amman, *Boulos & Jallad* 8145 (K000318258); Madaba, d.1929 (HUJ); Karak, Qaser, 500 m after the finishing of Wadi Mujib toward Karak, May 2006 (NCARE).

Iris palaestina (Baker) Boiss., Fl. Orient. 5 (1): 120. 1882.
Syn. *Xiphion palaestinum* Baker (1871).
Ras Naqab, 19 May 1974, *Boulos & Jallad* (AMM, E).

Iris petrana Dinsm., Fl. Syria ed. 2, 2: 599. 1933.
Ma'an, Al-Fujaij rangeland reserve (MSB-K 186357); Tafila, Lahda spring, road to Dana Reserve, BOT ERA project (NHJ 2013.0464).

Iris postii Mouterde, Nouv. Fl. Liban 1: 309. 1966.
Azraq, *Gillett* 15697 (K00031826).

Iris regis-uzziae Feinbrun, Notes Roy. Bot. Gard. Edinburgh 37(1): 75. 1978.
Maan Aqaba, *Birkinshaw* s.n. (K000318288).

Iris vartanii Foster, Gard. Chron. 1: 438. 1885.
North of Salt, *Meyers & Dinsmore* (E00333054).

Moraea sisyrinchium (L.) Ker Gawl., Ann. Bot. 1(2): 241. 1804.
Syn. *Iris sisyrinchium* L. (1753); *Gynandriris sisyrinchium* (L.) Parl. (1854).
Tafila, on the road from Qadeseyeh to Shoubak (MSB-K 339946); Tal Al-Rumman, *Al-Eisawi, Abu Sondos, Abu Anza & Shishani* (NHJ 2006.0059).

Romulea bulbocodium (L.) Sebast. & Mauri, Fl. Roman. Prodr. 17. 1818.
Syn. *Crocus bulbocodium* L. (1753).
Near the village Makhraba (ca 20 km W of Irbid), *Böcker* RB 989 (B10 0520734).

Romulea petraea Ref. Al-Eisawi Arab Gulf J. Sci. Res. 19: 168. 2001.
Petra (RSCN database).

Romulea phoenicia Mouterde, Bull. Soc. Bot. France 100: 349. 1953.
Ref. Balqa, Danin (2004).

Ixiolirionaceae

Ixiolirion tataricum (Pall.) Schult. & Schult.f., Syst. Veg. 7: 752. 1829.
Syn. *Amaryllis tatarica* Pall. (1776); *Amaryllis montana* Labill. (1791); *Ixiolirion montanum* (Labill.) Schult. & Schult.f. (1821); *Ixiolirion pallasii* Fisch. & C.A.Mey. ex Ledeb. (1852).
Jordan, *Cope & Al-Eisawi* 381 (K000286348); Tafila, Dana Reserve, Wadi Ahmer, BOT ERA project (NHJ 2013.0437).

Juncaceae

Juncus acutus L., Sp. Pl.: 325. 1753.
Azraq, 8.iii.1981, *Al-Oqlah* (YAR-H).

Juncus articulatus L., Sp. Pl.: 327. 1753.
Ma'an, 2005 (NCARE).

Juncus bufonius L., Sp.Pl. 1: 328. 1753.
6–7 km S Rum Rest House, 23 Mar. 1975, *Boulos et al.* (AMM); Ain Hesban, 17 Apr. 1911, *Meyers & Dinsmore* (E).

Juncus capitatus Weigel, Observ. Bot. 28. 1772.
Ref. Jordan Valley, Danin (2004).

Juncus effusus L., Sp. Pl.: 326. 1753.
Ain Musa, Petra, 1886 (AUB).

Juncus fontanesii J.Gay ex Laharpe, Essai Monogr. Jonc. 42. 1825.
Ma'an, Ain Sader spring, on the way from Wadi-Musa to Taybeh (MSB-K 277943, NCARE).

Juncus inflexus L., Sp. Pl.: 326. 1753.
Syn. *Juncus glaucus* Sibth. (1794) nom. illegit.; *Juncus tenax* Banks & Sol. (1794) nom. illegit.; *Juncus paniculatus* Hoppe ex J.A. & J.H.Schult (1829); *Juncus warakensis* Nab. (1929).
Al Ghour (YAR-H); Ajloun, RSCN database.

Juncus maritimus Lam., Encycl. 3: 264. 1789.
Azraq, around pools (FOR-H).

Juncus punctorius L.f., Suppl. Pl.: 208. 1782.
Ma'an, road from Dilagha to Wadi Araba, just after the end of Dilagha village houses (MSB-K 277965).

Juncus rigidus Desf., Fl. Atlant. 1: 312. 1798.
Syn. *Juncus maritimus* Lam. var. *arabicus* Asch. & Buchenau ex Boiss. (1882); *Juncus arabicus* (Asch. & Buchenau ex Boiss.) Adamson (1935).
Balqa, in gully on mountainside, in Wadi Al Zara (MSB-K 176655).

Juncus subulatus Forssk., Fl. Aegypt.-Arab.: 75. 1775.
Syn. *Juncus multiflorus* Desf. (1798) nom. illegit.
Dead Sea Area (RSCN database).

Juncaginaceae

Triglochin maritima L., Sp. Pl. 105. 1753.
Karak (Moab) (B100167666).

Triglochin palustris L., Sp. Pl.: 338. 1753.
Ref. RSCN database; Al-Eisawi 115 (2013).

Lamiaceae

Ajuga chamaepitys (L.) Schreb., Pl. Verticill. Unilab. Gen. Sp. 24. 1774.
Syn. *Teucrium chamaepitys* L. (1753).
Jerash, *Townsend* (K000318314).

Ajuga chamaepitys (L.) Schreb. subsp. **chia** (Schreb.) Arcang., Comp. Fl. Ital. 560. 1882.
Syn. *Ajuga chia* Schreb. (1774)
Tall al Birah, *Davies* 8621 (K000318321); Tal Al-Rumman, *Taifour & Aldamen* (NHJ 2009.0160).

Ajuga chamaepitys (L.) Schreb. subsp. **tridactylites** (Benth.) P.H.Davis, Notes Roy. Bot. Gard. Edinburgh 38: 29. 1980.
Syn. *Ajuga tridactylites* Ging. ex Benth. (1835).
Ma'an, Al Naqab, 23 Apr. 1954 (FOR-H).

Ajuga iva (L.) Schreb., Pl. Verticill. Unilab. Gen. Sp. 15. 1774.
Syn. *Teucrium iva* L. (1753.); *Bulga iva* (L.) Kuntze (1891).
Ajloun, Halaweh, 24 Feb.1982 (YAR-H).

Ajuga orientalis L., Sp. Pl. 2: 561. 1753.
Syn. *Bulga orientalis* (L.) Kuntze (1891).
Amman, To Eraq Amir Wadi Esseir, down to the wadi near Fugahaá street, near the old water lefting bridge, site no 10 (MSB-K 496195).

Ballota philistaea Bornm., Repert. Spec. Nov. Regni Veg. 17: 284. 1921.
Petra, Jebal Madhbah, 26.5.1997 (72324-HUJ).

Ballota saxatilis Sieber ex C.Presl, Delic. Prag. 81. 1822.
Syn. *Ballota rugosa* Benth. (1834).
Ajloun, Ajloun city centre, toward Irbid (MSB-K 185729); Dibeen Forest, *RBG-Ire & RBG-Jo* (NHJ 2008.0060).

Ballota undulata (Sieber ex Fresen.) Benth., Labiat. Gen. Spec. 595. 1834.
Syn. *Marrubium undulatum* Sieber ex Fresen. (1834).
Tal Al-Rumman, *Al-Eisawi* (NHJ 2006.0210).

Clinopodium graveolens Kuntze subsp. **rotundifolium** (Pers.) Govaerts, World Checkl. Seed Pl. 3(1): 17. 1999.
Syn. *Acinos rotundifolius* Pers. (1806); *Calamintha neorotundifolia* Mateo (1992).
E of Medaba, 25 Apr. 1911, *Meyers & Dinsmore* (E); Hammamat Al-Omoush village (IBBU-H).

Clinopodium insulare (Candargy) Govaerts, World Checkl. Seed Pl. 3(1): 17. 1999.
Syn. *Calamintha incana* (Sm.) Boiss. (1879); *Thymus incanus* Sm. (1809).
Amman, Site 50, 1 km from city of Wadi al Seer toward Iraq Al Ameenby (MSB-K 395715).

Clinopodium serpyllifolium (M.Bieb.) Kuntze subsp. **fruticosum** (L.) Bräuchler, Taxon 55: 980. 2006.
Syn. *Melissa fruticosa* L. (1753); *Micromeria fruticosa* (L.) Druce (1914).
Upper Jordan Valley (RSCN database).

Eremostachys molucelloides Bunge, Fl. Altaic. 2: 415. 1830.
Shobek to Gebel Aneiza (Aqaba), 21 Apr. 1945, *Davis* (E).

Eremostachys transjordanica Eig, Palestine J. Bot., Jerusalem Ser. 4: 173. 1948.
2.5 km S of Safawi-Ruwaishid rd. (BRDC-H).

Lallemantia iberica Fisch. & C.A.Mey., Index Seminum (LE) 6: 53. 1840.
Zarqa, 2 km from the Cross of Jarash (MSB-K 199692).

Lallemantia royleana (Benth.) Benth., Prodr. 12: 404. 1848.
Rum, 2016, *Fragman-Sapir* (NHJ).

Lamium amplexicaule L., Sp. Pl. 2: 579. 1753.
Tal Al-Rumman, *Al-Eisawi* (NHJ 2006.0311).

Lamium garganicum L., Sp. Pl. ed. 2, 808. 1763.
Ajloun, Halaweh (YAR-H).

Lamium moschatum Mill., Gard. Dict. ed. 8, 4. 1768.
Dibeen National Park, 2005, *Mahklouf* s.n. (AMM); Jarash, Thagret Asfoor (YAR-H).

Lamium orientale (Fisch. & C.A.Mey.) E.H.L.Krause, Deutschl. Fl. ed. 2, 11: 137. 1903.
Syn. *Wiedemannia orientalis* Fisch. & C.A.Mey. (1838).
Amman, Wadi Seer, Al-Bahatharea, 1 km after the main water spring in the wadi, site no 9 (MSB-K 496162).

Lavandula coronopifolia Poir., Encycl. 3: 308. 1789.
Syn. *Lavandula stricta* Delile (1813).
Aqaba, Site 36, Wadi Al-Yutum (MSB-K 558095); Wadi Yutum, *El-Oqlah et al.* (NHJ 2013.0616).

Lavandula dentata L., Sp. Pl. 2: 572. 1753.
Jordan, *Baierle & Kürschner* 86–455 (E00106959).

Lavandula pubescens Decne., Ann. Sci. Nat., Bot. sér. 2, 2: 246. 1834.
Aqaba, Wadi Rum (MSB-K 228185); Tafila, Wadi Hassa, near the spring BOT ERA project (NHJ 2013.0551).

Marrubium alysson L., Sp. Pl. 2: 582. 1753.
Lower Jordan Valley (RSCN database).

Marrubium cuneatum Banks & Sol., Nat. Hist. Aleppo ed. 2, 2: 255. 1794.
Ajloun, Safsafa village, *Taifour, Al Khawaldah & Alshwaer* (NHJ 2013.0666).

Marrubium globosum subsp. **libanoticum** (Boiss.) P.H. Davis, Kew Bull. 6: 71. 1951.
Syn. *Marrubium libanoticum* Boiss. (1853).
Rum (http://flora.org.il/en/plants/MARLIB/).

Marrubium vulgare L., Sp. Pl. 2: 583. 1753.
Ajloun, from Ajloun Castle about 5 km toward the valley (MSB-K 211875.); Dana Reserv, RBG-Ire & RBG-Jo (NHJ 2008.0090).

Melissa officinalis L., Sp. Pl. 2: 592. 1753.
Ref. Al-Eisawi 116 (2013).

Mentha aquatica L., Sp. Pl. 2: 576. 1753.
Ajloun, Abu Zaton (YAR-H).

Mentha longifolia (L.) L., Fl. Monsp. 19. 1756.
Syn. *Mentha longifolia* (L.) Huds. (1762); *Mentha spicata* L. var. *longifolia* L. (1753); *Mentha sylvestris* L. (1753).
Amman, Wadi Seer, towards Adasseyeh, Bahhath & Buhaira villages (MSB-K 212090); Tal Al-Rumman bridge, *Al Shudiefat & Khalidi* (NHJ 2013.0007).

Mentha pulegium L., Sp. Pl. 2: 577. 1753.
Upper and Lower Jordan valley (RSCN database).

Mentha suaveolens Ehrh., Beitr. Naturk. 7: 149. 1792.
Ref. RSCN database; Al-Eisawi 117 (2013).

Micromeria danaensis Danin, Willdenowia 27: 161. 1997.
Aqaba, Dana Reserve, Wadi Barra area, below forest ranger station, 2 km SE of the visitor center; in

crevices and near smooth-faced white sandstone outcrops, *Danin* 963013 (B, type); Tafila, near Dana Reserve, type locality, BOT ERA project (NHJ 2013.0492).

Micromeria graeca (L.) Benth. ex Rchb., Fl. Germ. Excurs. 1: 311. 1831.
Syn. *Satureja graeca* L. (1753).
Salt, iv.1888 (AUB).

Micromeria myrtifolia Boiss. & Hohen., Diagn. Pl. Orient. ser. 1, 5: 19. 1844.
Balqa, Umm-Jozeh to Rmaimein, inside olive orchard (MSB-K 243779).

Micromeria nervosa (Desf.) Benth., Labiat. Gen. Spec. 376. 1834.
Syn. *Satureja nervosa* Desf. (1798).
Irbid, Ishtafeina to Deir Abu Said, (MSB-K 114297); Tal Al-Rumman, *Al-Eisawi* (NHJ 2006.0217).

Micromeria sinaica Benth., Labiat. Gen. Spec. 380. 1834.
Karak, Mujib valley (MSB-K 339751); Tafila, Dana Reserve, plateau near top of Rumana mountain, BOT ERA project (NHJ 2013.0443).

Moluccella laevis L., Sp. Pl. 2: 587. 1753.
Irbid, 2 km before NCARTT station at Maru (MSB-K 200567); Mafraq, Erhab, *Al Shudiefat & Khalidi* (NHJ 2013.0605).

Moluccella spinosa L., Sp. Pl. 2: 587. 1753.
Ma'an, Shoubak & Al Mansurah (YAR-H).

Nepeta cilicica Boiss. ex Benth., Prodr. 12: 388 .1848.
Syn. *Nepeta pycnantha* Boiss. ex Benth. (1848)
Mafraq, Umm Jemal, Sep. 1875. (AUB).

Nepeta congesta Fisch. & C.A.Mey., Ann. Sci. Nat., Bot. 4, 1: 34. 1854.
Syn. *Nepeta cryptantha* Boiss. & Hausskn. (1879.); *Nepeta involucrata* (Bunge) Bornm. (1912); *Oxynepeta involucrata* Bunge (1878).
Tafila, after Hamad bin Jazi forest toward Rumaneh Camp & Ain Lahtha (MSB-K 200198).

Nepeta curviflora Boiss., Diagn. Pl. Orient. ser. 1, 5: 22. 1844.
Amman, Site 50, 1 km from city of Wadi Al Seer toward Iraq Al Ameenby (MSB-K 395704); Rum, *Taifour* (NHJ 2013.0416).

Nepeta italica L., Sp. Pl.: 571. 1753.
Petra, Falahat (1999).

Nepeta stricta (Banks & Sol.) Hedge & Lamond, Notes Roy. Bot. Gard. Edinburgh 38(1): 45. 1980.
Syn. *Satureja stricta* Banks & Sol. (1794); *Nepeta calycina* Fenzl (1843).
Ref. Al-Eisawi 117 (2013) as *Nepeta calycina*.

Origanum dayi Post, Bull. Herb. Boissier 1: 405. 1893.
Maan, Wadi Al Naqab (FOR-H).

Origanum jordanicum Danin & Künne, Willdenowia 25(2): 603. 1996.
Endemic. Type: Jordan, Aqaba, *I. Künne* 94-09–19 (HUJ).

Origanum petraeum Danin, Willdenowia 19(2): 401. 1990.
Endemic. Type: Jordan, Aqaba, *I. Künne* 21.8.4 (HUJ).

Origanum punonense Danin, Willdenowia 19(2): 402. 1990.
Endemic. Type: Jordan, Aqaba (Karak area), *Baierle & Kürschner* 86–193 (holo. BSB).

Origanum ramonense Danin, Israel J. Bot. 17: 101. 1968.
Ras Naqab, 1963 (HUJ).

Origanum syriacum L., Sp. Pl. 2: 590. 1753.
Syn. *Majorana syriaca* (L.) Raf. (1840).
Karak, Karak–Jordan Valley road, site 45 (MSB-K 395597); Tal Al-Rumman, *Taifour & Aldamen* (NHJ 2009.0063).

Otostegia fruticosa (Forssk.) Schweinf. ex Penzig, Atti Congr. Bot. Genova 356. 1893.
Syn. *Clinopodium fruticosum* Forssk. (1775).
Wadi Rum, 1999 (RSCN database); Ref. Qishawi, M. (1999). Wadi Rum Nature Reserve Flora Survey. Royal Society for the Conservation of Nature.

Phlomis aurea Decne., Ann. Sci. Nat., Bot. sér. 2, 2: 251. 1834.
Ref. Al-Eisawi 118 (2013).

Phlomis brachyodon (Boiss.) Zohary ex Rech.f., Oesterr. Bot. Z. 89: 290. 1940.
Syn. *Phlomis armeniaca* Willd. var. *brachyodon* Boiss. (1853).
Amman, Wadi Al Sheta (MSB-K 243735; Tal Al-Rumman, *Al-Eisawi, Abu Sondos, Abu Anza & Shishani* (NHJ 2006.0061).

Phlomis floccosa D.Don, Edwards's Bot. Reg. 15: t. 1300. 1830.
Syn. *Phlomis samia* L. var. *bicolor* Viv. (1824); *Phlomis bicolor* (Viv.) Benth. (1834).
Ref. Al-Eisawi 118 (2013) as *Phlomis bicolour*.

Phlomis fruticosa L., Sp. Pl. 2: 584. 1753.
Ref. Al-Eisawi 118 (2013).

Phlomis herba-venti L. subsp. **pungens** (Willd.) Maire ex De Fillipps, Bot. J. Linn. Soc. 64(2): 233. 1971.
Syn. *Phlomis pungens* Willd. (1800).
Karak, from desert road to Karak, way to Mohta University (MSB-K 277840).

Phlomis kurdica Rech.f., Oesterr. Bot. Z. 89: 274. 1940.
Salt, (AUB); Tafila Province, 2014, *Oran* (AMM); Irbid, Zobia (YAR-H).

Phlomis nissolii L., Sp. Pl. 2: 585. 1753.
Jordan, 17.iii.1895 (AUB).

Phlomis orientalis Mill., Gard. Dict. ed. 8. 9. 1768.
Wadi Lagsein&Salt, Deir Alla (FOR-H).

Phlomis platystegia Post, Bull. Herb. Boissier 1: 407. 1893.
Tafila, outside Dana near Spring (MSB-K 151854); Dana Reserve, *RBG-Ire & RBG-Jo* (NHJ 2008.0076)

Phlomis syriaca Boiss.,Diagn. Pl. Orient. ser. 1, 12: 89. 1853.
Tal Al-Rumman RBG, Sector No. 20, *El-Oqlah, Taifour & Al Khawaldah* (NHJ 2013.0229).

Phlomis viscosa Poir., Encycl. 5: 271. 1804.
Jebel Siaghah, Madaba, 30.iv.1877, 216 (AUB); Ajloun, on way from Sakib to Rajib, *Taifour* (NHJ); Dibeen National Park, 2005, *Mahklouf* s.n. (AMM).

Phlomoides laciniata (L.) Kamelin & Makhm., Bot. Zhurn. (Moscow & Leningrad) 75: 249. 1990.
Syn. *Phlomis laciniata* L. (1753); *Eremostachys laciniata* (L.) Bunge (1830).
Jordan, *Boulos & Al-Eisawi* 6424 (E00387842); Tal Al-Rumman, *Al-Eisawi* (NHJ 2006.0274); Karak, 1 km after Wadi Al Mougeb toward Ariha & Al Karak city (MSB-K 199795).

Pleudia aegyptiaca (Raf.) Will, N. Schmalz, Class.-Bockh., Turkish J. Bot. 39(4): 701. 2015.
Syn. *Salvia aegyptiaca* L. (1753).
Tafila, Southern Ghor Municipality (MSB-K 413510); Ein trab, *Taifour, Al Khawaldah, Alshwaer & Narita* (NHJ 2013.0710).

Prasium majus L., Sp. Pl. 2: 601. 1753.
Ajloun, 2 km north of Sakeb towards Ainjana on Jarash-Ajloun road (MSB-K 158738); Tal Al-Rumman, *Al-Eisawi* (NHJ 2006.0126).

Salvia africana-caerulea L., Sp. Pl. 26. 1753.
Syn. *Salvia acetabulosa* L (1767).
Salt, Abu Dali, 12 Apr. 1890 (AUB).

Salvia bracteata Banks & Sol., Nat. Hist. Aleppo ed. 2, 2: 242. 1794.
Amman, Yaduda (FOR-H); Ajloun (AUB).

Salvia ceratophylla L., Sp. Pl.: 2.7 1753.
Tafila Province, 2014, *Oran* (AMM); Ma'an, 20 km S of Petra *Frey & Kürschner*.

Salvia deserti Decne., Ann. Sci. Nat., Bot. sér. 2, 2: 248. 1834.
Jordan, *Davis* 9244 (E00209652).

Salvia dominica L., Sp. Pl.: 25. 1753.
Syn. *Salvia graveolens* Vahl (1804).
Irbid, Banikanah, Hammat Elmukhabeh, 29.ii.1983 (YAR-H).

Salvia eigii Zohary, Beih. Bot. Centralbl. 48(2): 543. 1931.
Upper Jordan Valley (YAR-H).

Salvia fruticosa Mill., Gard. Dict., ed. 8. n. 5, 1768.
Irbid, Alkoura, Bergish, Rujum el husa, with *Quercus coccifera*, 2015, *El-Oqlah* 2015; Almazar-Irbid (L1074458) (http://data.gbif.org/occurrences/299071472).

Salvia hierosolymitana Boiss., Diagn. Pl. Orient. 12: 61. 1853.
Tafila, Dana village, inside Dana Nature Reserve (MSB-K 199809).

Salvia indica L., Sp. Pl.: 26. 1753.
Syn. *Salvia brachycalyx* Boiss., 1879, nom. illegit.
Ajloun, 15 km north of Ajloun to Irbid at 3 km after Ibbin junction at Ras Munif (MSB-K 151991); Ajloun district, outside Ajloun reserve near Elynabi village, BOT ERA project (NHJ 2013.0530).

Salvia judaica Boiss., Diagn. Pl. Orient. ser. 1, 12: 61. 1853.
Al-Mafraq- Nadrah- between Rehab and Jerash, (IBBU-H); Jordan Valley (RSCN database).

Salvia lanigera Poir., Encycl. Suppl. 5: 49. 1817.
Tafila, Dana reserve (MSB-K 339935).

Salvia multicaulis Vahl, Enum. Pl. Obs. 1: 225. 1804.
Syn. *Salvia pinardii* Boiss. (1853).
Amman, 5 km from University of Applied Science toward Alouk (MSB-K 276670); Tafila (FOR-H).

Salvia napifolia Jacq., Hort. Bot. Vindob. 2: 71. 1773.
Ref. RSCN database; Al-Eisawi 119 (2013).

Salvia palaestina Benth., Labiat. Gen. Spec. 718. 1835.
Amman, Atlad Al Ali, opposite Jordan University (MSB-K 278320); Tal Al-Rumman, *Al-Eisawi, Abu-Sondos, Abu Anza & Shishani* (NHJ 2006.0063).

Salvia peratica Paine, Palest. Expl. Soc. 3: 118. 1875.
Amman (RSCN database).

Salvia pinnata L., Sp. Pl.: 27. 1753.
Ref. Al-Eisawi 119 (2013).

Salvia rubifolia Boiss., Diagn. Pl. Orient. 12: 56. 1853.
(YAR-H); [http://flora.org.il/en/plants/SALRUB/].

Salvia samuelssonii Rech.f., Ark. Bot. ser. 2, 1: 318. 1950.
Felsensteppe, Berg Nebo (W175433) (http://data.gbif.org/occurrences/367082842).

Salvia sclarea L., Sp. Pl. 27. 1753.
Tal Al-Rumman, *Taifour & Aldamen* (NHJ 2009.0154).

Salvia spinosa L., Mant. Pl. Altera 511. 1771.
Azraq, *Gillett* 15621 (K000318308); Azraq, Wadi Bottom (South west the Castle Amra), *El-Oqlah* (NHJ 2013.0587).

Salvia syriaca L., Syst. Nat. ed. 10 2: 854. 1759.
Tafila Province, 2014, *Oran* (AMM); Al Mafraq, Irbid-Mafraq street, 30 May 1985 (YAR-H).

Salvia verbenaca L., Sp. Pl.: 25. 1753.
Syn. *Salvia controversa* Ten. (1831).
Tafila, 20 km west of Jorf el Darawish to Tafila (MSB-K 151795).

Salvia virgata Jacq., Hort. Bot. Vindob. 1: 14. 1770.
Salt, v.1886 (AUB).

Salvia viridis L., Sp. Pl.: 24. 1753.
Syn. *Salvia horminum* L. (1753).
Adasein, *Al-Eisawi* 7833 (K000315811); Tal Al-Rumman, *Al-Eisawi* (NHJ 2006.0028).

Salvia viscosa Jacq., Misc. Austriac. 2: 328. 1781.
(YAR-H); [http://flora.org.il/en/plants/SALVIS/].

Satureja nabateorum Danin & Hedge, Willdenowia 28(1–2): 135. 1998.
Endemic. Type: Jordan, Aqaba, *Danin* 97021 (holo. HUJ; iso. B, E, K).

Satureja thymbra L., Sp. Pl. 2: 567. 1753.
Tafila Province, 2014, *Oran* (AMM); Dibeen (FOR-H).

Satureja thymbrifolia Hedge & Feinbrun, Israel J. Bot. 17: 213. 1969.
El-Hesma, *Eig, Zohary & Feinbrun* s.n. (K000193589).

Scutellaria brevibracteata Stapf., Denkschr. Kaiserl. Akad. Wiss., Wien. Math.-Naturwiss. Kl. 1: 99. 1885.
Syn. *Scutellaria peregrina* L. (1753).
Irbid, Irbid to Umm Gais main road, Site no 1 (MSB-K 496058).

Scutellaria galericulata L., Sp. Pl.: 599. 1753.
Al-Mafraq, Rawdhet Al-Ameera Bassma (IBBU-H).

Scutellaria rubicunda Hornem., Hort. Bot. Hafn. 2: 968. 1815.
Dead Sea Area (RSCN database).

Scutellaria tomentosa Bertol., Misc. Bot. 2: 11. 1843.
Syn. *Scutellaria fruticosa* Pers. (1806) nom. illegit.
Amman, from Wasfi Al Tal street opposite Altarbia Alvideieh school, right street toward Khalda main centre (MSB-K 228392).

Scutellaria utriculata Labill., Pl. Syr. 4: 11. 1812.
(YAR-H). [http://flora.org.il/en/plants/SCUUTR/].

Sideritis perfoliata L., Sp. Pl. 2: 575. 1753.
Dibeen Forest Reserve, 2004, *Mahklouf* (AMM); Ajloun Forest Reserve (RSCN database).

Sideritis pullulans Vent., Descr. Pl. Nouv. t. 98. 1803.
Amman, Khalda, on the way to Sweleh (MSB-K 185796); Irbid, Korah, *Taifour et al.* (NHJ 2013.0674).

Stachys aegyptiaca Pers., Syn. Pl. 2(1): 124. 1806.
Tafila Province, 2014, *Oran* (AMM); 6–7 km South Rum rest house, 23 Mar. 1975, *Boulos et al.* (AMM, E).

Stachys arabica Hornem., Hort. Bot. Hafn. 554. 1815.
Amman, 29.IV.1886 (AUB).

Stachys cretica L., Sp. Pl. 2: 581. 1753.
Jarash, Souf, 5 km after oak forest, site no 33 (MSB-K 497686).

Stachys distans Benth., Prodr. 12: 472. 1848.
Kufur Yasi, 29.iv.1929, *Dinsmore* (YAR-H).

Stachys libanotica Benth., Prodr. 12: 462. 1848.
Ref. RSCN database; Al-Eisawi 119 (2013).

Stachys longispicata Boiss. & Kotschy, Fl. Orient. 4: 725. 1879.
Upper Jordan Valley (RSCN database).

Stachys neurocalycina Boiss., Diagn. Pl. Orient. 12: 85. 1853.
After guava orchard near Yarmouk river (MSB-K 419062).

Stachys nivea Labill., Icon. Pl. Syr. 3: 5. 1809.
Jabal Rum, iv.1907, *Dinsmore* 5963 (AUB-H).

Stachys palaestina L., Sp. Pl. ed. 2, 1674. 1763.
(YAR-H); [http://flora.org.il/en/plants/STAPAL/]

Stachys paneiana Mouterde, Saussurea 4: 25. 1973.
Ref. RBG Jordan checklist (unpubl.)

Stachys spectabilis Choisy ex DC., Mém. Soc. Phys. Genève 1(2): 457. 1823.
(YAR-H).

Stachys viticina Boiss., Diagn. Pl. Orient. 12: 77. 1853.
Dibbin Forest Reserve, *Abuhammour, Al Mfadi & Al Khateeb* (NHJ).

Stachys zoharyana Eig, Palestine J. Bot., Jerusalem Ser. 4: 172. 1948.
(YAR-H). [http://flora.org.il/en/plants/STAZOH/].

Teucrium capitatum L., Sp. Pl. 2: 566. 1753.
Al Karak, Wadi Ibn Hammad SCA (RSCN database); Qatar Reserve Herbarium, *Abu Yahya, Eid & Ananbeh* (2010). Flora Baseline Survey, Qatar Protected Area.

Teucrium creticum L., Sp. Pl. 563. 1753.
(YAR-H).

Teucrium decaisnei C.Presl, Abh. Königl. Böhm. Ges. Wiss. ser. 5, 3: 530. 1845.
Syn. *Teucrium pilosum* Asch. & Schweinf. (1887).
Jordan, *McDonald* s.n. (K000286307).

Teucrium divaricatum Sieber ex Heldr., Exsicc. (Autogr. Herb. Graec. Norm.) 1856: 290. 1857.
Amman, on the way to the dead sea, just before Na'our bridge junction (MSB-K 341365).

Teucrium lamiifolium D'Urv., Mém. Soc. Linn. Paris 1: 320 1822.
(YAR-H).

Teucrium leucocladum Boiss., Diagn. Pl. Orient. 5: 44. 1844.
Syn. *Teucrium leucocladum* Boiss. subsp. *jordanicum* Danin, Willdenowia 27: 172 (1997), Type: Jordan, Dana Reserve, Wadi Barra, *Danin* 63601 (holo. HUJ; iso. B, E, K).

Teucrium montbretii Benth., Ann. Sci. Nat., Bot. II, 6: 56. 1836.
Aqaba, Dana Reserve, Wadi Barra area, below the forest-ranger station, 2 km SE of the visitor center (HUJ).

Teucrium orientale L., Sp. Pl. 2: 562 1753.
Ref. RSCN database; Al-Eisawi 120 (2013).

Teucrium parviflorum Schreb., Pl. Verticill. Unilab. Gen. Sp. 31. 1774.
Upper Jordan valley (RSCN database).

Teucrium polium L., Sp. Pl. 566. 1753.
Irbid, first junction to Jarash, beside Al Faqua nurseries (MSB-K 176703); Tal Al-Rumman, *Al-Eisawi* (NHJ 2006.0233)

Teucrium procerum Boiss. & Blanch, Diagn. Pl. Orient. 2, 4: 56. 1859.
Ma'an, Rajif-Ras Al-Naqab road, 5 km after Rajif junction on the left side of road, 6 Nov. 2008 (NCARE).

Teucrium pruinosum Boiss., Fl. Orient. 4(2): 808. 1879.
Ma'an, Rajif-Ras Al-Naqab road, 5 km after Rajif junction, site no 31 (MSB-K 497594).

Teucrium rigidum Benth., Prodr. 12: 578. 1848.
Ref. RSCN database; Al-Eisawi 120 (2013).

Teucrium scordium L. subsp. **scordioides** (Schreb.) Arcang., Comp. Fl. Ital. 559. 1882.
Syn. *Teucrium scordioides* Schreb. (1774); *Teucrium lanuginosum* Hoffmanns. & Link (1809); *Teucrium petkovii* Urum. (1913).
Zarqa River, *Dinsmore* 3527 (AUB); Upper Jordan valley (RSCN database).

Teucrium spinosum L., Sp. Pl. 2: 566. 1753.
Dinsmore 5605 (AUB); (YAR-H).

Thuspeinanta persica (Boiss.) Briq., Nat. Pflanzenfam. 4(3a): 229. 1896
Aqaba (RSCN database).

Thymbra capitata (L.) Cav., Elench. Pl. Horti Matr. 37. 1803.
Syn. *Satureja capitata* L. (1753); *Thymus capitatus* (L.) Hoffmanns. & Link (1809); *Coridothymus capitatus* (L.) Rchb.f. (1857).
Amman, from Wasfi Al Tal street opposite Altarbia Alvideieh school (MSB-K 244318).

Thymbra spicata L., Sp. Pl. 2: 569. 1753.
Jarash, Tell Arruman site (MSB-K 228381); Tal Al-Rumman, *Royal Botanic Garden Team*, (NHJ 2008.0018).

Thymus bovei Benth., Labiat. Gen. Spec. 342. 1834.
Al-Mafraq, Al-Mdawara village (IBBU-H); Wadi Musa (YAR-H).

Vitex agnus-castus L., Sp. Pl. 2: 638. 1753.
Irbid, Al Yormouk river (MSB-K 341332); Northeastern Badia, Al-Dafyanah, Al Shudiefat & Khalidi (NHJ 2013.0008).

Ziziphora capitata L., Sp. Pl.: 21. 1753.
Ajloun, Souf-Ibbein road, forest on right side of road (MSB-K 276902);Tal Al-Rumman, *Taifour & Aldamen* (NHJ 2009.0066).

Ziziphora tenuior L., Sp. Pl.: 21. 1753.
Jordan, *Mooney* 4371 (K000509401).

Lauraceae

Laurus nobilis L., Sp. Pl.: 369. 1753.
Jordan (E00310131).

Liliaceae

Fritillaria persica L., Sp. Pl.: 304. 1753.
Syn. *Theresia libanotica* Boiss. (1854); *Fritillaria libanotica* (Boiss.) Baker (1874).
Jordan, *Roth* s.n. (K000900875); Tal Al-Rumman, *Taifour & Aldamen* (NHJ 2009.0052).

Gagea chlorantha (M.Bieb.) Schult. & Schult.f., Syst. Veg. 7: 551. 1829.
Syn. *Ornithogalum chloranthum* M.Bieb. (1808); *Gagea damascena* Boiss. et Gaill. (1859); *Gagea caucasica* Stapf (1885).
Karak, Wadi Dana, between Dana and Lahta, *Frey & Kürschner* (B).

Gagea commutata K.Koch, Linnaea 22: 227. 1849.
Al-Mafraq, Borqu' (IBBU-H).

Gagea dayana Chodat & Beauverd, Pl. Post. 1: 8. 1932.
Reported from the Naqab Highlands by Danin (http://flora.org.il/en/plants/GAGDAY/).

Gagea fibrosa (Desf.) Schult. & Schult.f., Syst. Veg. 7: 552. 1829.
Syn. *Ornithogalum fibrosum* Desf. (1798).
Ma'an, Rajif-Ras Al-Naqab road (MSB-K 416304).

Gagea libanotica (Hochst.) Greuter, Israel J. Bot. 19: 158. 1970.
Syn. *Lloydia libanotica* Hochst. (1845).
25 km W of Ma'an, 14 Apr. 1980, *Frey & Kürschner* (B).

Gagea minima (L.) Ker Gawl., J. Sci. Arts 1: 180. 1816.
Syn. *Gagea arvensis* (Pers.) Dumort. (1827); *Ornithogalum minimum* L. (1753); *Anacamptis sancta* (L.) R.M.Bateman (1997).
Ref. Jordan valley, 6.iii.1953, *El Khayat*.

Gagea reticulata (Pall.) Schult. & Schult.f., Syst. Veg. 7: 542. 1829.
Syn. *Ornithogalum reticulatum* Pall. (1776); *Ornithogalum circinnatum* L.f. (1782); *Gagea reticulata* (Pall.) J.A. & J.H.Schultes var. *tenuifolia* Boiss. (1882); *Gagea tenuifolia* (Boiss.) Fomin (1909);
Jordan, 17.02.1974, *Al-Eisawi* 1642 (B100190924).

Gagea rigida Boiss. & Spruner, Diagn. Pl. Orient. 7: 108. 1846.
Madaba, 17 Feb. 1954 (FOR-H).

Gagea villosa (M.Bieb.) Sweet, Hort. Brit. 418. 1826.
Syn. *Ornithogalum villosum* M.Bieb. (1808).
Irbid, Al Ramtha, 20 Mar. 1983 (YAR-H).

Lilium candidum L., Sp. Pl.: 302. 1753.
RSCN database; Ref. Al-Eisawi 124 (2013).

Tulipa agenensis DC., Liliac. 1: t. 6.0 1804.
Jarash, Tall Rumman, Garden site (MSB-K 419327); Tal Al-Rumman, *Taifour & Aldamen* (NHJ 2009.0216).

Tulipa biflora Pall., Reise Russ. Reich. 3: 727. 1776.
Syn. *Tulipa polychroma* Stapf (1885).
1–4 km S Ras en Naqb, 19 Mar. 1975, *Boulos & Jallad* (AMM, E).

Tulipa systola Stapf, Denkschr. Akad. Wien I. 17. 1885.
Syn. *Tulipa montana* Boiss. (1827)
Tal Al-Rumman, *Al-Eisawi* (NHJ 2006.0140).

Linaceae

Linum album Kotschy ex Boiss., Diagn. Pl. Orient. 6: 27. 1846.
2.5 km S of the Safawi-Ruwaishid rd. (BRDC-H).

Linum bienne Mill., Gard. Dict. ed. 8, no. 8. 1768.
Majdal, 9.5.1927 (30655-HUJ).

Linum corymbulosum Rchb.,Fl. Germ. Excurs. 2: 834. 1832.
Syn. *Linum liburnicum* (non Scop.) auctt. mult. incl. Rawi *Linum maritimum* L.
Eastern upper slopes of the Jordan valley (NCARE).

Linum mucronatum Bertol., Misc. Bot. 1: 18. 1842.
Ras en Naqb, *Boulos, Jallad & Lahham* 6546 (K000286303).

Linum nodiflorum L., Sp. Pl.: 280. 1753.
Dibeen National Park, 2005, *Mahklouf* s.n. (AMM).

Linum peyronii Post, Pl. Postianae 3: 6. 1892.
Amman to Zizia, 15.iv.1929 (30859-HUJ).

Linum pubescens Banks & Sol., Nat. Hist. Aleppo ed. 2, 2: 268. 1794.
Salt, *Al-Eisawi* 3525 (K000286276); Tal Al-Rumman, *Al-Eisawi* (NHJ 2006.0325).

Linum strictum L., Sp. Pl.: 279. 1753.
Balqa, Al-Rumman hills (MSB-K 340690); Tal Al-Rumman, *Al-Eisawi* (NHJ 2006.0030).

Linum toxicum Boiss., Fl. Orient. 1: 854. 1867.
Environs of Alma, 21.7.1942 (31013-HUJ).

Linum utatissimum L., Sp. Pl.: 277. 1753.
Amman, 1953, *Kasapligil* (FOR-H). Found as an escape.

Loranthaceae

Plicosepalus acaciae (Zucc.) Wiens & Polhill, Nordic J. Bot. 5: 221. 1985.
Syn. *Loranthus acaciae* Zucc. (1843).
Tal Al-Rumman, *HRH Princess Basma bint Ali & Nawash* (NHJ 2011.0020).

Viscum cruciatum Sieber ex Boiss., Voy. Bot. Espagne 2: 274. 1840.
Wadi Orjan, *Royal Botanic Garden Team* (NHJ 2011.0021).

Lythraceae

Ammannia baccifera L., Sp. Pl.: 120. 1753.
Upper Jordan Valley, Danin (2005); RSCN database.

Lythrum borysthenicum (M.Bieb. ex Schrank) Litv., Fl. Sredn. Rossii ed. 5: 209. 1917.
Madaba Governorate, Hidan (NHJ).

Lythrum hyssopifolia L., Sp. Pl.: 447. 1753.
Balqa, Rummaimin area, just beside the waterfall (MSB-K 278238).

Lythrum junceum Banks & Sol., Nat. Hist. Aleppo ed. 2, 2: 253. 1794.
Tell el Kadi, *Meyers & Dinsmore* 2915 (E00332166).

Lythrum salicaria L., Sp. Pl.: 446. 1753.
Irbid, Al-Adassiyeh (IBBU-H).

Lythrum tribracteatum Salzm. ex Spreng., Syst. Veg. 4(2): 190. 1827.
Upper Jordan Valley RSCN database & (YAR-H).

Malvaceae

Abutilon fruticosum Guill. & Perr., Fl. Seneg. Tent. 70 .1831.
Aqaba, Wadi Yutum, *Boulos et al.* 7520 (E00437591).

Abutilon hirtum (Lam.) Sweet, Hort. Brit. 53. 1826.
Syn. *Sida hirta* Lam., Encycl. (1783).
Wadi Araba (YAR-H).

Abutilon indicum (L.) Sweet, Hort. Brit. 54 .1826.
Syn. *Sida indica* L. (1756).
Dead Sea area, Wadi Araba (YAR-H).

Abutilon pannosum (G.Forst.) Schltdl., Bot. Zeitung (Berlin) 9: 828. 1851.
Syn. *Sida pannosa* G.Forst. (1789); *Abutilon glaucum* (Cav.) Sweet (1826); *Abutilon muticum* (Delile ex DC.) Sweet (1830).
Karak, Ghor Hadeetheh, the first entrance to the south most shore of the Dead Sea, Site no 27 (MSB-K 497505).

Abutilon theophrasti Medik., Malvenfam. 28. 1787.
N Jordan, adventive weed in irrigated fields (YAR-H).

Acaulimalva acaulis (Dombey ex Cav.) Krapov, Darwiniana 19: 15. 1974.
Syn. *Malva acaulis* Dombey ex Cav. (1786).
Along Zarqa River, 2004, *Al-Dwiri* s.n. (AMM).

Alcea acaulis (Cav.) Alef., Oesterr. Bot. Z. 12: 251. 1862.
Syn. *Althaea acaulis* Cav. (1786).
Jordan, Salt, *Al-Eisawi* 3533 (K000286265) as var. *albiflora*; Shoubak, Experimental Farm, *Gillett* 15902 (K000286270) as var. *brachyantha*; Tal Al-Rumman, *Al-Eisawi* (NHJ 2006.0022).

Alcea apterocarpa Boiss., Fl. Orient. 1: 830. 1867.
Karak, and Irbid, 2.iv.1984 (YAR-H).

Alcea chrysantha (Sam.) Zohary, Bull. Res. Council Israel Sect. D, Bot. 11: 221. 1963.
Syn. *Althaea chrysantha* Sam. (1935).
Amman, *Rechinger* 12849 (K000286272).

Alcea digitata Alef., Oesterr. Bot. Z. 12: 252. 1862.
Irbid, Jordan Valley at Baqora, 5 km north of Shuna (MSB-K 151751).

Alcea dissecta (Baker f.) Zohary, Israel J. Bot. 12: 16. 1963.
Syn. *Althaea disecta* Baker f. (1890).
Irbid area (YAR-H).

Alcea galilaea Zohary, Bull. Res. Council Israel, Sect. D, Bot. 11: 224. 1963.
Umm Qays area (YAR-H); Ajloun, 15 Mar. 1996, *Danin* (HUJ).

Alcea setosa (Boiss.) Alef., Oesterr. Bot. Z. 12: 255. 1862.
Irbid, first junction to Jarash (MSB-K 176699); Tal Al-Rumman, *Al-Eisawi* (NHJ 2006.0297).

Alcea striata Alef., Oesterr. Bot. Z. 12: 252. 1862.
Madaba, 39 km S of Madaba (MSB-K 114448).

Alcea striata Alef. subsp. **rufescens** (Boiss.) Cullen, Notes Roy. Bot. Gard. Edinb. 27: 216. 1967.
Syn. *Alcea rufescens* Boiss. (1867).
Wadi Heidan (Moab), *Davis* 8903 (E00431563).

Althaea hirsuta L., Sp. Pl. 2: 687. 1753.
Balqa, Tell Rumman, Botanic Garden site (MSB-K 496771); Tal Al-Rumman, *Taifour & Aldamen* (NHJ 2009.0062).

Althaea ludwigii L., Mant. Pl. 98. 1767.
Ma'an, El-Jafr, in the vicinity of the Police Station, *Boulos & Jallad* 7924 (E00465000).

Corchorus olitorius L., Sp. Pl.: 529. 1753.
Meyers & Dinsmore 1112 (E00437436).

Corchorus trilocularis L., Mant. Pl. 77. 1767.
Al Ghour, Baqora, 1.xi.1954 (YAR-H).

Grewia tenax (Forssk.) Fiori, Agric. Colon. 1911, v. Suppl., 23. 1912.
Syn. *Chadara tenax* Forssk. (1775).
Ref. Al-Eisawi 170 (2013).

Grewia villosa Willd., Neue Schriften Ges. Naturf. Freunde Berlin 4: 205. 1803.
Lower Jordan Valley, Dead Sea area (YAR-H).

Hibiscus micranthus L.f., Suppl. Pl. 308. 1782.
10–12 km N of Aqaba, 13.3.1974, *Boulos, Al-Eisawi & Jallad* (AMM).

Hibiscus trionum L., Sp. Pl. 2: 697. 1753.
Tal Al-Rumman, *Al-Eisawi* (NHJ 2006.0009).

Lavatera punctata All., Auct. Fl. Pedem. 26. 1789.
Syn. *Malva punctata* (All.) Alef. (1862).
Mts. above Wadi Kafamga, *Trought* (E00454058).

Lavatera trimestris L., Sp. Pl. 2: 692. 1753.
Um Rehan, 11.iv.1955 (YAR-H).

Malva aegyptia L., Sp. Pl. 2: 690. 1753.
Syn. *Malva aegyptia* L. subsp. *leiocarpa* Iljin.
800 m W of Safawi-Azraq rd., 23 Apr. 1992 (IBBU-H).

Malva multiflora (Cav.) Soldano, Banfi & Galasso, Atti Soc. Ital. Sci. Nat. Mus. Civico Storia Nat. Milano 146(2): 230. 2005.
Syn. *Lavatera cretica* L. (1753); *Malope multiflora* Cav. (1786).
Deir Alla farm, *Trought* (E00454012); Irbid, Dear Abu Said, 08.05.1983 (YAR-H).

Malva neglecta Wallr., Syll. Pl. Nov. 1: 140. 1824.
Ma'an, Site 13. Shoubak-Qadeseyeh road (MSB-K 554318); Dana, *Taifour* (NHJ 2013.0399).

Malva nicaeensis All., Fl. Pedem. 2. 40. 1785.
Syn. *Malva flexuosa* I.Riedl (1815).
Karak, Wadi Al Arab, Wadi Zeglab (MSB-K 185648); Tal Al-Rumman, *Al-Eisawi* (NHJ 2006.0268).

Malva parviflora L., Demonstr. Pl. 18. 1753.
Tafila, after crossing at Afra at 2–3 km from Tafila (MSB-K 186313);Tal Al-Rumman, *Shishani & Alkoran* (NHJ 2011.0029).

Malva pusilla Sm., Engl. Bot. 4: t. 241, 1794.
Syn. *Malva rotundifolia* L. (1753).
Almafraq, Manshiyeh (YAR-H).

Malva sylvestris L., Sp. Pl. 2: 689. 1753.
Tal Al-Rumman, *Al-Eisawi* (NHJ 2006.0359).

Malvella sherardiana (L.) Jaub. & Sp., Ill. Pl. Orient. 5: 46. 1854.
Syn. *Malva sherardiana* L. (1763).
Jordan, *Al-Eisawi* 9992 (K000286274).

Sida acuta Brum.f., Fl. Indica 147. 1768.
Jordan, *sin. loc.* (YAR-H). Jordan velley (http://flora.org.il/en/plants/SIDACU/).

Menispermaceae

Cocculus pendulus (J.R.Forst. & G.Forst.) Diels, Pflanzenr. IV, 94: 237. 1910.
Syn. *Epibaterium pendulum* J.R.Forst. & G.Forst. (1775).
Wadi Yutum, *Taifour, Al Khawaldah & Al Mogherbi* (NHJ 2013.0701).

Molluginaceae

Glinus lotoides L., Sp. Pl.: 463. 1753.
Syn. *Glinus dictamnoides* Burm.f. (1768); *Mollugo hirta* Thunb. (1974).
Jarash, Tel Rumman, beside King Talal Dam (MSB-K 341491);Tell Rumman, *Taifour & Aldamen* (NHJ 2009.0060).

Moraceae

Ficus carica L., Sp. Pl. 2: 1059. 1753.
Petra, vi.1885 (AUB); Petra 29.3.1936 (31159-HUJ). Cultivated and naturalised in some localities.

Ficus palmata Forssk., Fl. Aegypt.- Arab. 179. 1775
Syn. *Ficus pseudosycomorus* Decne. (1834).
Aqaba, 6–7 km south Rum Rest House, *Boulos et al.* 7666 (E00405338); Rum, *Taifour* (NHJ 2013.0413).

Moringaceae

Moringa peregrina (Forssk.) Fiori, Agric. Colon. 4: 59 1911.
Syn. *Hyperanthera peregrina* Forssk. (1775); *Moringa aptera* Gaertn. (1791).
Wadi Zerka, *Davis* 8824 (E00448590);Tafila, Wadi Finan, RSCN reserve (MSB-K 278319); Dead Sea, *Nuaimat, Al Khawaldah & Alshwaer* (NHJ 2013.0660).

Myrtaceae

Myrtus communis L., Sp. Pl.: 471. 1753.
Karak Governorate, Alkaser, Ibn Hammad, *Taifour* (NHJ).

Nitrariaceae

Nitraria retusa (Forssk.) Asch., Verh. Bot. Vereins Prov. Brandenburg 18: 94. 1876.
Wadi Araba, *Hunting Aero Survey* (E00436002436002); Ghor Al Safi, Ghor Al Safi main road toward Potash factory (MSB-K 412616).

Nitraria schoberi L., Syst. Nat., ed. 10. 2: 1044. 1759.
Azraq Wetland Reserve, *Al-Eisawi* 19414 (K000318311).

Peganum harmala L., Sp. Pl.: 444. 1753.
Syn. *Peganum harmala* L. var. *stenophyllum* Boiss.
Azraq, Azraq Al-Shamali (MSB-K 176297); Azraq, near Shomari, *Taifour, Al Khawaldah.&.Alshwaer* (NHJ 2013.0728).

Nyctaginaceae

Boerhavia coccinea Mill., Gard. Dict. ed. 8, 4. 1768.
Aqaba, inside Aqaba City in a water canal for runoff rain water (MSB-K 340416).

Boerhavia repens L., Sp. Pl.: 3. 1753.
Syn. *Boerhavia diffusa* L. (1753).
9 km NE Aqaba, Boulos, *Jallad & Lahham Aqaba* AMM); in runoff canal, 31 May 2006 (NCARE).

Boerhavia sinuata (Meikle) Greuter & Burdet, Willdenowia 16: 448. 1987.
Syn. *Commicarpus sinuatus* Meikle (1974).
Lower Jordan valley, Danin (http://flora.org.il/en/plants/COMSIN/).

Commicarpus helenae (Roem. & Schult.) Meikle, Hooker's Icon. Pl. 7(4): t. 3694, p. 1. 1971.
Syn. *Boerhavia helenae* Roem. & Schult. (1822).
Araba valley, Danin (http://flora.org.il/en/plants/COMHEL/).

Commicarpus plumbagineus (Cav.) Standl., Contr. U.S. Natl. Herb. 18: 101. 1916
Syn. *Boerhavia plumbaginea* Cav., Icon. (1793); *Commicarpus africanus* (Lour.) Dandy (1950).
Irbid, coming from Baqora towards Al-Shoneh Al-Shamaleyeh (MSB-K 340438).

Nymphaeaceae

Nuphar lutea (L.) Sm., Fl. Graec. Prodr. 1: 361. 1809.
Syn. *Nymphaea lutea* L. (1753).
Hsileh, *Meyers & Dinsmore* (E); Upper Jordan Valley (RSCN database).

Oleaceae

Fraxinus angustifolia Vahl subsp. **syriaca** (Boiss.) Yalt., Fl. Turkey 6: 152. 1978.
Syn. *Fraxinus syriaca* Boiss. (1849).
Sweileh, *RBG-Ire & RBG-Jo* (NHJ 2008.0027).

Olea europaea L., Sp. Pl.: 8. 1753.
Amman, Site 54. Tal Al-Rumman (MSB-K 395841); Wadi Orjan, Royal Botanic Garden Team (NHJ 2011.0027).

Phillyrea latifolia L., Sp. Pl.: 8. 1753.
Syn. *Phillyrea media* L. (1759).
Dair Abusaid Bergish hunting reserve, *El-Oqlah* (NHJ 2013.0052).

Onagraceae

Epilobium hirsutum L., Sp. Pl.: 347. 1753.
Jarash, opposite the main building of Al Mearad municipality (MSB-K 211990).

Epilobium parviflorum Schreb, Spic. Fl. Lips. 155. 1771.
Ajloun, Soos Rasoon, 31 Oct. 1983 (YAR-H).

Epilobium tetragonum L., Sp. Pl.: 348. 1753.
Syn. *Epilobium tournefortii* Freyn
Irbid, Ain El Turab, 17 Nov. 1983 (YAR-H).

Ludwigia adscendens (L.) H.Hara subsp. **diffusa** (Forssk.) P.H.Raven, Kew Bull. 15: 476. 1962.
Syn. *Jussiaea diffusa* Forssk.(1775); *Jussiaea repens* Boiss. (1753); *Jussiaea stolonifera* Guill. & Perr. (1833); *Jussiaea repens* L. var. *diffusa* (Forssk.) Brenan (1894); *Ludwigia stolonifera* (Guill. & Perr.) P.H.Raven (1964).
Upper Jordan Valley (YAR-H).

Orchidaceae

Anacamptis collina (Banks & Sol. ex Russell) R.M.Bateman, Pridgeon & M.W.Chase, Lindleyana 12(3): 120. 1997.
Syn. *Orchis collina* Banks & Sol. ex Russell (1794).
Aman, opposite Jordan University (NCARE).

Anacamptis laxiflora (Lam.) R.M.Bateman, Pridgeon & M.W.Chase, Lindleyana 12(3): 120. 1997.
Syn. *Orchis laxiflora* Lam. (1799).
Karak, Merj, 2 Apr. 1877 (AUB).

Anacamptis palustris (Jacq) R.M.Bateman, Pridgeon & M.W.Chase, Lindleyana 12(3): 120. 1997.
Syn. *Orchis palustris* Jacq. (1786).
Azraq, 16 Apr. 1964, *Townsend* (K000718232).

Anacamptis papilionacea (L.) R.M.Bateman, Pridgeon & M.W.Chase, Lindleyana 12(3): 120. 1997.
Syn. *Orchis papilionacea* L. (1753); Orchis caspia Trautv. (1873).
Ref. Balqa, Karak, Danin (2004); Balqa, Al-Rumman hills (MSB-K 340704); Tell Ruman (NHJ 2013–2006).

Anacamptis pyramidalis (L.) Rich., De Orchid. Eur. 33. 1817.
Jarash, *Burma* 396 (AUB); Dibeen Forest Reserve, May 2006, *Taifour & Othman* (DFR-H).

Anacamptis sancta (L.) R.M.Bateman, Pridgeon & M.W.Chase, Lindleyana 12(3): 120. 1997.
Balqa, Tal Al-Rumman, on the way to the water (MSB-K 338282); Tal Al-Rumman RBG, Sector No. 23, *Al Mfadi* (NHJ 2013.0091).

Cephalanthera longifolia (L.) Fritsch, Oesterr. Bot. Z. 38: 81. 1888.
Syn. *Serapias helleborine* L. var. *longifolia* L. (1753).
Irbid, Barkash, from main road of Zabia down to the valley (MSB-K 276935); Dibeen, 18 Apr. 2012, *Taifour* (NHJ 2013–2006).

Epipactis veratrifolia Boiss. & Hohen., Diagn. Pl. Orient. 13: 11. 1854.
Amman, Dead Sea. 2 km after Dead Sea Panorama junction towards Aqaba, site no 17 (MSB-K 496782); Wadi Hassa, Afra hot spring, BOTERA project (NHJ 2013.0501).

Himantoglossum caprinum (M.Bieb.) Spreng., Syst. Veg. (ed. 16) [Sprengel] 3: 694. 1826.
Amman (YAR-H); also recorded by Danin (http://flora.org.il/en/plants/HIMCAP/).

Limodorum abortivum (L.) Sw., Nova Acta Regiae Soc. Sci. Upsal. 6: 80. 1799.
Jarash, site 24, Dibbin reserve (MSB-K 556127).

Neotinea maculata (Desf.) Stearn, Ann. Mus. Goulandris 2: 79. 1975.
Dibeen Forest Reserve, May, 2006, *Taifour & Othman* (FOR-H).

Neotinea tridentata (Scop.) R.M.Bateman, Pridgeon & M.W.Chase, Lindleyana 12(3): 122. 1997.
Syn. *Orchis tridentata* Scop., Fl. Carniol., ed. 2. 2: 190. 1772.
Ajloun, 11 Apr. 2012, *Abuhammour, Al Mfadi & Al Khateeb* RBG Jordan Herbarium Specimens list 2006–2013.

Ophrys apifera Huds., Fl. Angl. (Hudson) 340. 1762.
Syn. *Ophrys oestrifera* M.Bieb., Fl. Taur.-Caucas. 2: 369. 1808.
Jerash, Dibeen, *El-Oqlah, Taifour & Said* (NHJ 2013.0104).

Ophrys fusca Link, J. Bot. (Schrader) 1799 (2): 324. 1800.
Dibeen National Park, 2005, *Mahklouf* s.n. (AMM).

Ophrys lutea Icon. [Cavanilles] ii. 46. t. 160. 1793.
Jerash, Dibeen, *El-Oqlah, Taifour & Said* (NHJ 2013.0103).

Ophrys sphegodes Mill., Gard. Dict., ed. 8. n. 8. 1768.
Ajloun, Amman, (YAR-H).

Ophrys transhyrcana Czerniak., Not. Syst. Herb. Hort. Petrop. 4. 1. 1923.
Syn. *Ophrys sintenisii* Fleischm. & Bornm. (1923).
Dibeen National Park, 2005, *Mahklouf* s.n. (AMM).

Ophrys umbilicata Desf., Ann. Mus. Hist. Nat. 10: 227. 1807.
Syn. *Ophrys carmeli* Fleischm. & Bornm. (1923).
Tal Al-Rumman, *Taifour & Aldamen* (NHJ 2009.0099).

Orchis anatolica Boiss., Diagn. Pl. Orient. ser. 1, 5: 56. 1844.
Jerash, Dibeen, *El-Oqlah, Taifour & Said* (NHJ 2013.0105).

Orchis galilaea (Bornm. & M.Schulze) Schltr., Repert. Spec. Nov. Regni Veg. 19: 47. 1923.
Ajloun Rorest Reserve, RSCN report (Impact of Ajloun Forest Reserve on Plant Biodiversity Conservation and Supporting Local People Livelhood).

Orchis punctulata Stev. ex Lindl., Gen. Sp. Orchid. Pl. 273. 1835.
Sin. loc. (YAR-H); Ajloun, Amman (RSCN).

Orchis quadripunctata Cirillo ex Ten., Fl. Napol. Prodr.: 53. 1812.
Irbid, 04 Nov. 1983 (YAR-H).

Orchis simia Lam., Fl. Franç. (Lamarck) 3: 507. 1779.
Dibeen National Park, 2005, *Mahklouf* s.n. (AMM).

Platanthera clavellata (Michx.) Luer, Native Orchids Florida 148. 1972.
Syn. *Orchis tridentata* Muhl. ex Willd. (1805).
Dibeen National Park, 2005, *Mahklouf* s.n. (AMM).

Orobanchaceae

Bartsia trixago L., Sp. Pl. 2: 602. 1753.
Syn. *Bellardia trixago* (L.) All. (1785).
Salt, IV.1895, *Fox* 701 (AUB).

Cistanche salsa (C.A.Mey.) Beck, Nat. Pflanzenfam. [Engler & Prantl] iv. III b. 129. 1893.
Syn. *Phelipaea salsa* C.A.Mey. (1830).
Jordan, East of Jordan, Paine10 (K000283208).

Cistanche tubulosa (Schenk) Wight ex Hook.f., Fl. Brit. India, 4. 2: 324. 1884.
Syn. *Phelypaea tubulosa* Schenk (1840.).
Aqaba, Site 1. Wadi Arab (MSB-K 552554); Mafraq Hazim, *Taifour, Al Mfadi & Said* (NHJ 2013.0048).

Orobanche aegyptiaca Pers., Syn. Pl. [Persoon] 2(1): 181. 1806.
Tabaget Fahel, 20.IV.1967 (AUB).

Orobanche cernua Loefl., Iter Hispan. 152. 1758.
Ma'an, Wadi Musa territory, Al Taybeh & Um'Qaser area (MSB-K 228141); Wadi Hassa, Borbeta hot spring, BOT ERA project (NHJ 2013.0497).

Orobanche crenata Forssk., Fl. Aegypt.-Arab.: 113. 1775.
Irbid, Kofour As Said, 22 Feb.1982 (YAR-H).

Orobanche grisebachii Reut., Prodr. [A. P. de Candolle] 11: 28. 1847.
Ref. RSCN database; Al-Eisawi 133 (2013).

Orobanche lavandulacea Rchb., Iconogr. Bot. Pl. Crit. 7: 48 1829.
Al Ghour, Ghour Feefa, 04 Apr.1984 (YAR-H).

Orobanche minor Sm., Engl. Bot. t. 422. 1797.
Ref. RSCN database; Al-Eisawi 133 (2013).

Orobanche mutelii F.W. Schultz, Fl. Franç. (Mutel) 2: 353, t. 43. 1835.
Dibeen Forest Reserve, 2004, *Mahklouf* s.n. (AMM).

Orobanche nana (Reut.) Beck, Biblioth. Bot. 19: 91. 1890.
Ref. RSCN database; Al-Eisawi 133 (2013).

Orobanche palaestina Reut., Prodr. [A. P. de Candolle] 11: 718. 1847.
Moab, RSCN database.

Orobanche ramosa L., Sp. Pl. 2: 633. 1753.
Ref. RSCN database; Al-Eisawi 133 (2013).

Orobanche schultzii Mutel, Fl. Franç. (Mutel) 2: 352. 1835.
Amman, on way to Bahath (MSB-K 416348).

Parentucellia latifolia (L.) Caruel, Fl. Ital. 6: 480. 1885.
Syn. *Euphrasia latifolia* L. (1753.)
Amman, opposite Jordan University directly opposite stadium (MSB-K 275994).

Parentucellia viscosa (L.) Caruel, Fl. Ital. 6: 482. 1885.
Syn. *Bartsia viscosa* L. (1753).
Jebel Siaghah, Madaba, 12.IV.1877 (AUB–132).

Oxalidaceae

Oxalis corniculata L., Sp. Pl.: 435. 1753.
Rum, 17 Apr. 2012, *Taifour et al.* (RBG Herbarium Specimens list 2006–2013-HM).

Oxalis pes-caprae L., Sp. Pl.: 434. 1753.
Tal Al-Rumman, *Taifour & Aldamen* (NHJ 2009.0135).

Papaveraceae

Argemone mexicana L., Sp. Pl. 508 1753.
Ref. Al-Gharaibeh (2005).

Ceratocapnos turbinata (DC.) Lidén, Opera Bot. 88: 38. 1986.
Syn. *Ceratocapnos palaestina* Boiss., Diagn. Pl. Orient. ser. 1, 8: 12. 1849; *Fumaria turbinata* Sm. ex DC., Syst. Nat. [Candolle] 2: 132. 182.
Ajloun district, Wadi Orjan, BOT ERA project (NHJ 2013.0543).

Fumaria asepala Boiss., Fl. Orient. 1: 135. 1867.
Syn. *Fumaria parviflora* Lam. var. *latisecta* (non Hausskn.) Blakelock
Petra (YAR-H); RSCN database.

Fumaria bracteosa Pomel, Nouv. Mat. Fl. Atl. 2: 239. 1875.
Syn. *Fumaria micrantha* Lag. var. *parlatoriana* Boiss.; *Fumaria densiflora* DC. var. *parlatoriana* (Boiss.) Hausskn.; *Fumaria densiflora* DC. var. *bracteosa* (Pomel) Batt. & Trabut; *Fumaria parlatoriana* (Boiss.) Dinsm. (1898).
Salt, Dir Alla, 1955 (FOR-H).

Fumaria capreolata L., Sp. Pl. 2. 701. 1753.
Tell Ar Rumman, 28 Feb.2012, *Al Mfadi & Al Khateeb* (NHJ).

Fumaria densiflora DC., Cat. Pl. Horti Monsp. 113. 1813.
Syn. *Fumaria micrantha* Lag. (1816).
Madaba, Ayun Musa, *Meyers & Dinsmore* (E00410959); Tal Al-Rumman, *Taifour & Aldamen* (NHJ 2009.0145).

Fumaria judaica Boiss., Diagn. Pl. Orient. ser. 1, 8: 15. 1849.
Jordan, *Meyers & Dinsmore* (E00162178).

Fumaria macrocarpa Parl., Pl. Nov. 5. 1842.
Aqraba, 1983 (YAR-H).

Fumaria officinalis L., Sp. Pl. 2: 700. 1753.
Ref. RSCN survey list; Al-Eisawi 134 (2013).

Fumaria parviflora Lam., Encycl. 2(2): 567. 1788.
Syn. *Fumaria vaillantii* Lois.; *Fumaria parviflora* Lam. var. *persica* Pugsl.; *Fumaria parviflora* Lam. var. *indicoides* Pugsl.; *Fumaria indica* (Non Pugsl.) Blakelock
Ma'an, 30 km southwest of Ma'an, *Dinsmore* (E00443067);.Tal Al-Rumman, *Al-Eisawi, Abu-Sondos, Abu Anza & Shishani* (NHJ 2006.0094)

Fumaria petteri Rchb. subsp. **thuretii** (Boiss.) Pugsley, Icon. Fl. Germ. Helv. 3: 1. 1838.
Syn. *Fumaria thuretii* Boiss. (1854).
Irbid, 04 Apr. 1983 (YAR-H).

Glaucium aleppicum Boiss. & Hausskn., Fl. Orient. 1: 122. 1867.
Irbid, road from Umm Qeis down to Yarmouk Valley, *Boulos, Loutfy & Hauke* 5824 (E00402455); Tal Al-Rumman, *Taifour & Aldamen* (NHJ 2010.0066).

Glaucium arabicum Fresen., Mus. Senckenberg. 1: 174. 1833.
Amman, Site 20, Hisha property, near the water reservoir (MSB-K 557205).

Glaucium corniculatum (L.) Curtis, Fl. Londin. 6: t. 32. 1789.
Jarash, near Rumman (MSB-K 114138).

Glaucium flavum Crantz, Stirp. Austr. Fasc. 2: 131. 1763.
Irbid, Almogam Al Deam (YAR-H).

Glaucium grandiflorum Boiss. & A.Huet, Diagn. Pl. Orient. ser. 2, 5: 15. 1856.
Shoubak, Experimental Farm, *Gillett* 15828 (K000283201).

Glaucium leiocarpum Boiss., Fl. Orient. 1: 122. 1867.
Irbid, Al Mogam Aldeam, 20 May 1984 (YAR-H).

Hypecoum aegyptiacum (Forssk.) Asch. & Schweinf., Mém. Inst. Égypt. 2: 37. 1887.
Aqaba, rd. to Rum, 30 Nov.1955 (FOR-H); Mafraq, Al Mdawara s.n. (Badia).

Hypecoum dimidiatum Delile, Fl. Arabie Pétrée 84. 1830.
Wadi Ibn Hammad SCA, 2011 (Wadi Ibn Hammad SCA baseline survey 2011).

Hypecoum imberbe Sm., Fl. Graec. Prodr. 1(1): 107. 1806.
Syn. *Hypecoum grandiflorum* Benth., Cat. Pl. Pyr. 91 (1826.)
Amman, Bahath, after *Zizyphus* field ends (MSB-K 417725); Tal Al-Rumman, *Al- Eisawi, Abu- Sondos, Abu Anza & Shishani* (NHJ 2006.0066).

Hypecoum littorale Wulfen, Collectanea 2: 205. 1788.
Syn. *Hypecoum geslinii* Coss. & Kralik (1857).
Wadi Rum, 1996 (Eisawi & von Lowen in Informatore Botanico Italiano-Volume 28, 1996).

Hypecoum pendulum L., Sp. Pl.: 124. 1753.
Ma'an, Site 25, Shoubak, opposite the secondary school (MSB-K 557227).

Hypecoum procumbens L., Sp. Pl.: 124. 1753.
Tafila, Dana spring near Dana village, BOT ERA project (NHJ 2013.0468).

Papaver argemone L., Sp. Pl.: 506. 1753.
Salt, 20 km NW of Amman (MSB-K 114116); Tal Al-Rumman, *Al- Eisawi, Abu- Sondos, Abu Anza & Shishani* (NHJ 2006.0347).

Papaver carmeli Feinbrun, Israel J. Bot. 12: 83. 1963.
Irbid, Yarmuk University campus, 20 May 1984 (YAR-H).

Papaver decaisnei Hochst & Steud ex Elkan, Tent. Monogr. Papver 26. 1839.
Ma'an, 5 km NW Rum railway station, *Frey & Kürschner* (TUB).

Papaver glaucum Boiss. & Hausskn., Fl. Orient. 1: 116. 1867.
Syn. *Papaver rogersii* Exell (1930).
Wadi Musa to Bin Dabaghat (Aqaba), *Davis* 8948 (E00403618).

Papaver humile Fedde, Bull. Herb. Boissier 2, 5: 446. 1905.
Medaba, *Meyers & Dinsmore* 598 (E00402422).

Papaver hybridum L., Sp. Pl.: 506. 1753.
Salt, Wadi Shuaib (MSB-K 114389); Tal Al-Rumman, *Taifour & Aldamen* (NHJ 2009.0071).

Papaver polytrichum Boiss. & Kotschy, Diagn. Pl. Orient. ser. 2, 5: 14. 1856.
Azraq, Wadi Bottom (South west the Castle Amra), *El-Oqlah* (NHJ 2013.0582).

Papaver rhoeas L., Sp. Pl.: 507. 1753.
Syn. *Papaver rhoeas* L. var. *syriacum* Boiss. sensu Anth.; *Papaver bipinnatum* (non C.A.Mey.) Sirj.
Madaba, SW of Ma'een (MSB-K 113980).

Papaver subpiriforme Fedde, Bull. Herb. Boissier II, 5: 169. 1905.
Tal Al-Rumman, *Taifour & Aldamen* (NHJ 2009.0065).

Papaver syriacum Boiss. & C.I.Blanche, Diagn. Pl. Orient. ser. 2, 6: 8. 1859.
Tal Al-Rumman, *Shishani & Alkoran* (NHJ 2011.0063).

Papaver umbonatum Boiss., Diagn. Pl. Orient. ser. 1, 8: 11. 1849.
Upper slopes of Jordan Valley, 15 May 2007 (NCARE).

Roemeria hybrida (L.) DC., Syst. Nat. [Candolle] 2: 92. 1821.
Amman, *Boulos* 8758 (K000283205).

Roemeria procumbens Aarons. & Oppenh., Bull. Soc. Bot. Genève 1930, Ser. II. xxii. 305. 1931.
Karak, Wadi Araba, *Frey & Kürschner* (TUB); Mafraq, Al Mansoorah (IBBU-H).

Plantaginaceae

Anarrhinum forskaohlii (J.F.Gmel.) Cufod., Bull. Jard. Bot. État Bruxelles 33(Suppl.): 891. 1963.
Syn. *Simbuleta forskaohlii* J.F.Gmel. (1791).
Amman, Site 50, 1 km from city of Wadi el seer toward Iraq Al Ameenby (MSB-K 395726).

Bacopa monnieri (L.) Wettst., Proc. Acad. Nat. Sci. Philadelphia xcviii. 94. 1946.
Al Mazra'ah, *Mooney* 4617(K000283211).

Bornmuellerantha aucheri (Boiss.) Rothm., Mitth. Thüring. Bot. Vereins n.f., 50: 226. 1943.
Syn. *Odontites aucheri* Boiss. (1844).
Near the village Makhraba (ca 20 km W of Irbid), *Böcker* RB 984 (B100520741).

Chaenorhinum calycinum (Banks & Sol.) P.H.Davis, Notes Roy. Bot. Gard. Edinburgh 36(1): 4. 1978.
Syn. *Chaenorhinum rytidospermum* (Fisch. & C.A.Mey.) Kuprian. (1955).
Ref. RSCN database; Al-Eisawi 163 (2013).

Globularia arabica Jaub. & Spach, Ill. Pl. Orient. 3: 75. 1848.
Tafila, Dana Reserve, BOT ERA project (NHJ 2013.0456).

Kickxia acerbiana (Boiss.) Täckh. & Boulos, Publ. Cairo Univ. Herb. 5: 98. 1974.
Syn. *Linaria acerbiana* Boiss. (1897).
Ma'an, 5 km NW Rum railway station, *Frey & Kürschner* (E); nr. Aqaba, *Boulos & Lahham* 7414 (AMM, BM, E).

Kickxia aegyptiaca (L.) Nábělek, Spisy Prír. Fak. Masarykovy Univ. 70: 31. 1926.
Syn. *Kickxia aegyptiaca* (L.) Nábělek subsp. *aegyptiaca* (1926); *Antirrhinum aegyptiacum* L. (1753).
Zarqa, Alouk (MSB-K 227926); Tal Al-Rumman, *Al-Eisawi* (NHJ 2006.0207).

Kickxia azraqensis Boulos & Lahham, Candollea 32(3): 277. 1977.
Jordan, *Täckholm, El-Hadidi, Lahham & Boulos* 8981 (K000979879; iso B); Azraq, Wadi Bottom (South west the Castle Amra), *El-Oqlah* (NHJ 2013.0592).
This name is unresolved.

Kickxia cirrhosa (L.) Fritsch, Excursions fl. Oesterreich 492. 1897.
Wadi Ajloun, 17 Jun. 1981 (YAR-H).

Kickxia commutata (Bernh. ex Rchb.) Fritsch, Excursions fl. Oesterreich 492. 1897.
Syn. *Linaria commutata* Bernh. ex Rchb. (1831).
10–12 km N of Aqaba, 13.iii.1974, *Boulos, Eisawi & Jallad* (E).

Kickxia elatine (L.) Dumort. subsp. **crinita** (Mabille) Greuter, Boissiera 13: 106. 1967.
Syn. *Linaria sieberi* Rchb. (1831); *Kickxia sieberi* (Rchb.) Dörfl. & Allan (1940); *Linaria crinita* Mabille, (1867).
Ref. Balqa, Danin (2004).

Kickxia floribunda (Boiss.) Täckh. & Boulos, Publ. Cairo Univ. Herb. 5: 97. 1974.
Syn. *Linaria floribunda* Boiss. (1853).
Aqaba, Wadi Araba, site no 20 (MSB-K 496829).

Kickxia heterophylla (Schousb.) Dandy, Fl. Pl. Sudan 3. 137. 1956.
Syn. *Linaria spartioides* Brouss. ex Buch (1828); *Kickxia spartioides* (Brouss. ex Buch) Janch. (1933).

Kickxia judaica Danin, Israel J. Bot. 26(1): 23. 1977.
Ref. Karak, Danin (2004).

Kickxia kneuckeri (Bornm.) Tackholm & Boulos, Publ. Cairo Univ. Herb. 5: 97. 1972. publ. 1974.
Tal Al-Rumman, *Taifour & Aldamen* (NHJ 2009.0152).
This name is unresolved.

Kickxia lanigera (Desf.) Hand.-Mazz., Ann. K. K. Naturhist. Hofmus. 27: 403. 1913.
Syn. *Linaria lanigera* Desf. (1798).
Petra, 22.vii.1907 (AUB).

Kickxia petrana Danin, Willdenowia 21(1–2): 181. 1991.
Endemic. Type: Jordan, Petra, *I. Künne* 3488 (holo B (B 100296152); iso. HUJ; B (B100296151)).

Kickxia spuria (L.) Dumort., Fl. Belg. 35. 1827.
Syn. *Antirrhinum spurium* L. (1753).
Upper Jordan Valley, RSCN survey record.

Kickxia spartioides (Brouss. ex Buch) Janch., Oesterr. Bot. Z. 82: 152. 1933.
Syn. *Linaria spartioides* Brouss. ex Buch (1828).
Nr. Aqaba, *Boulos & Lahham* 7437 (AMM, BM, E).
This name is unresolved.

Linaria albifrons (Sibth. & Sm.) Spreng., Syst. Veg. (ed. 16) [Sprengel] 2: 793. 1825.
Syn. *Antirrhinum albifrons* Sibth. & Sm. (1809).
Irbid, 5.iv.1983 (YAR-H).

Linaria chalepensis (L.) Mill., Gard. Dict. ed. 8, 12. 1768.
Syn. *Antirrhinum chalepense* L. (1753).
Dibeen National Park, 2005, *Mahklouf* s.n. (AMM).

Linaria damascena Boiss. & Gaill., Diagn. Pl. Orient. ser. 2, 6: 130. 1859.
Shoubak, RSCN record; Ref. Al-Eisawi 164 (2013).

Linaria haelava Delile, Descr. Egypte, Hist. Nat. 66. 1813.
Syn. *Antirrhinum haelava* Forssk. (1775).
Ajloun Forest Reserve, *Böcker* RB 901 (B100520887).

Linaria joppensis Bornm., Verh. K. K. Zool.-Bot. Ges. Wien 48: 612. 1898.
Al Maghrar, 29 Mar. 1994, Borzatti & Barsotti (Studies on the Flora of Jordan 14. The species of the Hisma Basin (Wadi Rum Desert). This name is unresolved.

Linaria micrantha (Cav.) Hoff. & Link., Fl. Portug. 1: 258. 1811.
Syn. *Antirrhinum micranthum* Cav. (1791).
Zarqa, Site 4. Alouk (MSB-K 552646); Tal Al-Rumman, *Al-Eisawi* (NHJ 2006.0442).

Linaria simplex Desf., Tabl. École Bot. 65. 1804.
Syn. *Antirrhinum simplex* Willd. (1800).
Tal Al-Rumman, *Taifour & Aldamen* (NHJ 2010.0060).

Linaria tenuis (Viv.) Spreng., Syst. Veg. 2: 795. 1825.
Syn. *Antirrhinum tenue* Viv. (1824).
Jebal Hiswa, 30 Mar. 1994, Borzatti & Barsotti (Studies on the Flora of Jordan 14. The species of the Hisma Basin (Wadi Rum Desert).

Linaria triphylla Mill., Gard. Dict., ed. 8. n. 2. 1768.
40 km E of Amman, nr. Zarqa, *Frey & Kürschner* (E).

Lindenbergia indica (L.) Vatke, Oesterr. Bot. Z. 25(1): 10. 1875.
Syn. *Lindenbergia sinaica* Benth. (1835).
Ref. RSCN record; Al-Eisawi 164 (2013).

Misopates orontium (L.) Raf., Autik. Bot. 158. 1840.
Syn. *Antirrhinum orontium* L. (1753).
Jarash, Dibbeen (MSB-K 27673).

Plantago afra L., Sp. Pl., ed. 2. 1: 168. 1762.
Syn. *Plantago psyllium* L. (1753); *Plantago stricta* Schousb. (1800); *Plantago stricta* Schousb. var. *divaricata* (Zucc.) Bameoud.
Irbid, Wadi Al Arab Dam (MSB-K 199636); Tal Al-Rumman, *Al-Eisawi* (NHJ 2006.0327)

Plantago albicans L., Sp. Pl.: 114. 1753.
Syn. *Plantago cylindrical* Zohary subsp. *mesopotamica*; *Plantago boissieri* Hausskn. & Bornm. var. *tenera* Pilger
Irbid, 6.V.1886 (AUB); Irbid (YAR-H).

Plantago amplexicaulis Cav., Icon. 2: 22. 1793.
Wadi Mudeisisat, S of Azraq, *Frey & Kürschner* (E); Al Mafraq, Badia.

Plantago arenaria Waldst. & Kit., Descr. Icon. Pl. Hung. 1: 51. 1801.
Tal Al-Rumman, *Al-Eisawi* (NHJ 2006.0177).

Plantago bellardii All., Fl. Pedem. 1: 82. 1785.
Balqa, 500 m after Rumman hills (MSB-K 340667); Tal Al-Rumman, *Al-Eisawi* (NHJ 2006.0405).

Plantago ciliata Desf., Fl. Atlant. 1: 137, t. 39. 1798.
Ma'an, 20 km from Ma'an to Modawara 12 Jun. 1996 (NCARE).

Plantago coronopus L., Sp. Pl.: 115. 1753.
Amman, Dead Sea area (MSB-K 275112); Tal Al-Rumman, *Al-Eisawi* (NHJ 2006.0427).

Plantago cretica L., Sp. Pl.: 114. 1753.
Tal Al-Rumman, *Taifour & Aldamen* (NHJ 2009.0138).

Plantago cylindrica Forssk., Fl. Aegypt.-Arab.: 31. 1775.
Aqaba, Wadi Araba, 6 km from Rahmeh resthouse & gas station to Amman (MSB-K 228200).

Plantago lagopus L., Sp. Pl. 1: 114. 1753.
Irbid, close to Zahar (MSB-K 114334); Tal Al-Rumman, *Al-Eisawi* (NHJ 2006.0210).

Plantago indica L., Syst. Nat. ed. 10 2: 896. 1759.
Syn. *Plantago psyllium* L. (1753); *Psyllium indicum* Mirb. (1800).
Tal Al-Rumman, *Taifour & Aldamen* (NHJ 2009.0064).

Plantago lanceolata L., Sp. Pl.: 113. 1753.
Tafila, Dana village, beside Dana Natural Reserve (MSB-K 200202).

Plantago major L., Sp. Pl.: 112. 1753.
Tal Al-Rumman, *RBG-Ire & RBG-Jo* (NHJ 2008.0103).

Plantago maris-mortui Eig, Palestine J. Bot., Jerusalem Ser. 1: 231. 1938.
Amman, Bahath, after the *Zizyphus* field ends (MSB-K 41771.4).

Plantago notata Lag., Gen. Sp. Pl. [Lagasca] 7. 1816.
Jordan, *Paines*.n. (K000779627); Azraq, 16 Apr. 2012, Alkhalidi et al. RBGJ Herbarium Specimens list 2006–2013-HM.

Plantago orbignyana Steinh. ex Decne., Prodr. 13(1): 704. 1852.
Tafila, Dana Reserve, Barra, BOT ERA project (NHJ 2013.0484).

Plantago ovata Forssk., Fl. Aegypt.-Arab.: 31. 1775.
Syn. *Plantago decumbens* Forssk. (1775).
Tell Ar Rumman, 13 Apr. 2009, *Taifour et al.* RBGJ Herbarium Specimens list 2006–2013-HM.

Plantago phaeostoma Boiss. & Heldr., Diagn. Pl. Orient. ser. 2, 4: 71. 1859.
Irbid, Wadi Egam, 14 Mar.1983 (YAR-H).

Plantago sinaica Decne., Prodr. [A. P. de Candolle] 13(1): 733. 1852.
Syn. *Plantago arabica* Boiss., Diagn. Pl. Orient. ser. 1, 12: 94. 1853.
Ref. RSCN record; Al-Eisawi 135 (2013).

Plantago squarrosa Murray, Commentat. Soc. Regiae Sci. Gott. 4(Cl. Phys.): 38 1781 publ. 1782.
Sin. loc. (YAR-H).
Ref. Al-Eisawi 135 (2013).

Veronica agrestis L., Sp. Pl.: 13. 1753.
Ref. RSCN record; Al-Eisawi 166 (2013).

Veronica anagallis-aquatica L., Sp. Pl.: 12. 1753.
Syn. *Veronica lysimachioides* Boiss. (1856).
Ma'an, Wadi Musa, Ain Musa (MSB-K 496069); Dana spring near Dana village, BOT ERA project (NHJ 2013.0470).

Veronica anagalloides Guss., Pl. Rar. 5, t. 3. 1826.
Tafila, *Danin* 2014; sin loc. (YAR-H).

Veronica arvensis L., Sp. Pl.: 13. 1753.
Upper Jordan valley (RSCN database).

Veronica campylopoda Boiss., Diagn. Pl. Orient. ser. 1, 4: 80. 1844.
Ma'an, Rajif-Ras Al-Naqab Road, 0.5 km before the junction to the Desert Highway (MSB-K 413495).

Veronica cymbalaria Bodard, Mém. Véronique Cymb. 3. 1798.
Tal Al-Rumman, *Al-Eisawi, Abu-Sondos, Abu Anza & Shishani* (NHJ 2006.0074).

Veronica hederifolia L., Sp. Pl.: 13. 1753.
Along Zarqa River, 2004, *Al-Dwiri* s.n. (AMM).

Veronica macrostachya Vahl, Enum. Pl. 1. 71. 1804.
Petra (YAR-H).

Veronica orientalis Mill., Gard. Dict., ed. 8. n. 10. 1768.
Jebel Siaghah, Madaba, iv.1877 (AUB).

Veronica panormitana Tineo & Guss., Fl. Sic. Prodr. Suppl. (1): 4. 1832.
Zubia reserve, N of Ajloun, 16 Mar. 1996, *Danin* (HUJ).

Veronica persica Poir., Encycl. 8: 542. 1808.
Dibeen National Park, 2005, *Mahklouf* s.n. (AMM).

Veronica polita Fr., Novit. Fl. Suec. Alt. 1. 1828.
Tal Al-Rumman, *Al-Eisawi* (NHJ 2006.0300).

Veronica syriaca Roem. & Schult., Syst. Veg. 1: 116. 1817.
Tal Al-Rumman, *Taifour & Aldamen* (NHJ 2009.0030).

Veronica viscosa Boiss. Diagn. Pl. Orient. 12: 47. 1853.
Near the village Makhraba (ca 20 km W of Irbid), *Böcker* RB 988 (B 100520731).

Platanaceae

Platanus orientalis L., Sp. Pl. 2: 999. 1753.
Ajloun, Rajeb village, site no 6 (MSB-K 496139).

Plumbaginaceae

Limonium carnosum (Boiss.) O.Kuntze, Revis. Gen. Pl. 2: 395. 1891.
Syn. *Statice carnosa* Boiss. (1848).
Amman, Azraq, *Frey & Kürschner* 80–320 (E00453668).

Limonium globuliferum Kuntze, Revis. Gen. Pl. 2: 395. 1891.
Azraq, 01 Jul. 1953 (FOR-H).

Limonium lobatum (L.f.) Chaz., Revis. Gen. Pl. 2: 395. 1891.
Syn. *Statice lobata* L.f. (1782); *Statice aegyptiaca* Pers. (1805); *Statice thouinii* Viv. (1824); *Limonium thouinii* (Viv.) Kuntze (1891).
Azraq, Druz, *Townsend* 308 (K000283194); Tal Al-Rumman, *Taifour & Aldamen* (NHJ 2009.0095).

Limonium meyeri (Boiss.) Kuntze, Revis. Gen. Pl. 395. 1891.
Syn. *Statice meyeri* Boiss. (1848).
Lower Jordan Valley, RSCN record; Tafila Province, 2014, *Oran* (AMM).

Limonium narbonense Mill., Gard. Dict., ed. 8. n. 2, 1768.
Azraq, 14 Oct. 1953 (FOR-H)

Limonium pruinosum Kuntze, Revis. Gen. Pl. 2: 396. 1891.
Tafila, after site no. 40 up to Tafileh by 6 km (MSB-K 228266).

Limonium sinuatum (L.) Mill., Gard. Dict., ed. 8. n. 6. 1768.
Syn. *Statice sinuata* L. (1753).
Sin. loc. (YAR-H); Tafila Province, 2014, *Oran* (AMM).

Plumbago europaea L., Sp. Pl.: 151. 1753.
Amman, Wadi Al Sheta, after Amman municipality housing by 1 km (MSB-K 244581).

Psylliostachys spicata (Willd.) Nevski, Trudy Bot. Inst. Akad. Nauk S.S.S.R., Ser. 1, Fl. Sist. Vyssh. Rast. 4: 314. 1937.
Syn. *Statice spicata* Willd. (1798); *Limonium spicatum* (Willd.) Kuntze (1891).
Salt flats Azraq, *Hunting Aero Survey* 126A (E00453758).

Poaceae

Aegilops bicornis (Forssk.) Jaub. & Spach, Ill. Pl. Orient.4: 11. 1851.
Syn. *Triticum bicorne* Forssk. (1775).
Wadi Rum, *Davis* 10442 (E00363016).

Aegilops caudata L., Sp. Pl. 1051. 1753.
Wadi Rum, *Boulos & Lahham* 7825 (AMM, BM, E).

Aegilops columnaris Zhuk., Trudy Prikl. Bot. 18: 489. 1928.
Azraq (YAR-H).

Aegilops crassa Boiss., Diagn. Pl. Orient. 7: 129. 1846.
Al-Mafraq, Mfadha village (IBU-H).

Aegilops cylindrica Host, Icon. Descr. Gram. Austriac.2: 6. 1802.
2.5 km N of Safawi-Ruwaished rd. (YAR-H).

Aegilops geniculata Roth, Bot. Abh. Beobacht.: 45. 1787.
W of Wadi Rumeimin, *Meyers & Dinsmore* M724 (E00355598); Tal Al-Rumman, *Al-Eisawi* (NHJ 2006.0394).

Aegilops kotschyi Boiss., Diagn. Pl. Orient.7: 129. 1846.
On the road between Quweira and Naqb Ishtar, *Hunting Aero Survey* (E00363160); Tal Al-Rumman RBG, Sector No. 47, *El-Oqlah & Al Mfadi* (NHJ 2013.0333).

Aegilops longissima Schweinf. & Muschl., Man. Fl. Egypt1: 156. 1912.
Tafila, road to Dana from main Tafila-Shaubah road (MSB-K 151980).

Aegilops lorentii Hochst., Flora 28: 25. 1845.
Syn. *Aegilops biuncialis* Vis. (1851).
Tal Al-Rumman, *Al-Eisawi* (NHJ 2006.0319).

Aegilops neglecta Req. ex Bertol., Fl. Ital. 1: 787. 1834.
Syn. *Aegilops ovata* L. (1753).
3 km before Mafraq/Jarash junction, May 1981, Bioversity International Database [Online].

Aegilops peregrina (Hack.) Maire & Weiller, Fl. Afrique N.3: 358. 1955.
Syn. *Triticum peregrinum* Hack. (1907); *Aegilops variabilis* Eig (1929); *Triticum peregrinum* Hack. subsp. *cylindrostachys* (Eig & Feinbrun) Kimber & Feldman (1987).
Jordan, *Meyers & Dinsmore* 1629 (K000979174); Tal Al-Rumman RBG, Sector No. 23, *El-Oqlah & Taifour*, (NHJ 2013.0177).

Aegilops speltoides Tausch, Flora 20: 108. 1837.
24 km S of Site 7617, May, 1981/05, Bioversity International Database [Online].

Aegilops searsii Feldman & Kislev, Israel J. Bot.26: 191. 1978.
Syn. *Triticum searsii* Feldman & Kislev (1977); *Sitopsis searsii* (Feldman & Kislev) Á.Löve (1984).
Ma'an, in grounds of NCARTT station (MSB-K 151913).

Aegilops triuncialis L., Sp. Pl.: 1051. 1753.
Tal Al-Rumman, *RBG-Ire & RBG-Jo* (NHJ 2008.0035).

Aegilops umbellata Zhuk., Trudy Prikl. Bot. 18: 447, 483, f. 20. 1928.
Al Mafraq, 18 Apr. 1983 (YAR-H).

Aegilops vavilovii (Zhuk.) Chennav., Acta Horti Gothob. 23: 167. 1960.
Ma'an, in grounds of NCARTT station (MSB-K 151902).

Aeluropus lagopoides (L.) Trin. ex Thwaites, Enum. Pl. Zeyl. 374. 1864.
Jordan river, *Meyers & Dinsmore* 1315 (E00358316); Tell Qorma [Qarma], *Townsend* 278 (K000315641).

Aeluropus littoralis (Gouan) Parl., Fl. Ital. 1: 461. 1850.
Ain el Beida [Ayn al Bayda], *Townsend* 62 (K000315634).

Agrostis gigantea Roth, Tent. Fl. Germ. 1: 31. 1788.
Irbid, Wadi Alrain (YAR-H).

Agrostis stolonifera L., Sp. Pl.: 62. 1753.
Amman, Dead Sea area (YAR-H).

Aira elegantissima Schur, Verh. Mitth. Siebenbürg. Vereins Naturwiss. Hermannstadt 4: 85. 1853.
Ref. Al-Eisawi 137 (2013).

Alopecurus aequalis Sobol., Fl. Petrop. 16. 1799.
Salt, Wadi Shuaib (YAR-H).

Alopecurus arundinaceus Poir., Encycl. 8: 766. 1808.
Jerash, *Meyers & Dinsmore* G1703 (E00291990).

Alopecurus myosuroides Huds., Fl. Angl. 23. 1762.
Syn. *Alopeucurus agrestis* L. (1762).
Dibeen Forest Reserve, *Mahklouf* s.n. (AMM).

Alopecurus utriculatus Sol., Nat. Hist. Aleppo, ed. 2, 2: 243. 1794.
Wadi Sir, *Meyers & Dinsmore* M467 (E00398180); near the village Makhraba (ca 20 km W of Irbid), *Böcker* RB 978 (B 10 0520728).

Alopecurus utriculatus subsp. **anthoxanthoides** (Boiss.) Dogan, Notes Roy. Bot. Gard. Edinburgh 40: 86. 1982.
Syn. *Alopecurus anthoxanthoides* Boiss. (1854).
Tafila, Dana For. Reserve (FOR-H).

Ammochloa palaestina Boiss., Diagn. Pl. Orient.13: 52. 1854.
Ref. Al-Eisawi 137 (2013).

Andropogon distachyos L., Sp. Pl.: 1046. 1753.
Salt, *Meyers & Dinsmore* M706 (E00358909).

Anthephora laevis Stapf & C.E.Hubb., Fl. Trop. Afr. 9: 937. 1930.
Ref. Wadi Araba, Danin (2004).

Aristida adscensionis L., Sp. Pl.: 82. 1753.
Zerka Main (Moab),26-Apr–45, around hot springs, *Davis* 9431 (E00398832).

Aristida caerulescens Desf., Fl. Atlant. 1: 109. 1798.
Petra, Apr. 1966 (AUB); Jebal Rehebi, 31 Mar. 1994, *Borzatti* (AMM).

Aristida sieberiana Trin., Neue Entd. 2: 61. 1821.
N of Wadi Kefrein, *Meyers & Dinsmore* 1628 (E00398868).

Arrhenatherum album (Vahl) Clayton, Kew Bull. 16: 250. 1962.
Syn. *Avena alba* Vahl (1791).
Azraq, Wadi Ratam (FOR-H).

Arrhenatherum elatius (L.) P.Beauv. ex J.Presl & C.Presl., Fl. Cech.: 17. 1819.
Syn. Avena elatior L. (1753).
Irbid, Saham (YAR-H); Ajloun, RSCN database,

Arrhenatherum kotschyi Boiss., Diagn. Pl. Orient.7: 122. 1846.
Ajloun, *Meyers & Dinsmore* G1691 (E00356964); Tal Al-Rumman RBG, Sector No. 35, *El-Oqlah & Al Mfadi,* (NHJ 2013.0363).

Arrhenatherum palaestinum Boiss., Diagn. Pl. Orient.13: 51. 1854.
Jerash, *Meyers & Dinsmore* 9700 (K000315622); Tal Al-Rumman, *Al-Eisawi* (NHJ 2006.0175).

Arundo donax L., Sp. Pl. 81. 1753.
Karak, Ghor Al-Safi (MSB-K 199614).

Avena barbata Pott ex Link, J. Bot. (Schrader) 1799(2): 315. 1800.
Syn. *Avena hirtula* Lag. (1816); *Avena wiestii* Steud (1854).
Petra, Davis 9143 (E00356745); Dana Reserve, *RBG-Ire & RBG-Jo* (NHJ 2008.0038).

Avena clauda Durieu, Rev. Bot. Recueil Mens.1: 360. 1845.
Al-Mafraq, Rasm Al-Hisaan (IBBU-H).

Avena eriantha Durieu, Rev. Bot. Recueil Mens. 1: 360. 1845.
Al-Mafraq, Hosha (IBBU-H).

Avena fatua L., Sp. Pl.: 80. 1753.
Ziga (Moab), *Davis* 9628 (E00356805).

Avena longiglumis Durieu, Rev. Bot. Recueil Mens.1: 359. 1845.
Al Balqa, Site 008. S. of Naqb Ishtar, *Hunting Aero Survey* 49C (E00356813); Tal Al-Rumman, *Al-Eisawi* (NHJ 2006.0027).

Avena sativa L., Sp. Pl. 79. 1753.
Irbid (FOR-H).

Avena sterilis L., Sp. Pl. ed. 2: 118. 1762.
Above Zarqa river, *Meyers & Dinsmore* G17 (E00356889); Tal Al-Rumman, *Al-Eisawi* (NHJ 2006.0296).

Avena sterilis subsp. **ludoviciana** (Durieu) Gillet & Magne, Nouv. Fl. Franç. ed. 3: 352. 1873.
Syn. *Avena ludoviciana* Durieu (1855).
Irbid, Al Ramtha (YAR-H).

Boissiera squarrosa (Banks & Sol.) Nevski, Trudy Sredne-Aziatsk. Gosud. Univ., Ser. 8b, Bot.17: 30. 1934.
Syn. *Pappophorum squarrosum* Banks & Sol. (1794).
Tafila, road from Dana reserve to Tafileh, 50m after the junction to Rummana camp (MSB-K 277862).

Bothriochloa ischaemum (L.) Keng, Contr. Biol. Lab. Sci. Soc. China, Bot. Ser. 10: 201. 1936.
Syn. *Andropogon ischaemum* L. (1753).
Makhaibi (FOR-H).

Brachiaria mutica (Forssk.) Stapf, Fl. Trop. Afr. 9: 526. 1919.
Syn. *Urochloa mutica* (Forssk.) T.Q.Nguyen (1966).
Lowe Jordan Valley (http://flora.org.il/en/plants/UROMUT/).

Brachypodium distachyon (L.) P.Beauv., Ess. Agrostogr.: 101, 155. 1812.
Syn. *Bromus distachyos* L. (1756).
Tal Al-Rumman, *Al-Eisawi* (NHJ 2006.0003).

Brachypodium sylvaticum (Huds.) P.Beauv., Ess. Agrostogr.: 101, 155. 1812.
Syn. *Festuca sylvatica* Huds. (1762).
Ref. (FOR-H); Al-Eisawi 138 (2013).

Briza maxima L., Sp. Pl. 70. 1753.
Balqa, Tal Al-Rumman, Royal Botanic Garden (MSB-K 341217); Tal Al-Rumman, *Al-Eisawi* (NHJ 2006.0238).

Briza minor L., Sp. Pl.: 70. 1753.
Between Hesban and Wadi Sir, 28-Apr–11, *Meyers & Dinsmore*, M1048 (E00407520).

Bromus aegyptiacus Tausch, Flora 20: 124. 1837.
Ref. RSCN database; Al-Eisawi 138 (2013).

Bromus alopecuros Poir., Voy. Barbarie 2: 100. 1789.
Syn. *Bromus sericeus* Ten. (1811).
Ajloun, *Meyers & Dinsmore* G1429 (E00363928); Tal Al-Rumman, *Al-Eisawi* (NHJ 2006.0147); Karak, Wadi Araba, Feinan, Wadi Khalid, *Frey & Kürschner* (B); Maan, 10 km NE Wadi Musa, 13.iv.1987, *Baierle* & Todt. 87.281.

Bromus brachystachys Hornung, Flora 16: 417. 1833.
Syn. *Bromus tigridis* Boiss. & Noë (1859); *Bromus aegyptiacus* subsp. *palaestinus* Melderis (1952); *Bromus palaestinus* (Melderis) Mouterde (1956).
Tell al Kadi, *Meyers & Dinsmore* (E).

Bromus catharticus Vahl, Symb. Bot. 2: 22. 1791.
Syn. *Bromus willdenowii* Kunth (1829).
Ref. Wadi Araba, Danin (2004).

Bromus danthoniae Trin., Verz. Pfl. Casp. Meer.: 24. 1831.
Jordan, *Hunting Aero Survey* 218A (E00169395); Irbid, Bani Knana. *El-Oqlah & Bani Said* (NHJ 2013.0146).

Bromus diandrus Roth, Bot. Abh. Beobacht.: 44. 1787.
Ajloun district, Wadi Orjan, BOT ERA project (NHJ 2013.0541).

Bromus fasciculatus C.Presl, Cyper. Gramin. Sicul.: 39. 1820.
Azraq, *Townsend* 318 (K000386957); Tal Al-Rumman, *Al-Eisawi* (NHJ 2006.0204).

Bromus haussknechtii Boiss., Fl. Orient.5: 648. 1884.
Karak, Wadi Araba, (NE from Ain Finan), 13.iv.1986, *Baierle* 86–1821.
Ref. Danin (http://flora.org.il/en/plants/BROHAU/).

Bromus hordeaceus L., sp.Pl.: 77. 1753.
Baqla and Amman, Danin 2005; RSCN database.

Bromus intermedius Guss., Fl. Sicul. Prodr. 1: 114. 1827.
1.4 km from Duhni Qa'a, 2 km N of Safawi-Ruwaished rd. (BRDC-H).

Bromus japonicus Thunb., Syst. Veg. ed. 14 119. 1784.
Above Zarqa River, 2.v.1911 (44021-HUJ); Ajloun Forest Reserve (RSCN database).

Bromus lanceolatus Roth, Catal. Bot.1: 18. 1797.
Syn. *Bromus macrostachys* Desf. (1798).
Karak, Qasser-Karak road, just outside Qasser town (MSB-K 276522); Wadi Sir, *Meyers & Dinsmore* M1412 (E00368469); Tal Al-Rumman, *Al-Eisawi* (NHJ 2006.0006).

Bromus madritensis L., Cent. Pl. 1,5. 1755.
Vaqla and Amman, Damin 2005; RSCN database.

Bromus pseudobrachystachys H.Scholz, Bot. Jahrb. Syst. 91: 462. 1972.
(FOR-H).

Bromus rigidus Roth, Bot. Mag. (Römer & Usteri) 4(10): 21. 1790.
Al Mafraq, Borqu (IBBU-H).

Bromus rubens L., Cent. Pl. I 5. 1755.
Syn.: *Bromus madritensis* (L.) Husn. ssp. *rubens* (L.) Husn. (1899).
Wadi Rum, *Davis* 9004 (E00368488); Tal Al-Rumman, *Al-Eisawi* (NHJ 2006.0242).
Dana Reserve, *RBG-Ire & RBG-Jo* (NHJ 2008.0077); Amman, opposite to University of Jordan (MSB-K 340427).

Bromus scoparius L., Cent. Pl. I 6. 1755.
Al-Mafraq, Borqu (IBBU-H); 800 m W of the Safawi-Azraq rd. (BRDC-H); Irbid (YAR-H).

Bromus squarrosus L., Sp. Pl.: 76. 1753.
Ref. RSCN database; Al-Eisawi 139 (2013).

Bromus scoparius L., Sp. Pl.: 6. 1753.
Al Karak, midway between Qatrana and Jurf ed Darawish Station, *Boulos & Jallad* 7858 (E00368668); Tal Al-Rumman, *Al-Eisawi* (NHJ 2006.0019).

Bromus sterilis L., Sp. Pl.: 77. 1753.
Along Zarqa River, 2004, *Al-Dwiri* s.n. (AMM); Irbid (YAR-H); Madaba (YAR-H); Ajloun (RSCN database).

Bromus syriacus Boiss. & Blanche, Diagn. Pl. Orient. ser. 2, 4: 139. 1859.
Irbid, W Um Qais (IBBU-H).

Bromus tectorum L., Sp. Pl.: 77. 1753.
Madaba (MSB-K 114068); Tal Al-Rumman, *Al-Eisawi* (NHJ 2006.0396)

Bromus tomentellus Boiss., Diagn. Pl. Orient.7: 126. 1846.
Ma'an, site 20, Hisha property, near the water reservoir (MSB-K 557847).

Castellia tuberculosa (Moris) Bor, Indian Forester 74: 90. 1948.
Syn. *Catapodium tuberculosum* Moris (1841).
Jordan (AUB); Dead Sea area, *Danin* (http://flora.org.il/en/plants/CASTUB/), not recorded from Jordan by Danin.

Catabrosa aquatica (L.) P.Beauv., Ess. Agrostogr.: 97. 1812.
Syn. *Aira aquatica* L. (1753); *Catabrosa capusii* Franch. (1884).
Jerash, *Meyers & Dinsmore* G1744 (E00403165); Azraq, *Frey & Kürschner* (E00448744).

Catapodium rigidum (L.) C.E.Hubb., Fl. Bedfordshire: 437. 1953.
Syn. *Poa rigida* L. (1755); *Sclerochloa rigida* (L.) Link (1821); *Desmazeria rigida* (L.) Tutin (1952).
Ajloun, Za'atari road, about 15 km to Zai area, (MSB-K 276784); Tal Al-Rumman, *Al-Eisawi* (NHJ 2006.0042); above Zarqa river, *Meyers & Dinsmore* G529 (E00407603).

Cenchrus ciliaris L., Mant. Pl. 2: 302. 1771.
Syn. *Pennisetum ciliare* (L.) Link (1827).
Aqaba, site 001, nr H.A.S. Quweira camp, *Hunting Aero Survey* 17C (E00365516); Ghor As Safi (HUJ); Mujib River (HUJ).

Cenchrus echinatus L., Sp. Pl.: 1050. 1753.
Aqaba, Aqaba-Saudi border highway, hill just opposite to ship of Potash company (MSB-K 412649).

Centropodia forsskalii (Vahl) Cope, Kew Bull. 37: 658. 1983.
Syn. *Danthonia forsskalii* (Vahl) Trin. (1827).
Nr. Aqaba, *Boulos & Lahham* 7440 (AMM, BM, E); Ma'an, in wadi west of El Inab, *Hunting Aero Survey* 79A (E00357115).

Chloris barbata Sw., Fl. Ind. Occid. 1: 200. 1797.
Wadi Zarqa, *Danin & Hedge* (E).

Chloris gayana Kunth, Révis. Gramin. 1: 89. 1829.
Irbid (YAR-H).

Chloris virgata Sw., Fl. Ind. Occid. 1: 203. 1797.
Irbid (YAR-H).

Chrysopogon gryllus (L.) Trin., Fund. Agrost. (Trinius) 188. 1820.
Ref. RSCN database; Al-Eisawi 140 (2013).

Coix lacryma-jobi L., Sp. Pl.: 972. 1753.
Ref. RSCN database; Al-Eisawi 140 (2013).

Cornucopiae alopecuroides L., Mant. Pl. 1: 28. 1767.
Salt, Wadi Shuaib, 1955, Kaspligil (FOR-H).

Cornucopiae cucullatum L., Sp. Pl.: 54. 1753.
Castle of Lubebeh, *Meyers & Dinsmore* (E).

Corynephorus divaricatus (Pourr.) Breistr., Procès-Verbaux Soc. Dauphin. Études Biol., Bio-Club sér. 3, 17: 3. 1950.
Syn. *Aira divaricata* Pourr. (1788).
Majdal to Jarash, 9 May 1927 (HUJ).

Crithopsis delileana (Schult.) Roshev., Zlaki: 319. 1937.
Syn. *Elymus delileanus* Schult. (1824).
Ain Hesban, *Meyers & Dinsmore* M773 (E00355810); Tal Al-Rumman, *Al-Eisawi* (NHJ 2006.0032).

Crypsis schoenoides (L.) Lam., Tabl. Encycl. 1: 166. 1791.
Syn.: *Phleum schoenoides* L. (1753).
Tal Al-Rumman, *Royal Botanic Garden Team* (NHJ).

Cutandia dichotoma (Forssk.) Trab., Fl. Algérie, Monocot.: 237. 1895.
Syn. *Festuca dichotoma* Forssk. (1775).
About 40 km S of Ma'an, 18 Apr. 1929 (HUJ).

Cutandia maritima (L.) Barbey, Fl. Sard. Comp.: 72. 1885.
Syn. *Triticum maritimum* L. (1762).
RSCN database; (YAR-H).

Cutandia memphitica (Spreng.) Benth., J. Linn. Soc., Bot.19: 118. 1881.
Syn. *Dactylis memphitica* Spreng. (1801).
Petra, Regione del Sinai, *Figari* (E00403221).

Cymbopogon commutatus (Steud). Stapf, Bull. Misc. Inform. Kew 1907: 211. 1907.
Syn. *Andropogon commutatus* Steud. (1854).
Ref. RSCN database.

Cynodon dactylon (L.) Pers., Syn. Pl. 1: 85. 1805.
Syn.: *Panicum dactylon* L. (1753).
Bank of Zarqa river, *Meyers & Dinsmore* G3, (E00357428); nr. Aqaba, *Boulos & Lahham* 5377 (AMM, BM, E).

Cynosurus coloratus Lehm. ex Steud., Fl. Afr. Austral. Ill.: 439. 1841.
Syn. *Cynosurus callitrichus* Barb.-Boiss. & Barbey (1882).
Madaba, Ghernata, 5 km NE of Medaba (MSB-K 113887); Medaba, *Meyers & Dinsmore* M7 (E00403010).

Cynosurus echinatus L., Sp. Pl.: 72. 1753.
Ref. Balqa, Danin (2004).

Cynosurus elegans Desf., Fl. Atlant. 1: 82. 1798.
Syn. *Cynosurus effusus* Link (1799).
Tafila, outside Dana village on the way to MOA guesthouse (MSB-K 228129); Tal Al-Rumman, *Al-Eisawi* (NHJ 2006.0243); Ajloun, South of Ajloun, *Hunting Aero Survey* 111B (E00403035).

Dactylis glomerata L., Sp. Pl.: 71. 1753.
Jordan, *Täckholm, El-Hadidi, Lahham, & Boulos* 8921 (K001064706); Tal Al-Rumman, *Al-Eisawi* (NHJ 2006.0165).

Dactyloctenium aegyptium (L.) Willd., Enum. Pl.: 1029. 1809.
Syn. *Cynosurus aegyptius* L. (1753).
Ref. Al-Eisawi 141 (2013).

Desmostachya bipinnata (L.) Stapf, Fl. Cap.7: 632. 1900.
Syn. *Briza bipinnata* L. (1756). *Eragrostis bipinnata* (L.) Muschl. (1895).
Karak, *Frey & Kürschner* 80–789 (E00358140); Wadi Araba, 09.iii.1955, *Kapsligil* (FOR-H).

Dichanthium annulatum (Forssk.) Stapf, Fl. Trop. Afr.9: 178. 1917.
Syn. *Andropogon annulatus* Forssk., Fl. Aegypt.-Arab.: 173 1775.
Safawi, *Cope & Al-Eisawi* D. 351 (K000315614).

Dichanthium foveolatum (Delile) Roberty, Boissiera 9: 163. 1960.
Syn. *Andropogon foveolatus* Delile (1813).
Wadi Araba, Danin (2004).

Digitaria sanguinalis (L.) Scop., Fl. Carniol. ed. 2, 1: 52 .1771.
Syn. *Panicum sanguinale* L. (1753).
W Ras en Naqb, 26 Oct. 1999, *Borzatti & Barsotti* (AMM).

Dinebra retroflexa (Vahl) Panz., Denkschr. Königl. Akad. Wiss. München4: 270. 1813 publ. 1814.
Syn. *Cynosurus retroflexus* Vahl (1791).
Irbid, Al Hammah (YAR-H).

Echinaria capitata (L.) Desf., Fl. Atlant.2: 385. 1799.
Syn. *Cenchrus capitatus* L. (1753).
Hunting Aero Survey 70D (E00448481).

Echinochloa colona (L.) Link, Hort. Berol.2: 209. 1833.
Syn. *Panicum colonum* L. (1759).
Reseifeh, Wadi Jerba, *Harding* (E00365067).

Echinochloa crus-galli (L.) Beauv., Ess., Agrostogr.: 53, 1812.
Syn. *Panicum crus-galli* L. (1753).
Balqa and Dead sea, Danin (2004).

Eleusine indica (L.) J.Gaertn., Fruct. Sem. Pl. 1: 8. 1788.
Araba valley (http://flora.org.il/en/plants/ELEIND/).

Elymus elongatus (Host) Runemark, Hereditas (Lund) 70: 156. 1972.
Syn. *Elytrigia elongata* (Host ex P.Beauv.) Nevski (1933).
Lower Jordan Valley, Jordan banks near Allenby Bridge, *Zohary* (HUJ).

Enneapogon desvauxii P.Beauv., Ess. Agrostogr.: 82. 1812.
Syn. *Enneapogon brachystachyus* (Jaub. & Spach) Stapf (1900).
Karak, Wadi Araba, Feinan, Khirbet el-Ghuwebe, *Frey & Kürschner* (B); Wadi Yutum, on the slopes along the road, 9 km N of Aqaba, *Boulos & Jallad* (AMM).

Enneapogon persicus Boiss., Diagn. Pl. Orient.5: 71. 1844.
RSCN database; (YAR-H).

Eragrostis aegyptiaca (Willd.) Delile, Descr. Egypte, Hist. Nat. 157. 1813.
Ref. RSCN database; Al-Eisawi 142 (2013).

Eragrostis barrelieri Daveau, J. Bot. (Morot) 8: 289. 1894.
23 km SW of Ma'an, 9 Oct. 1941 (HUJ).

Eragrostis cilianensis (All.) Janch., Mitt. Naturwiss. Vereins Univ. Wien 5: 110. 1907.
Syn. *Poa cilianensis* All. (1785).
Ma'an, Petra, Wadi Musa, *Frey & Kürschner* (B).

Eragrostis echinchloidea Stapf, Fl. Cap. 7: 627. 1900.
Ref. Wadi Araba, Danin (2004).

Eragrostis minor Host, Fl. Austriaca1: 135. 1827.
Ghor es Safi, Nov. 1883, *Hart* (HUJ).

Eragrostis palmeri S.Wats., Proc. Amer. Acad. Arts 18: 182. 1883.
Al Karak, Ghour Al Mazrah (YAR-H).

Eragrostis pilosa (L.) P.Beauv., Ess. Agrostogr.: 71. 1812.
Syn. *Poa pilosa* L. (1753).
Ref. RSCN database; Al-Eisawi 142 (2013).

Eragrostis virescens J.Presl, Reliq. Haenk. 1: 276. 1830.
Tafila, Ein Fidan, 13 km E of Ein, *Danin* (http://flora.org.il/en/plants/ERAVIR/).

Eremopoa persica (Trin.) Roshev, Fl. URSS 2: 430. 1934.
Syn. *Poa persica* Trin. (1831).
Ajloun, *Meyers & Dinsmore* (E); Burgu pipe station (FOR-H).

Eremopyrum bonaepartis (Spreng.) Nevski, Trudy Bot. Inst. Akad. Nauk S.S.S.R., Ser. 1, Fl. Sist. Vyssh. Rast.1: 18. 1933.
Syn. *Triticum bonaepartis* Spreng. (1880); *Eremopyrum confusum* Melderis (1964).
Ma'an, Shoubak area, on the way to Wadi Musa (MSB-K 277921).

Eremopyrum distans (K.Koch) Nevski, Trudy Bot. Inst. Akad. Nauk S.S.S.R., Ser. 1, Fl. Sist. Vyssh. Rast.1: 18. 1933.
Syn. *Agropyron distans* K.Koch (1848).
Tafila, Tafila to Ghor road, 10 km after the junction toward Ghor (MSB-K 496818).

Eremopyrum orientale (L.) Jaub. & Spach, Ill. Pl. Orient.4: 26. 1851.
Syn. *Secale orientale* L. (1753).
Azraq, *Frey & Kürschner* (B).

Festuca arundinacea Schreb. subsp. **arundinacea**, Spic. Fl. Lips.: 57. 1771.
Amman, main road from NCARTT toward Abu Nsair, beside the water spring (Ain Abu Nsair) (MSB-K 200420).

Gastridium ventricosum (Gouan) Schinz & Thell., Vierteljahrsschr. Naturf. Ges. Zürich 58: 39. 1913.
Syn. *Gastridium scabrum* C.Presl, Fl. Sicul. xliv 1826.
Jarash, Magdal, *Meyers & Dinsmore* (E).

Gaudinia fragilis (L.) P.Beauv., Ess. Agrostogr.: 95. 1812.
Syn. *Avena fragilis* L. (1753).
Jarash, 10 May 1927 (HUJ); Lyera, *Meyers & Dinsmore* (E).

Halopyrum mucronatum (L.) Stapf, Icon. Pl. 25: t. 2448, 1896.
Syn. *Uniola mucronata* L. (1762); *Desmazeria unioloides* Deflers (1889).
Ref. RSCN database; Al-Eisawi 142 (2013).

Henrardia pubescens (Bertol.) C.E.Hubb., Blumea, Suppl.3: 19. 1946.
Syn. *Rottboellia pubescens* Bertol. (1842).
Ma'an, Aqaba, 2 km W of Bir ed-Dabbaghat (SSW Shoubak), *Baierle* 87–842 (E00365819).

Heteranthelium piliferum (Sol.) Hochst. ex Jaub. & Spach, Ill. Pl. Orient.4: 24. 1850.
Syn. *Elymus pilifer* Sol. (1794).
Musa to Petra (HUJ); 10 km NE of the Safawi-Mafraq rd. (BRDC-H).

Hordeum bulbosum Sieber ex Kunth, Enum. Pl.1: 449. 1833.
Madaba, Mlaih to Makawer, in school yard (Baqe Al Kabqi) (MSB-K 185682); Tal Al-Rumman RBG, Sector No. 98, *El-Oqlah, Taifour & Al Khawaldah* (NHJ 2013.0271).

Hordeum distichon L., Sp. Pl. 85. 1753.
Karak, Bioversity International Database. (http://bioversityinternational.org).

Hordeum marinum subsp. **gussoneanum** (Parl.) Thell., Vierteljahrsschr. Naturf. Ges. Zürich 52: 441.1908.
Syn. *Hordeum geniculatum* All. (1785).
Azraq Oasis, in the vicinity of the hot spring, *Jallad, Lahham & Ismail* 7032 (E00363682).

Hordeum murinum L. subsp. **glaucum** (Steud.) Tzvelev, Novosti Sist. Vyssh. Rast.8: 67. 1971.
Syn. *Hordeum glaucum* Steud. (1854).
Tal Al-Rumman, *Taifour & Aldamen* (NHJ 2009.0207) & *Al-Eisawi* (NHJ 2006.0046).

Hordeum murinum subsp. **leporinum** (Link) Arcang., Comp. Fl. Ital. 805. 1882.
Syn. *Hordeum leporinum* Link (1834).
Tabaget Fahel, May 1887 (AUB).

Hordeum spontaneum K.Koch, Linnaea 21: 430. 1848.
Syn. *Hordeum ithaburense* Boiss. (1854).
40 km E of Amman, *Dinsmore* 14685 (E00363813); Tal Al-Rumman, *Al-Eisawi* (NHJ 2006.0014).

Hyparrhenia hirta (L.) Stapf, Fl. Trop. Afr.9: 315. 1918.
Syn. *Andropogon hirtus* L. (1753).
Wadi Heidan (Moab), *Davis* 9561 (E00361135); Tal Al-Rumman, *RBG-Ire & RBG-Jo* (NHJ 2008.0112).

Imperata cylindrica (L.) Raeusch., Nomencl. Bot., ed. 3: 10. 1797.
Syn. *Lagurus cylindricus* L. (1759).
Ma'an, near Nafeh in Arabia petraea, Schimper 114 (E00358504); Tal Al-Rumman, *Shishani & Alkoran* (NHJ 2011.0081).

Lagurus ovatus L., Sp. Pl.: 81. 1753.
Irbid, Al Ramtha (YAR-H).

Lamarckia aurea Moench, Methodus 201. 1794.
Mt. Hor, east side, *Davis* 8577 (E00369986); Tal Al-Rumman, *Al-Eisawi* (NHJ 2006.0295); Ajloun forest Reserve, *Bocker* RB 899 (B 100520885).

Lasiurus scindicus Henrard, Blumea4: 514. 1941.
Aqaba, 5 km after Beer Maz Kor rest house, site no 3 (MSB-K 496070).

Leptochloa fusca (L.) Kunth, Révis. Gramin.1: 91. 1829.
Syn. *Diplachnefusca*(L.) Stapf (1900); *Leptochloa fusca* subsp. *uninervia* (J.Presl) N.Snow (1998). *Leptochloa uninervia* (J.Presl) Hitchc. & Chase (1917).
Ref. Al-Eisawi 141 (2013) as *Diplachnefusca.*

Loliolum subulatum (Banks & Sol.) Eig, J. Bot.75: 189. 1937.
Syn. *Triticum subulatum* Banks & Sol. (1794); *Nardurus subulatus* (Banks & Sol.) Bor (1965).
Ma'an, Site 15. Qadeseyeh-Shoubak road, Fujaij Satation (MSB-K 554374).

Lolium multiflorum Lam., Fl. Franç. 3: 621. 1778.
Ajloun, Bergesh, Irbid (YAR-H).

Lolium perenne L., Sp. Pl.: 83. 1753.
Wadi Malih, *Dinsmore* 5064 (E00407657).

Lolium persicum Boiss. & Hohen., Diagn. Pl. Orient. 13: 66. 1854.
Dibeen Forest Reserve, 2004, *Mahklouf* (AMM).

Lolium rigidum Gaudin, Agrost. Helv.1: 334. 1811.
Syn. *Lolium loliaceum* (Bory & Chaub.) Hand.-Mazz. (1914).
Zarqa, by 1 km edge of barley field (MSB-K 341321); Tal Al-Rumman, *Al-Eisawi* (NHJ 2006.0005).

Lolium temulentum L., Sp. Pl.: 83. 1753.
Al-Mafraq- Al-Zubaidiyeh (IBBU-H); 1999); Ma'an, Ammab, Balqa, Danin (2004).

Lygeum spartum Loefl. ex L., Gen. Pl. ed. 5 522. 1754.
Irbid, Jordan Univ. Science & Tech. (YAR-H).

Melica cupanii Guss., Fl. Sicul. Prodr. Suppl.: 17. 1832.
Ajloun Forest Reserve (RSCN database).

Melica minuta L., Mant. Pl.1: 32. 1767.
Aqaba (RSCN database); (YAR-H).

Melica persica Kunth Révis. Gramin. 1: 351. 1830.
Syn. *Melica canescens* (Regel) Lavrenko ex Nevski (1934).
Aqaba, 1.5 km E of Haj (near Wadi Musa), *Baierle* 87–772 (E00410076); Tafila, before the main road down toward Dana village (MSB-K 228277).

Milium pedicellare (Bornm.) Roshev. ex Melderis, Ark. Bot., n.s., 2: 291. 1952.
Syn. *Milium vernale* M.Bieb. var. *pedicellare* Bornm. (1914).
East of Suf, 4 May 1911 (HUJ); Ajloun (RSCN database).

Milium vernale M.Bieb., Fl. Taur.-Caucas.1: 53. 1808.
E of Salt, 24 Apr. 1911, *Meyers & Dinsmore* (E).

Oloptum miliaceum (L.) Röser &H.R.Hamasha, Pl. Syst. Evol. 298: 365. 2012.
Syn. *Agrostis miliacea* L. (1753); *Piptatherum miliaceum* (L.) Coss. (1812); *Piptatherum thomasii* (Duby) Kunth (1833).
Ma'an, Petra, *Dinsmore* 10146 (E00443905); Tafila, on the way to Dana from Ghour, *Taifour, Al Khawaldah & Al Mogherbi* (NHJ 2013.0687).

Panicum antidotale Retz., Observ. Bot. 4: 17. 1786.
Ref. Wadi Araba, Danin (2004).

Panicum miliaceum L., Sp. Pl. 58. 1753.
Ref. Wadi Araba, Danin (2004).

Panicum repens L., Sp. Pl. ed. 2: 87. 1762.
Lower Jordan Valley (RSCN database); (FOR-H).

Panicum turgidum Forssk., Fl. Aegypt.-Arab.: 18. 1775.
In regione Wadi Hebran, Arabiae petr. ad rupes, *Schimper* 152 (E00355503); Ma'an: 15 km S von Aqaba, *Frey & Kürschner* 5811 (E00355500).

Parapholis filiformis (Roth) C.E.Hubb., Blumea, Suppl. 3: 14. 1946.
Mafraq, Al Al-Bayt University (IBBU-H).

Parapholis incurva (L.) C.E.Hubb., Blumea, Suppl.3: 14. 1946.
Syn. *Aegilops incurva* L. (1753).
Along Zarqa River, 2004, *Al-Dwiri* s.n. (AMM).

Parapholis marginata Runemark, Bot. Not. 115: 8, 14. 1962.
Al Ramtha, Al Shajarah (YAR-H).

Paspalidium geminatum (Forssk.) Stapf, Fl. Trop. Afr.9: 583. 1920.
Syn. *Panicum geminatum* Forssk. (1775).
Umm Qeis, *Boulos, Jallad & Lahham* 8418 (K000315583).

Paspalum dilatatum Poir.Encycl. 5: 35. 1804.
(YAR-H); [http://e-monocot.org/taxon/urn:kew.org:wcs:taxon:431483].

Pennisetum alopecuroides (L.) Spreng., Syst. Veg. 1: 303. 1824.
Wadi Hassa, Borbeta hot spring, BOT ERA project (NHJ 2013.0496).

Pennisetum divisum (Forssk. ex J.F.Gmel.) Henrard, Blumea 3: 162. 1938.
Syn. *Panicum divisum* Forssk. ex J.F.Gmel. (1791).
Madaba, Amman-Dead Sea highway, before Zara Springs (hot springs) (MSB-K 227797).

Pennisetum glaucum (L.) R.Br., Prodr. Fl. Nov. Holl.: 195. 1810.
Syn. *Setaria glauca* (L.) P.Beauv. (1812); *Setaria lutescens* F.T.Hubb.(1916) nom. illeg.
Jerash, 29 Oct. 1926 (HUJ); Wadi Suhaib (YAR-H).
[http://e-monocot.org/taxon/urn:kew.org:wcs:taxon:432725]

Pennisetum orientale Rich., Syn. Pl.1: 72. 1805.
Ref. Al-Eisawi 145 (2013).

Pennisetum purpureum Schumach., Beskr. Guin. Pl. 44. 1827.
Jarash, King Talal Dam (YAR-H).

Pennisetum setaceum (Forssk.) Chiov., Boll. Soc. Bot. Ital.1923: 113. 1923.
Syn. *Phalaris setacea* Forssk. (1775).
Irbid, Asahknh (YAR-H).

Phacelurus digitatus (Sm.) Griseb., Spic. Fl. Rumel. 2: 424. 1846.
Ref. Al-Eisawi 145 (2013).

Phalaris aquatica L., Cent. Pl. I 4. 1755.
Syn. *Phalaris tuberosa* L (1771); *Phalaris nodosa* L. (1774), nom. illegit.
Amman, Khalda, from the Al Waha Circle toward Sweleh (MSB-K 2003.05).

Phalaris brachystachys Link, Neues J. Bot.1(3): 134. 1806.
Al Mafraq, Al Khanasri (YAR-H).

Phalaris canariensisL, Sp. Pl. 54. 1753.
Salt, Dier Alla (YAR-H); Amman (http://flora.org.il/en/plants/PHACAN/).

Phalaris minor Retz., Observ. Bot. 3: 8. 1783.
Al Karak, 4 km south Qatrana, along the road to Aqaba, *Boulos, Jallad & Lahham* 6587 (E00369602); Amman, Dabala site on the desert highway (MSB-K 554293); Tal Al-Rumman, *Al-Eisawi* (NHJ 2006.0176).

Phalaris paradoxa L., Sp. Pl. ed. 2: 1665. 1763.
Madaba, above Zarqa River, *Meyers & Dinsmore* G771 (E00291880); Tal Al-Rumman, *Al-Eisawi* (NHJ 2006.0320).

Phleum subulatum (Savi) Asch. & Graebn., Syn. Mitteleur. Fl.2(1): 154. 1899.
Syn. *Phalaris subulata* Savi (1798).
Between Salt and Rumman, *Meyers & Dinsmore* M1408 (E00398421); Tal Al-Rumman, *Taifour & Aldamen* (NHJ 2009.0076).

Phragmites australis (Cav.) Trin. ex Steud., Nomencl. Bot. ed. 2, 2: 324. 1841.
Jordan river, *Meyers & Dinsmore* 4288 (E00443993).

Piptatherum blancheanum É.Desv. ex Boiss., Diagn. Pl. Orient. ser. 2, 4: 127. 1859.
Irbid, Kufur Asad Forest (IBBU-H).

Piptatherum coerulescens (Desf.) P.Beauv., Ess. Agrostogr. 18. 1812.
Ref. Al-Eisawi 146 (2013).

Piptatherum holciforme (M.Bieb.) Roem. & Schult., Syst. Veg. 2: 328. 1817.
Syn. *Agrostis holciformis* M.Bieb. (1808).
Tal Al-Rumman, *Taifour & Aldamen* (NHJ 2009.0024).

Piptatherum miliaceum (L.) Coss., Notes Pl. Crit. 129. 1812.
Syn. *Oryzopsis miliacea* (L.) Benth. & Hook. ex Asch. & Schweinf. (1887).
Jarash, Jarash ruins, *Davis* 9685 (E00443908).

Poa annua L., Sp. Pl. 68. 1753.
Irbid, Ain Hammam, 08.v.1983, *Karim* (YAR-H).

Poa bulbosa L., Sp. Pl. 70 1753.
Syn. *Poa eigii* Feinbrun (1941).
Midway between Amman and Madaba, *Boulos, Al-Eisawi & Jallad* 6066 (E00367629); Tafila, Dana reserve, Al Hussein youth camp (MSB-K 176493); Tal Al-Rumman, *Al-Eisawi* (NHJ 2006.0318)

Poa compressa L., Sp. Pl.: 69. 1753.
Ref. Al-Eisawi 146 (2013).

Poa hackelii Post, Bull. Herb. Boissier 5: 760. 1897.
Amman, South of Ziza, *Dinsmore* 4571 (E00367525).

Poa infirma Kunth, Nov. Gen. Sp.1: 158. 1816.
Along Zarqa River, 2004, *Al-Dwiri* s.n. (AMM); Upper Jordan Valley (RSCN database).

Poa moabitica Bor, Notes Roy. Bot. Gard. Edinburgh 31: 396. 1972.
17 km E of Wadi Musa (HUJ).

Poa pratensis L., Sp. Pl.: 67. 1753.
40 km E of Amman, *Dinsmore* (E); Petra (YAR-H).

Poa sinaica Steud., Syn. Pl. Glumac.1: 256. 1854.
Azraq, Shishan [Azraq ash Shishan], *Townsend* 135 (K000315610).

Poa timoleontis Heldr. ex Boiss., Fl. Orient. 5: 607. 1884.
Irbid, Malka (YAR-H); Jarash (YAR-H).

Poa trivialis L., Sp. Pl.: 67. 1753.
Tal Al-Rumman, *RBG-Ire & RBG-Jo* (NHJ 2008.0107).

Polypogon fugax Nees ex Steud., Syn. Pl. Glumac. 1: 184. 1854.
Ref. Al-Eisawi 147 (2013).

Polypogon maritimus Willd. subsp. **maritimus**, Ges. Nat. Freunde Berlin Mag. Neue Schr. 3: 422 (1801).
Western bank area, near the N shore of the Dead Sea, just West of the mouth of the Jordan River, *Kramer* (E); Ajloun (YAR-H); Balqa Governorate, *Al-Sowaimeh, Abuhammour Al Mfadi & Al Khateeb* (NHJ).

Polypogon monspeliensis (L.) Desf., Fl. Atlant.1: 66. 1798.
Syn. Alopecurus monspeliensis L., (1753).
Madaba, Madaba-Karak road, Wadi Walleh, just beside the bridge over stream (MSB-K 276511); Tal Al-Rumman, *RBG-Ire & RBG-Jo* (NHJ 2008.0053).

Polypogon viridis (Gouan) Breistr., Bull. Soc. Bot. France110(Sess. Extraord.): lxxxix 1966.
Syn. *Agrostis viridis* Goun (1762).
Meyers & Dinsmore 6189 (E00369763).

Psilurus incurvus (Gouan) Schinz & Thell., Vierteljahrsschr. Naturf. Ges. Zürich 58: 40. 1913.
Above Zarqa river, *Meyers & Dinsmore* G1047 (E00407876); Tal Al-Rumman, *Al-Eisawi* (NHJ 2006.0154).

Puccinellia convoluta (Hornem.) Fourr., Ann. Soc. Linn. Lyon n.s., 17: 184. 1869.
Amman on way to Azraq (FOR-H).

Puccinellia distans (Jacq.) Parl., Fl. Ital.1: 367. 1850.
Syn. *Poa distans* Jacq. (1764).
Ref. Al-Eisawi 146 (2013).

Rhizocephalus orientalis Boiss., Diagn. Pl. Orient. 5: 69. 1844.
18 km N of the Safawi-Ruwaished rd., 20 km from Safawi (BRDC-H); Amman to Ziza, 15 Apr. 1929 (HUJ).

Rostraria berythea (Boiss. et Blanche) Holub., Folia Geobot. Phytotax. 9: 271. 1974.
Ain Hesban, *Meyers & Dinsmore* M779 (E00449639); Tal Al-Rumman, *Al-Eisawi* (NHJ 2006.0035).

Rostraria cristata (L.) Tzvelev, Novosti Sist. Vyssh. Rast. 7: 47. 1970 publ. 1971.
Syn. *Rostraria smyrnaea* (Trin.) H.Scholz (1989).
Jebel Attarus, Moab, *Davis* 8721 (E00449592); Tal Al-Rumman, *Al-Eisawi* (NHJ 2006.0141).

Rostraria obtusiflora (Boiss.) Holub, Folia Geobot. Phytotax.9: 27.1 1974.
Syn. *Koeleria obtusiflora* Boiss. (1846).
Majdal, 9 May 1927 (HUJ); RSCN database.

Rostraria pumila (Desf.) Tzvelev, Novosti Sist. Vyssh. Rast.7: 48. 1970 publ. 1971.
Syn. *Avena pumila* Desf. (1798).
Ref. Al-Eisawi 148 (2013); [http://e-monocot.org/taxon/urn:kew.org:wcs:taxon:439538].

Saccharum ravennae (L.) L., Syst. Veg. ed. 13, 2: 88. 1774.
Syn. *Andropogon ravennae* L. (1763).
Karak, Dead Sea area, Zarqa Ma'in, *Frey & Kürschner*.

Saccharum spontaneum L., Mant. Pl. 2: 183. 1771.
Wadi Rum (Qishawi, 1999: Wadi Rum Nature Reserve Flora Survey).

Schismus arabicus Nees, Fl. Afr. Austral. Ill.: 422. 1841.
Jordan, *Hepper* 3123 (K000714985); Tal Al-Rumman, *Al-Eisawi* (NHJ 2006.0113); Ajloun forest reserve, *Bocker* RB 900 (B 10 0520886).

Schismus barbatus (L.) Thell., Bull. Herb. Boissier, sér. 2, 7: 391. 1907.
Syn. *Festuca barbata* L. (1753).
Wadi Rum, *Gillett* 16057 (K000315599); nr. Aqaba, *Boulos & Lahham* 7384 (AMM, BM, E).

Schoenefeldia gracilis Kunth, Révis. Gramin. 1: 283. 1830.
(FOR-H).

Sclerochloa dura (L.) P.Beauv., Ess. Agrostogr.: 98. 1812.
Syn. *Cynosurus durus* L. (1753).
Al-Mafraq, Borqu' (IBBU-H); RSCN database.

Sclerochloa woronowii (Hack.) Tzvelev, Bot. Mater. Gerb. Bot. Inst. Komarova Akad. Nauk S.S.S.R. 18: 27. 1957.
Syn. *Scleropoa woronowii* Hack. (1912).
Wadi Rum, nr. rest house, 14 Apr. 1974, *Boulos* (AMM).

Scolochloa festucacea (Willd.) Link, Hort. Berol. 1: 136. 1827.
Syn. *Arundo festucacea* Willd. (1809); *Schedonorus arundinaceus* Roem. & Schult. (1817).
Aqaba (YAR-H). [http://flora.org.il/en/plants/SCHARU/].

Secale cereale L., Sp. Pl. 84. 1753.
Syn. *Secale strictum* C.Presl (1826).
(YAR-H). [http://e-monocot.org/taxon/urn:kew.org:wcs:taxon:441740].

Setaria parviflora (Poir.) M.Kerguelen, Lejeunia 120: 161. 1987.
(YAR-H); [http://e-monocot.org/taxon/urn:kew.org:wcs:taxon:442600].

Setaria pumila (Poir.) Roem. & Schult., Syst. Veg. 2: 891. 1817.
Ajloun, RSCN database.

Setaria verticillata (L.) P.Beauv., Ess. Agrostogr.: 51. 1812.
Syn. *Panicum verticillatum* L. (1762); *Setaria adhaerens* (Forssk.) Chiov. (1919).
Jebel Amman, Amman, in UNDP Garden, 3rd circle, *Boulos* 5821 (E00365198).

Setaria viridis (L.) P.Beauv., Ess. Agrostogr. 51. 1812.
As Salt, Wadi Shuaib (YAR-H).

Sorghum halepense (L.) Pers., Syn. Pl.1: 101. 1805.
Syn. *Holcus halepensis* L. (1753).
Aied Al-Shbeeb Nursery (Private) (IBBU-H); Ajloun, Rasoon (YAR-H).

Sorghum virgatum (Hack.) Stapf, Fl. Trop. Afr. 9: 111. 1917.
Aied Al-Shbeeb Nursery (Private) (IBBU-H); Irbid, Al Hamma (YAR-H).

Sphenopus divaricatus (Gouan) Rchb., Fl. Germ. Excurs.1: 45. 1830.
Syn. *Poa divaricata* Gouan (1773).
Azraq, *Frey & Kürschner* (E); [http://e-monocot.org/taxon/urn:kew.org:wcs:taxon:443897].

Spodiopogon pogonanthus (Boiss. & Balansa) Boiss., Fl. Orient. 5: 457. 1884.
Syn. *Andropogon pogonanthus* Boiss. & Balansa (1858).
Ref. Al-Eisawi 149 (2013).

Sporobolus indicus (L.) R.Br., Prodr. Fl. Nov. Holl.: 170. 1810.
Syn. *Agrostis indica* L. (1753).
(YAR-H); [http://e-monocot.org/taxon/urn:kew.org:wcs:taxon:444243]. Introduced.

Sporobolus pungens (Schreb.) Kunth, Révis. Gramin. 1: 68. 1829.
(YAR-H); [http://e-monocot.org/taxon/urn:kew.org:wcs:taxon:444418].

Stipa arabica Trin. & Rupr., Sp. Gram. Stipac.: 77. 1842.
(YAR-H). [http://e-monocot.org/taxon/urn:kew.org:wcs:taxon:444697]

Stipa barbata Desf., Fl. Atlant. 1: 97. 1798.
Amman, Site 1, Dabala site on the desert highway (MSB-K 558017).

Stipa bromoides (L.) Dörfl., Exsicc. (Herb. Norm.) 34: 129, no. 3386 1897.
Ajloun Forest Reserve, RSCN database.

Stipa ehrenbergiana Trin. & Rupr., Sp. Gram. Stipac. 75. 1842.
[http://e-monocot.org/taxon/urn:kew.org:wcs:taxon:444996]; Ref. Al-Eisawi 149 (2013).

Stipa hohenackeriana Trin. & Rupr., Sp. Gram. Stipac.: 80. 1842.
RSCN database: Qishawi, El-Moghrabi & Fagir, Shaumari Reserve Herbarium 2011; Hasa to Manzel. 7 May 1927, (HUJ).

Stipa holosericea Trin., Mém. Acad. Imp. Sci. St.-Pétersbourg, Sér. 6, Sci. Math. 1: 81. 1831.
Syn. *Stipa fontanesii* Parl (1850).
Dhiban to Wadi Themid (Moab), *Davis* 9178 (E00410457).

Stipa letourneuxii Trab., Bull. Soc. Bot. France 36: 405. (1889).
Syn. *Stipa pellita* (Trin. & Rupr.) Tzvelev (1966).
[(YAR-H); http://e-monocot.org/taxon/urn:kew.org:wcs:taxon:445297]

Stipa lagascae Roem. & Schult., Syst. Veg. 2: 333. 1817.
Ma'an, 10 km N of Petra, 22 Jul. 1980 (HUJ); Al-Mafraq, Bwadhah (IBBU-H).

Stipagrostis acutiflora de Winter, Kirkia 3: 133. 1963.
[http://e-monocot.org/taxon/urn:kew.org:wcs:taxon:445934].

Stipagrostis obtusa (Delile) Nees, Linnaea 7: 293. 1832.
Syn. *Aristida obtusa* Delile (1813).
Below Naqb Ishtar, 27 Mar. 1955, *Hunting Aero Survey* (E); Aqaba, Wadi Yutum (FOR-H).

Stipagrostis ciliata (Desf.) de Winter, Kirkia 3: 133. 1963.
Aqaba, Ghor, Ram NE of Aqaba, *Hunting Aero Survey* 25A (E00402033).

Stipagrostis drarii (Täckh.) de Winter, Kirkia3: 134. 1963
Syn. *Aristida drarii* Täckh. (1942).
Aqaba, Wadi Araba, 3 km SW of Gharandal, *Baierle & Pimpl* 87–701 (E00402042).

Stipagrostis hirtigluma (Steud. ex Trin. & Rupr.) de Winter, Kirkia 3: 134. 1963.
Syn. *Aristida ciliata* Steud. & Hochst. (1840).
Karak, Wadi Araba, Feinan, Khirbet en-Nahas, *Frey & Kürschner*.

Stipagrostis lanata (Forssk.) de Winter, Kirkia3: 135. 1963.
Syn. *Aristida lanata* Forssk. (1775).
RSCN database; (YAR-H); [http://e-monocot.org/taxon/urn:kew.org:wcs:taxon:445969].

Stipagrostis multinerva H.Scholz, Osterr.Bot.Z. 117: 289. 1969.
Ref. Al-Eisawi 149 (2013). Not recorded fro Jordan but possibly found there [http://e-monocot.org/taxon/urn:kew.org:wcs:taxon:445976].

Stipagrostis obtusa (Delile) Nees, Linnaea 7: 293. 1832.
Syn. *Aristida obtusa* Delile (1813).
Aqaba, Ghor Ram, NE of Aqaba, *Hunting Aero Survey* 24B (E00440021).

Stipagrostis plumosa (L.) Munro ex T.Anderson, J. Proc. Linn. Soc., Bot.5 (Suppl. 1): 40. 1860.
Syn. *Aristida plumosa* L. (1763); *Aristida brachypoda* Tausch (1836).
Aqaba, Aqaba-Sudi border highway (MSB-K 412661).

Stipagrostis pungens (Desf.) De Winter, Kirkia 3: 135. 1963.
Zarqa, Azraq, *Dinsmore* 1806 (E00440062).

Stipagrostis raddiana (Savi) de Winter, Kirkia3: 135. 1963.
Syn. *Aristida raddiana* Savi (1837).
Jebel Uweinid [Qasr Uweinid], *Townsend* 174 (K000315604).

Stipagrostis scoparia (Trin. & Rupr.) De Winter, Kirkia 3: 136. 1963.
(YAR-H); [http://e-monocot.org/taxon/urn:kew.org:wcs:taxon:446009].

Stipagrostis uniplumis (Licht.) de Winter, Kirkia3: 136. 1963.
Syn. *Aristida uniplumis* Licht. (1817).
Ref. Al-Eisawi 149 (2013); [http://e-monocot.org/taxon/urn:kew.org:wcs:taxon:446016].

Stipellula capensis (Thunb.) Röser & Hamasha, Schlechtendalia 24: 92. 2012.
Syn. *Stipa capensis* Thunb. (1794); *Stipella capensis* (Thunb.) Röser & Hamasha (2012); *Stipa retorta* Cav. (1795); *Stipa tortilis* Desf. (1798).
Madaba, Dhiban to Wadi Themed, *Davis* 9614 (E00410314); Tal Al-Rumman, *Al-Eisawi* (NHJ 2006.0016).

Stipellula parviflora (Desf.) Röser & Hamasha, Schlechtendalia 24: 92. 2012.
Syn. *Stipa parviflora* Desf. (1798); *Stipella parviflora* (Desf.) Röser & Hamasha (2012).
Above Zarqa river, *Meyers & Dinsmore* G1731 (E00410490); Tal Al-Rumman RBG, Sector No. 80, *Taifour & Al Khawaldah* (NHJ 2013.0323).

Taeniatherum caput-medusae (L.) Nevski, Trudy Bot. Inst. Akad. Nauk S.S.S.R., Ser. 1, Fl. Sist. Vyssh. Rast.1: 22. 1933.
Syn. *Elymus caput-medusae* L. (1753); *Taeniatherum crinitum* (Schreb.) Nevski (1933).
Shoubak, Hisha, King property enclosed area (MSB-K 497561); Tal Al-Rumman RBG, Sector No. 20, *El-Oqlah, Taifour & Al Khawaldah* (NHJ 2013.0230).

Tetrapogon villosus Desf., Fl. Atlant.2: 388. 1799.
Tafila, S of Naqb Dahal, *Hunting Aero Survey* 199 (E00357585).

Tricholaena teneriffae (L.f.) Link, Handbuch1: 91. 1829.
Syn. *Saccharum teneriffae* L.f., Suppl. Pl.: (1782); *Panicum teneiffae* IL.f.) Spreng. (1824).
Wadi Tutm, *Hunting Aero Survey* 176 (E00365254).

Trisetaria glumacea (Boiss.) Maire, Bull. Soc. Hist. Nat. Afrique N. 33: 93. 1942.
Syn. *Trisetum glumaceum* Boiss. (1854).
Balqa Governorate, Salt, on the way to Wadi Shuaib, *Abuhammour, Al Mfadi* & *Al Khateeb* (NHJ).

Trisetaria koelerioides (Bornm. & Hack.) Melderis, Ark. Bot. n.s., 2: 292. 1952.
Syn. *Trisetum koelerioides* Bornm. & Hack. (1898).
Ref. RSCN database; Al-Eisawi 150 (2013).

Trisetaria linearis Forssk., Fl. Aegypt.-Arab.: 27. 1775.
Petra, 17 Apr. 1929 (HUJ); Dead Sea area (RSCN database).

Trisetaria macrochaeta (Boiss.) Maire, Bull. Soc. Hist. Nat. Afrique N. 33: 92. 1942.
Syn. *Trisetum macrochaetum* Boiss. (1854).
N Jordan, 2005, *Gharaibeh* (YAR-H).

Triticum boeoticum Boiss., Diagn. Pl. Orient. 13: 69. 1854.
Tafila, Tafileh-Qadeseyeh road, just opposite the cement factory (MSB-K 277851).

Triticum dicoccoides (Asch. & Graebn.) Schweinf., Ber. Deutsch. Bot. Ges. 26a (4): 309, 31. 1908.
Syn. *Triticum dicoccum* Schübl. var. *dicoccoides* Asch. & Graebn. (1901).
Irbid, Zoubia, Sheab Qatum, by roadside, 04.iv.2015, *El-Oqlah, Lahham & Khateeb* (YAR-H).

Triticum turgidum L., Sp. Pl.: 86. 1753.
Ajloun, Ajloun Reserve of RSCN (MSB-K 557984).

Triticum urartu Thumanjan ex Gandilyan, Bot. Zhurn. (Moscow & Leningrad) 57: 176. 1972.
Ref. Ma'an, Danin (2004).

Vulpia ciliata Dumort., Observ. Gramin. Belg.: 100. 1824.
Above 'Ain Fit, *Meyers & Dinsmore* 1048 (E00407486).

Vulpia fasciculata (Forssk.) Samp., Herb. Port.: 24 (1913).
Tal Al-Rumman, *Al-Eisawi* (NHJ 2006.0253).

Vulpia muralis (Kunth) Nees, Linnaea19: 694. 1847.
Syn. *Festuca muralis* Kunth (1822).
Ref. Al-Eisawi 150 (2013).

Vulpia myuros (L.) C.C.Gmel., Fl. Bad.1: 8. 1805.
Syn. *Festuca myuros* L. (1753).
Between Hesban and Wadi Sir, *Meyers & Dinsmore* M1048 (E00407509).

Vulpia persica (Boiss. & Buhse) V.I.Krecz. & Bobrov in V.L.Komarov (ed.), Fl. URSS 2: 535. 1934.
Ajloun, Eshtafenah (YAR-H).

Vulpia unilateralis (L.) Stace, Bot. J. Linn. Soc.76: 350. 1978.
Syn. *Triticum unilaterale* L. (1767); *Nardurus maritimus* (L.) Murb. (1900).
Bei Shoubak, *Frey & Kürschner* VO 5809 (E00407558).

Zingeria trichopoda (Boiss.) P.Smirn., Byull. Moskovsk. Obshch. Isp. Prir., Otd. Biol. n.s., 51(2): 67. 1946.
Syn. *Milium trichopodium* Boiss. (1854).
Ref. Al-Eisawi 151 (2013).

Polygalaceae

Polygala edumea Zohary, Palestine J. Bot. Ser. 2: 166. 1941.
Ma'an, Petra at foot of Mt. Hot, *Davis* 8644 (E00405099).

Polygala hohenackeriana Fisch. & C.A.Mey., Index Seminum Petropolitanus 4. 42. 1838.
El Heshma, 22 km N or Queira, 30.iii.1936 (29659-HUJ).

Polygala monspeliaca L., Sp. Pl. 2: 702. 1753.
Amman, opposite Jordan University (MSB-K 276223).

Polygala negevensis Danin, Israel J. Bot. 36(2): 67. 1987
Araba valley, Rum (http://flora.org.il/en/plants/POLNEG/).

Polygala sinaica Botsch., Novosti Sist. Vyssh. Rast. 368. 1964.
Ma'an, 20 km NE Queirah, *Frey & Kürschner* VO 5060 (E00406035).

Polygala sinaica Botsch. var. **glabrescens** (Zohary) Boulos, Fl. Egypt 2: 71. 2000.
Syn. *Polygala spinescens* Decne. var. *glabrescens* Zohary (1939).
Tafila, Dana Reserve, Rumana trail behind Rumana camp, BOT ERA Project (NHJ 2013.0454).

Polygonaceae

Atraphaxis billardieri Jaub. & Spach, Ill. Pl. Orient. 2: 14. 1844.
Ref. RSCN database; Al-Eisawi 151 (2013).

Atraphaxis spinosa L., Sp. Pl. 745. 1753.
Jordan, Ras en Naqb, *Boulos* 8510 (K000283175).

Calligonum comosum L'Hér., Trans. Linn. Soc. London 1: 180. 1791.
Syn. *Calligonum polygonoides* L. subsp. *Comosum* (L'Hér.) Soskov (1975).
Aqaba, Al-Diseh village (MSB-K 338226); Jordan, *Medp* 224 (K000286357); Wadi Araba, *Taifour* (NHJ).

Calligonum tetrapterum (Meisn. ex DC.) Jaub. & Spach, Ill. Pl. Orient. 5: 80. 1855.
Syn. *Pterococcus tetrapterus* Meisn., Prodr. [A. P. de Candolle] 14(1): 29. 1856
Ref. Al-Eisawi 151 (2013).

Emex spinosa (L.) Campd., Monogr. Rumex: 58. 1819.
Syn. *Rumex spinosus* L. (1753).
Petra, *Dinsmore* (E00475175); Tal Al-Rumman, *Al-Eisawi* (NHJ 2006.0406).

Fallopia convolvulus (L.) Á.Löve, Taxon 19: 200. 1970.
Karak, Lisan area, Danin 2014.

Persicaria acuminata (Kunth) M.Gómez, Anales Inst. Segunda Enseñ. 2: 278. 1896.
Syn. *Polygonum acuminatum* Kunth (1817).
Ajloun, Soos Rasoun (YAR-H).

Persicaria decipiens (R.Br.) K.L.Wilson, Telopea 3: 178. 1988.
Syn. *Polygonum salicifolium* Brouss. ex Willd. (1809).
Upper and Lower Jordan Valley (RSCN database).

Persicaria lanigera (R.Br.) & Soják, Preslia 46(2): 153. 1974.
Syn. *Polygonum lanigerum* R.Br. (1810).
Irbid, Ain Trab (YAR-H).

Persicaria lapathifolia (L.) Delarbre, Fl. Auvergne (Delarbre) ed. 2: 519. 1800.
Syn. *Polygonum lapathifolium* L. (1753).
Hulan, *Dinsmore* 12230 (E00472457).

Polygonum arygrocoleon Steud. ex Kunze, Linnaea (20) 1: 17 (1847).
Ref. Ras Naqab, Danin 2005; RSCN database.

Polygonum arenastrum Boreau, Fl. Centre France, ed. 3 [Boreau] 2: 559. 1857.
Amman, Abdali, *Boulos* 7802 (E00470866).

Polygonum aviculare L., Sp. Pl.: 362. 1753.
Amman, *Davis* (E00470879).

Polygonum bellardi All., Fl. Pedem. 2: 205. 1785.
Ref. Ajloun, Danin (2005); RSCN database.

Polygonum equisetiforme Sibth. & Sm., Fl. Graec. Prodr. 1(2): 266. 1809.
Amman, from the desert way to Thibham-Umm Al Rasas (MSB-K 186003).

Polygonum palaestinum Zohary, Fl. Palaest. 1: 54. 1966.
Tafila, Dana village, beside Dana Nature Reserve (MSB-K 200213); Tal Al-Rumman, *Taifour & Aldamen* (NHJ 2009.0077).

Polygonum patulum M.Bieb., Fl. Taur.-Caucas. 1: 304. 1808.
Tal Al-Rumman RBG, Sector No. 3, *El-Oqlah, Taifour & Al Khawaldah* (NHJ 2013.0215).

Rheum palaestinum Feinbrun, Palestine J. Bot., Jerusalem Ser. 3: 117. 1944.
Aqaba, west foot of Gebel Aneizi, *Davis* 8850 (E00473015).

Rheum ribes L., Sp. Pl.: 372. 1753.
About 20 km SW of Ma'an, *Dinsmore* 11746 (E00473006).

Rumex bucephalophorus L., Sp. Pl.: 336. 1753.
Balqa, Site 10. (MSB-K 554271); Tal Al-Rumman, *Al-Eisawi, Abu Sondos, Abu Anza & Shishani* (NHJ 2006.0060).

Rumex conglomeratus Murray, Prodr. Stirp. Gott.: 52. 1770.
Dibeen National Park, 2005, *Mahklouf* s.n. (AMM); Yarmuk River (YAR-H).

Rumex crispus L., Sp. Pl.: 335. 1753.
Al Shomary Wildlife Reserve (MSB-K 176769).

Rumex cyprius Murb., Acta Univ. Lund. n.s., ii. No. 14, 20. 1907.
Ajloun, from Ajloun towards Jordan Valley, site no 15 (MSB-K 496483); Tal Al-Rumman, *Al-Eisawi* (NHJ 2006.0188).

Rumex dentatus L., Mant. Pl. Altera 226. 1771.
Salt, Bellauna village (MSB-K 114378).

Rumex maritimus L., Sp. Pl.: 335. 1753.
Ajloun, Galaat El Rabadh (YAR-H).

Rumex nepalensis Spreng., Syst. Veg. (ed. 16) [Sprengel] 2: 159. 1825.
Ref. Al-Eisawi 152 (2013).

Rumex pictus Forssk., Fl. Aegypt.-Arab.: 76. 1775.
Jarash, Wadi Zarqa, 04.iv.1954, *Kaspligil* (FOR-H); Amman & Aqaba, Danin (2004).

Rumex pulcher L., Sp. Pl.: 336. 1753.
Above Zarqa River, *Meyers & Dinsmore* G346 (E00473282); Tal Al-Rumman, *Al-Eisawi* (NHJ 2006.0161).

Rumex roseus L., Sp. Pl.: 337. 1753.
Tabaget Fahel, 21.IV.1967, *Edgecombe* 21 (AUB); Ref. Moab, Danin (2004).

Rumex vesicarius L., Sp. Pl.: 336. 1753.
Zarqa, from Baireen Cross toward Al Kellah village (MSB-K 199670); Tal Al-Rumman, *Taifour & Aldamen* (NHJ 2009.0169).

Portulacaceae

Portulaca oleracea L., Sp. Pl.: 445. 1753.
Al-Mafraq, Al-Ba'ej (IBBU-H); Ajloun, Amman, Aqaba and Moab, *Danin* (2004).

Potamogetonaceae

Potamogeton compressus L., Sp. Pl. 127. 1753.
Jerash, 10 km N of Jerash to Irbid at 27 km from Irbid, 8 May 1996 (NCARE).

Potamogeton crispus L., Sp. Pl.: 126. 1753.
Ref. RSCN database; Al-Eisawi 152 (2013).

Potamogeton natans L., Sp. Pl.: 126. 1753.
Ref. RSCN database; Al-Eisawi 152 (2013).

Potamogeton pectinatus L., Sp. Pl.: 127. 1753.
Upper Jordan Valley (RSCN database).

Potamogeton perfoliatus L., Sp. Pl.: 126. 1753.
Burqu, 30 km North West of Ruweishid, *Woods* (E).

Zannichellia palustris L., Sp. Pl. 2: 969. 1753.
Azraq, *Frey & Kürschner* (BSB).

Primulaceae

Anagallis arvensis L. subsp. **foemina** (Mill.) Schinz & Thell., Bull. Herb. Boissier, ser. 2, 7: 497. 1907.
Syn. *Anagallis foemina* Mill. (1768).
Tal Al-Rumman, *Al-Eisawi* (NHJ 2006.0021).

Androsace maxima L., Sp. Pl.: 141. 1753.
Tafila, Dana reserve, 1 km down the village steppe road (MSB-K 176518); Tal Al-Rumman, *Taifour & Aldamen* (NHJ 2009.0177).

Cyclamen coum Mill., Gard. Dict. ed. 8, 6. 1768.
Salt, Wadi Shuaib, 17xii.1954, *Kasapligil* (FOR-H).

Cyclamen persicum Mill., Gard. Dict., ed. 8. n. 5, 1768.
Ajloun, *Al Mfadi et al.* (NHJ 2013.0025).

Lysimachia dubia Aiton, Hort. Kew. 1: 199. 1789.
Ref. Ajloun, Danin (2005); RSCN database.

Lysimachia linum-stellatum L., Sp. Pl.: 148. 1753.
Syn. *Asterolinon linum-stellatum* (L.) Duby (1844).
Tal Al-Rumman, *Taifour & Aldamen* (NHJ 2009.0178).

Samolus valerandi L., Sp. Pl.: 171. 1753.
Tafila Province, 2014, *Oran* (AMM); Wadi Rum, 17 Mar. 1978, *Al-Eisawi* (AMM).

Ranunculaceae

Adonis aestivalis L., Sp. Pl., ed. 2. 1: 771. 1762.
Irbid, c. 9 km NW Irbid, *Boulos, Jallad & Lahham* 8312 (E00440818); Tal Al-Rumman, *Taifour & Aldamen* (NHJ 2009.0020).

Adonis aleppica Boiss., Ann. Sci. Nat., Bot. sér. 2, 16: 350. 1841.
Jordan University of Science & Technology, 2005, *Al-Gharaibeh* s.n. (in citation only).

Adonis annua L., Sp. Pl.: 547. 1753.
Salt, iv.1886 (AUB); Dibeen Forest Reserve, May 2006, *Taifour & Othman* (DFR-H).

Adonis annua subsp. **cupaniana**(Guss.) C.Steinb., Webbia 25: 324. 1971.
Syn. *Adonis cupaniana* Guss. (1844).
Irbid, Wadi Al Arab, 14 Mar. 1983 (YAR-H).

Adonis dentata Delile, Descr. Egypte, Hist. Nat. 287. 1813.
Amman-Baghdad, 60 km W of H5, *Lamond* 44 (E0044088).

Adonis flammea Jacq., Fl. Austriac. 4: 29. 1776.
Tal Al-Rumman, *Al-Eisawi, Abu-Sondos, Abu Anza & Shishani* (NHJ 2006.0093).

Adonis microcarpa DC., Syst. Nat. [Candolle] 1: 222. 1817.
Between Wadi Musa village and The Siq (Petra), *Davis* 9482 (E00440924); Tal Al-Rumman, *Shishani & Alkoran* (NHJ 2011.0101).

Adonis palaestina Boiss., Diagn. Pl. Orient. ser. 1, 8: 1. 1849.
Jordan, *Davis* 9554 (E00051099); Dibeen Forest Reserve, May 2006, *Taifour & Othman* (DFR-H).

Anemone coronaria L., Sp. Pl.: 539. 1753.
Salt, Salt to Amman highway, on the way to Zai village (MSB-K 275101); Ajloun Ishtafina, *Al Mfadi, Al Khawaldah, Bani Said & Alshwaer* (NHJ 2013.0019).

Ceratocephala falcata (L.) Pers., Syn. Pl. 1: 341. 1805.
Al Mafraq, 20 km SE Mafraq, *Boulos, Jallad & Lahham* 6704 (E00444414).

Clematis cirrhosa L., Sp. Pl.: 544. 1753.
Ajloun, between Ajloun and Mar Elias, *Davis* 9438 (E00440319).

Clematis flammula L., Sp. Pl.: 544. 1753.
Ajloun, Al Sakheneh (YAR-H).

Consolida coelesyriaca Mouterde, Nouv. Fl. Liban 2: 23. 1970.
Ref. Aqaba, Danin (2004).

Consolida deserti-syriaci (Zohary) Munz, J. Arnold Arbor. 48. 187. 1967.
Syn. *Delphinium deserti-syriaci* Zohary (1941).
Ref. Al-Eisawi 154 (2013).

Consolida flava (DC.) Schrödinger, Ann. Naturhist. Mus. Wien 27(1): 43. 1913.
Syn. *Delphinium flavum* DC. (1818 publ. 1–15 Nov 1817]; *Delphinium deserti* Boiss. (1867).
Ma'an, Wadi Rum, *Frey & Kürschner* VO5580 (E00402660).

Consolida incana (E.D.Clarke) Munz, J. Arnold Arbor. xlviii. 181. 1967.
Syn. *Delphinium incanum* E.D.Clarke, Trav. Var. Eur. 2: 451. 1812; *Consolida rigida* (DC.) Bornm. (1914).
Jebel Siaghah, Madaba, 5.iv.1877 (AUB).

Consolida orientalis (J.Gay) Schrödinger, Zool.-Bot. Ges. Wien 4(5): 62. 1909.
Syn. *Consolida hispanica* (Willk. ex Costa) Greuter & Burdet (1989); *Delphinium hispanicum* Willk. ex Costa (1873).
Ajloun Forest Reserve, RSCN database; Aqaba, *Danin* (http://flora.org.il/en/plants/CONHIS/).

Consolida sclerocIada (Boiss.) Schrödinger, Ann. K. K. Naturhist. Hofmus. 27: 44. 1913.
Syn. *Delphinium sclerocladum* Boiss. (1849).
Shoubak, Hisha, King property enclosed area, site no 22 (MSB-K 497723); Tal Al-Rumman, *Al-Eisawi* (NHJ 2006.0429).

Consolida tomentosa (Aucher ex Boiss.) Schrödinger, Abh. K. K. Zool.-Bot. Ges. Wien 4(5): 9, 62. 1909.
Syn. *Delphinium tomentosum* Aucher ex Boiss. (1841).
Ref. Al-Eisawi 154 (2013).

Delphinium ithaburense Boiss., Diagn. Pl. Orient. ser. 1, 8: 9. 1849.
Tal Al-Rumman, *Al-Eisawi* (NHJ 2006.0430).

Delphinium peregrinum L., Sp. Pl.: 531. 1753.
Balqa, Tall Arrumman conserve (MSB-K 341343); Tal Al-Rumman, *Al-Eisawi* (NHJ 2006.0431).

Ficaria verna Huds., Fl. Angl. (Hudson) 214. 1762.
Syn. *Ranunculus ficaria* L. (1753).
Wadi Orjan, Royal Botanic Garden Team (NHJ 2011.0039); between Naur and Wadi Lir, 28 Apr. 1911, *Meyers & Dinsmore* (E).

Myosurus minimus L., Sp. Pl.: 284. 1753.
Ref. Al-Eisawi 154 (2013).

Nigella arvensis L., Sp. Pl. 2: 584. 1753.
Wadi Yutum, *Hunting Aero Survey* 174A (E00445127); Tal Al-Rumman RBG, Sector No. 55, *Taifour & Al Khawaldah* (NHJ 2013.0309).

Nigella ciliaris DC., Syst. Nat. [Candolle] 1: 327. 1817.
Tal Al-Rumman, *Al-Eisawi* (NHJ 2006.0168).

Nigella deserti Boiss., Ann. Sci. Nat., Bot. sér. 2, 16: 359. 1841.
Ref. Al-Eisawi 155 (2013).

Nigella latisecta P.H.Davis, Notes Roy. Bot. Gard. Edinburgh 26: 168. 1965.
Ref. Al-Eisawi 155 (2013).

Nigella orientalis L., Sp. Pl. 2: 584. 1753.
Ref. Al-Eisawi 154 (2013).

Nigella unguicularis (Poir.) Spenn., Monogr. Nigell. 12. 1829.
Syn. *Garidella unguicularis* Poir. (1812).
Amman, on the way to the dead sea, just before Na'our bridge junction, a hill on the right (MSB-K 341354); Dibeen Forest, *RBG-Ire & RBG-Jo* (NHJ 2008.0056).

Ranunculus aquatilis L., Sp. Pl.: 556. 1753.
Irbid, Al Brahemiyah (YAR-H).

Ranunculus arvensis L., Sp. Pl.: 555. 1753.
Jebel Siaghah, Madaba, 5.iv.1844 (AUB).

Ranunculus asiaticus L., Sp. Pl.: 552. 1753.
Aqaba, 4 km south of Rum Rest House, along the road to Rum Police Station, *Boulos, Jallad, Lahham & Abu Hmaidan* 7687 (E00442359);Tal Al-Rumman, *Al-Eisawi* (NHJ 2006.0211).

Ranunculus chius DC., Syst. Nat. [Candolle] 1: 299. 1817.
E of Suf, *Meyers & Dinsmore* (E).

Ranunculus cornutus DC., Syst. Nat. [Candolle] 1: 300. 1817.
Ajloun, Um Elyanabia (YAR-H).

Ranunculus damascenus Boiss. & Gaill., Diagn. Pl. Orient. ser. 2, 6: 5. 1859.
Jordan, *Boulos & Al-Eisawi* 6436 (E00018319).

Ranunculus macrorrhynchus Boiss., Diagn. Pl. Orient. ser. 1, 6: 5. 1846.
8 km E of Wadi Musa, 29.3.1936 (29679-HUJ).

Ranunculus marginatus d'Urv. Mém. Soc. Linn. Paris 1: 318. 1822.
Syn. *Ranunculus trachycarpus* Fisch. & C.A.Mey. (1837).
Jordan, *Rath* s.n. (K000692890); Salt, Damiya, 25 Mar. 1955 (FOR-H).

Ranunculus millefolius Banks & Sol., Nat. Hist. Aleppo ed. 2, 2: 254. 1794.
Above Ein Musa (Aqaba), *Davis* 8875 (E00375166);Tal Al-Rumman; *Al-Eisawi, Abu-Sondos, Abu Anza & Shishani* (NHJ 2006.0057).

Ranunculus muricatus L., Sp. Pl.: 555. 1753.
Castle of Shoubak, *Meyers & Dinsmore* (E).

Ranunculus neocuneatus C.C.Townsend, Taxon 16: 467. 1967.
Ref. RBG Jordan checklist (unpubl.).

Ranunculus oxyspermus Willd., Sp. Pl. 2: 1328. 1799.
Al-Tafila, Rashadiye (FOR-H).

Ranunculus paludosus Poir., Voy. Barbarie ii. 184. 1789.
Al-Mafraq, Bal'ama, Al-Jibal Al-Hamra (IBBU-H); Zeglab dam (YAR-H).

Ranunculus peltatus Schrank,Baier. Fl. 2: 103. 1789.
Maneura, *Meyers & Dinsmore* (E); Ajloun, Amman & Karak (Moab), Danin (2004).

Ranunculus scandicinus (Boiss.) P.H.Davis, Ark. Bot. n.s., 5: 143. 1960.
Jordan, Hesban, *Meyers & Dinsmore* M335 (E00444092).

Ranunculus sceleratus L., Sp. Pl.: 551. 1753.
Irbid, *Meyers, Fred & Dinsmore* 1803 (E00442980).

Ranunculus sphaerospermus Boiss. & Blanche, Diagn. Pl. Orient. II, 5: 6. 1856.
Between Lyera and Ain Aishe, *Meyers & Dinsmore* (E).

Thalictrum isopyroides C.A.Mey., Fl. Altaic. 2: 346. 1830.
Jordan, *Boulos & Jallad* 7803 (E00059156).

Resedaceae

Caylusea hexagyna (Forssk.) M.L.Green, Stand.-Sp. Nom. Conserv. 29, 63. 1926.
Syn. *Reseda hexagyna* Forssk. (1775).
Petra, *Boulos, Al-Eisawi & Jallad* 6307 (E00448682); Dana, *Taifour* (NHJ 2013.0406).

Ochradenus baccatus Delile, Descr. Egypte, Hist. Nat. 63. 1813.
Al Karak, Ghor Safi, southern tip of Dead Sea, *Boulos* 5816 (E00443401); Wadi Hassa, Borbeta hot spring, BOT ERA project (NHJ 2013.0498).

Oligomeris linifolius (Vahl ex Hornem.) J.F.Macbr., Contr. Gray Herb. 53: 13. 1918.
Road to Mudawwara, *Hunting Aero Survey* 41B (E00443377).

Oligomeris subulata Webb, Fragm. Fl. Aethiop. 26. 1854.
Al Karak, Env. of Dead Sea, 06-Apr–25, *Eig* (E00443382).

Reseda alba L., Sp. Pl.: 449. 1753.
Jordan, *Meyers & Dinsmore* 3431B (E00147995); Tal Al-Rumman, *Taifour & Aldamen* (NHJ 2010 .0004).

Reseda alopecuros Boiss., Diagn. Pl. Orient. ser. 1, 8: 55. 1849.
Along Zarqa River, 2004, *Al-Dwiri* s.n. (AMM); Irbid, Abu Logus (YAR-H).

Reseda arabica Boiss., Diagn. Pl. Orient. ser. 1, 1: 6. 1843.
Karak, Wadi Araba, 5 km E of Ain el-Fidan, *Frey & Kürschner* (BSB); Dead Sea area (YAR-H).

Reseda decursiva Forssk., Fl. Aegypt.-Arab.: 67. 1775.
Amman, Dead Sea area (MSB-K 275134); Al Mafraq, Um Alsarab (YAR-H).

Reseda globulosa Fisch. & C.A.Mey., Index Seminum (LE) 4: 45. 1838.
Al Zarqa, SW of Zarqa (FOR-H).

Reseda kahirina Boiss., Fl. Orient. 1: 430. 1867.
Burqu, Wadi Bottom, 24 Feb. 1955 (FOR-H).

Reseda lutea L., Sp. Pl.: 449. 1753.
Midway between Amman and Madaba, *Boulos, Al-Eisawi & Jallad* 6068 (E00448771); Tal Al-Rumman, *Al-Eisawi* (NHJ 2006.0294).

Reseda luteola L., Sp. Pl.: 448. 1753.
Irbid-Boshra, 6.V.1886 (AUB).

Reseda muricata C.Presl, Abh. Königl. Böhm. Ges. Wiss. V, 3: 438. 1845.
Jordan, *Jallad, Lahham & Abu Hmaidan* 7658 (E00151083); Tal Al-Rumman RBG, Sector No. 80, *Taifour &Al Khawaldah* (NHJ 2013.0324).

Reseda pruinosa Delile, Descr. Egypte, Hist. Nat. 63. 1813.
Wadi Lleie, *Borzatti & Barsotti* (AMM).

Reseda stenostachya Boiss., Diagn. Pl. Orient. ser. 1, 1: 5. 1843.
Aqaba, Wadi Yutum (RSCN database).

Reseda urnigera Webb, Fragm. Fl. Aethiop.-Aegypt. 25. 1854.
Syn. *Reseda boissieri* Müll.Arg. (1856).
Ref. Danin (2004); Lower Jordan valley (RSCN database).

Rhamnaceae

Rhamnus alaternus L., Sp. Pl.: 193. 1753.
Castle of Lubebeh, *Meyers & Dinsmore* (E).

Rhamnus disperma Ehrenb. ex Boiss., Fl. Orient. Suppl. 156. 1888.
Petra in the Sik, *Davis* 8756 (E00408088); Tafila, Dana spring near Dana village BOT ERA project (NHJ 2013.0472).

Rhamnus lycioides L., Sp. Pl., ed. 2. 1: 279. 1762.
Syn. *Rhamnus palaestina* Boiss. (1854).
Ma'an, 20 km NE Quweirah, *Frey & Kürschner* 80–659 (E00408093); Tal Al-Rumman, *Taifour & Aldamen* (NHJ 2009.0017).

Rhamnus punctata Boiss., Diagn. Pl. Orient. ser. 1, 2: 4. 1843.
Tafila Province, 2014, *Oran* (AMM).

Sageretia thea (Osbeck) M.C.Johnst., J. Arnold Arbor. 49: 378. 1968.
Syn. *Rhamnus thea* Osbeck (1757); *Sageretia brandrethiana* Aitch. (1865).
Jebel Rehebi, Wadi Rum, 27 Mar. 1997, *Borzatti & Barsotti* (AMM).

Ziziphus lotus (L.) Lam., Encycl. 3(1): 317. 1789.
Syn. *Rhamnus lotus* L. (1753).
Irbid, Al-Hemma, near River Yarmouk, *Boulos & Hauke* 5826 (E00405797); Tal Al-Rumman RBG, Sector No. 28, *El-Oqlah & Al Mfadi* (NHJ 2013.0374).

Ziziphus nummularia (Burm.f.) Wight & Arn., Cat. Ind. Pl. 31. 1833.
Al Ghour, Deir Alla (YAR-H).

Ziziphus spina-christi (L.) Desf., Fl. Atlant. 1: 201. 1798.
Syn. *Rhamnus spina-christi* L. (1753).
Balqa, Wadi Al Seer (MSB-K 176459); Wadi Hassa, Borbeta hot spring, BOT ERA project (NHJ 2013.0500).

Rosaceae

Cotoneaster nummularia Fisch. & C.A.Mey., Index Seminum Hort. Petrop. 2: 34. 1836.
Aqaba, near 'Ain el-Arja, *Baierle* 87–870 (E00408244).

Crataegus azarolus L., Sp. Pl.: 477. 1753.
Between Es-Salt and Rumman, *Meyers & Dinsmore* M673 (E00408889); Tal Al-Rumman, *Taifour & Aldamen* (NHJ 2009.0093)

Crataegus azarolus L. var. **aronia** L., Sp. Pl.: 477 .1753.
Syn. *Crataegus aronia* (L.) Bosc ex DC. (1825).
Tal Al-Rumman, *Taifour & Aldamen* (NHJ 2010.0049).

Crataegus monogyna Jacq., Fl. Austriac. (Jacquin) 3: 50, t. 292, f. 1, 1775.
Ref. Al-Eisawi 158 (2013).

Crataegus orientalis Pall. ex M.Bieb., Fl. Taur.-Caucas. 1: 387. 1808.
Ref. Al-Eisawi 158 (2013).

Neurada procumbens L., Sp. Pl.: 441. 1753.
Syn. *Neurada al-eisawii* Barsotti, Borzatti & Garbari (2000).
Aqaba, Quweira camp, Hunting Aero Survey(E00431608); nr. Aqaba, *Boulos & Lahham* 7292 (AMM, BM, E).

Potentilla geranioides Willd., Sp. Pl., ed. 4 2(2): 1101. 1799.
Road to Jerash, Wadi Al Zarqa, 15 Mar. 1954 (FOR-H).

Potentilla reptans L., Sp. Pl.: 499. 1753.
Ref. Gharaibeh (2005); (YAR-H).

Prunus amygdalus Batsch, Beytr. Entw. Gewächsreich 1: 30. 1801.
Syn. *Amygdalus communis* L. (1753).
Tal Al-Rumman, *RBG-Ire & RBG-Jo* (NHJ 2008.0061).

Prunus arabica (Olivier) Meikle, Kew Bull. 19(2): 229. 1965.
Syn. *Amygdalus arabica* Oliv. (1807).
Az Zarqa, Qasr Amra, *Hunting Aero Survey* 114 (E00406272); Amra, Wadi Bottom, *Taifour, Al Mfadi & Said* (NHJ 2013.0049).

Prunus korshinskyi Hand.-Mazz., Ann. Nat. Hofmus. Wien 27. 71. 1913.
Syn. *Amygdalus korshinskyi* (Hand.-Mazz.) Bornm. (1914).
10 km S Shoubak, *Frey & Kürschner* VO 5732 (E00406173); Shoubak, *Ananbeh, Al Khawaldah & Alshwaer* (NHJ 2013.0633).

Prunus microcarpa C.A.Mey., Verz. Pfl. Casp. Meer. (C.A. von Meyer). 166. 1831.
Syn. *Cerasus microcarpa* (C.A.Mey.) Boiss. (1872 or Jan 1873).
Amman, Bit Dabaghat, *Davis* 9231 (E00406560); Dana Reserve, *RBG-Ire & RBG-Jo* (NHJ 2008.0087).

Prunus prostrata Labill., Icones plantarum Syriae rariorum 15, t. 6, 1791.
Syn. *Cerasus prostrata* (Labill.) Ser. (1825).
Ref. Al-Eisawi 157 (2013) as *Cerasus prostrata*.

Prunus ursina Kotschy ex Boiss., Fl. Orient. 2: 652 (Dec 1872 or Jan 1873).
Tafileh, Shathem, Wadi Shathem Rawajfeh, 2015 (YAR-H).

Pyrus syriaca Boiss., Diagn. Pl. Orient. ser. 1, 10: 1. 1849.
Ajloun, *Davis* 9464 (E00408558).

Rosa canina L., Sp. Pl.: 492. 1753.
Jerash, Thagrat Asfour, *Taifour et al.* (NHJ 2013.0669).

Rosa phoenicia Boiss., Diagn. Pl. Orient. ser. 1, 10: 4. 1849.
Ajloun, Ajloun Reserve, *Taifour et al.* (NHJ 2013.0671).

Rosa pulverulenta M.Bieb., Flora Taur.-Cauc.: 399. 1808.
Ma'an-Shaubak Governorate, Juhir, *Taifour, Alkhawaldah & Almogherbi* (NHJ- NHJ.2013.0691).

Rubus sanctus Schreb., Icones descript. Plant. 15, t.18, 1766.
Syn. *Rubus sanguineus* Friv. (1835).
Irbid, Al-Hemma, near River Yarmouk, *Boulos & Hauke* 5827 (E00409061); Tal Al-Rumman, *Royal Botanic Garden Team* (NHJ 2008.0007).

Rubus tomentosus Borkh. var. **canescens** (DC.) Wirtq., Syn. Rub. Germ. 229. 1877.
Syn. *Rubus canescens* DC. (1813).
Jarash, Sd Al Malk Talal (YAR-H).

Sanguisorba minor Scop., Fl. Carniol., ed. 2. 1: 110. 1771.
Tal Al-Rumman RBG, Sector No. 71, *Taifour* (NHJ 2013.0207).

Sanguisorba verrucosa (Link ex G.Don) Ces., Stirpes ital. rar. 1840 ad tab.
Ma'an, Wadi Musa, near Petra, *Boulos & Jallad* 8137 (E00366224).

Sarcopoterium spinosum (L.) Spach, Ann. Sci. Nat., Bot. sér. 3, 5: 43. 1846.
Syn. *Poterium spinosum* L. (1753).
N von Salt, *Frey & Kürschner* 80–272 (E00441482); Tal Al-Rumman, *Al-Eisawi* (NHJ 2006.0180).

Rubiaceae

Asperula arvensis L., Sp. Pl.: 103. 1753.
Irbid, 15 km northwest of Irbid near Hibraz village (MSB-K151670).

Callipeltis cucullaris (L.) DC., Prodr. 4: 613. 1830.
Syn. *Valantia cucullaris* L. (1755); *Callipetis aptera* Boiss. & Buhse (1860).
Ma'an, site 19, Rajif-Ras-Naqab road (MSB-K 554503).

Callipeltis factorovskyi (Eig) Ehrend., Fl. Lowland Iraq: 569. 1964.
Syn. *Warburgina factorovskyi* Eig (1927).
Tal Al-Rumman, *Taifour & Aldamen* (NHJ 2009.0129).

Crucianella aegyptiaca L., Mant. Pl. 1: 38. 1767.
Syn. *Crucianella herbacea* Forssk. (1775).
Tal Al-Rumman RBG, Sector No. 16, *El-Oqlah & Al Mfadi* (NHJ 2013.0379).

Crucianella ciliata Lam., Encycl. 2: 217. 1786.
Shoubak (YAR-H).

Crucianella latifolia L., Sp. Pl.: 109. 1753.
Tal Al-Rumman, *RBG-Ire & RBG-Jo* (NHJ 2008.0048).

Crucianella macrostachya Boiss., Diagn. Pl. Orient. ser. 1, 3: 27. 1843.
Jarash, Rumman Hill (MSB-K 27820); Tal Al-Rumman, *Al-Eisawi* (NHJ 2006.0114).

Crucianella membranacea Boiss., Diagn. Pl. Orient. ser. 1, 3: 27. 1843.
Wadi Araba, 5 km E of Ain El-Fidan (IBBU-H).

Crucianella transjordanica Rech.f., Ark. Bot. ser. 2, 1: 324. 1950.
4 km W of Jerash (IBBU-H).

Cruciata articulata (L.) Ehrend., Notes Roy. Bot. Gard. Edinburgh 22: 396. 1958.
Syn. *Valantia articulata* L. (1753).
Tal Al-Rumman, *Taifour & Aldamen* (NHJ 2009.0125).

Cruciata taurica (Pall. ex Willd.) Ehrend., Notes Roy. Bot. Gard. Edinburgh 22: 393. 1958.
Syn. *Valantia taurica* Pall. ex Willd. (1806); *Galium coronatum* Sm. (1806).
Ref. FOR-H.

Galium adhaerens Boiss. & Balansa, Diagn. Pl. Orient. ser. 2, 5: 106. 1856.
Ref. RSCN database; Al-Eisawi 159 (2013).

Galium aparine L., Sp. Pl.: 108. 1753.
Tal Al-Rumman, *RBG-Ire & RBG-Jo* (NHJ 2008.0044).

Galium canum Req. ex DC., Prodr. 4: 602. 1830.
Petra, *Hepper* (K).

Galium cassium Boiss., Diagn. Pl. Orient. ser. 1, 10: 68. 1849.
Salt, 2014, *Danin* 317 (HUJ).

Galium chaetopodum Rech.f., Ark. Bot. ser. 2, 1: 324. 1950.
Zarqa, Azraq Al Amri (FOR-H).

Galium divaricatum Pourr. ex Lam., Encycl. 2(2): 580. 1788.
Ref. Karak, Danin (2004).

Galium elongatum C.Presl, Delic. Prag. 119. 1822.
Ref. RSCN database; Al-Eisawi 159 (2013).

Galium ghilanicum Stapf, Denkschr. Kaiserl. Akad. Wiss., Wien. Math.-Naturwiss. Kl. 50: 53. 1885.
Al Mafraq, 18 Apr. 1983 (YAR-H).

Galium hierochuntinum Bornm., Verh. Zool.-Bot. Ges. Wien 48: 596. 1898.
Tal Al-Rumman, *Shishani & Alkoran* (NHJ 2011.0106).

Galium hierosolymitanum L., Fl. Palaest. 13. 1756.
Ajloun Forest Reserve, RSCN database.

Galium humifusum M.Bieb., Fl. Taur.-Caucas. 1: 104. 1808.
Ma'an, Shoubak, on the way to Petra (MSB-K 243746).

Galium incanum Sm., Fl. Graec. Prodr. 1(1): 91. 1806.
Aqaba, RSCN database.

Galium incurvum Sibth. & Sm., Fl. Graec. Prodr. 1(1): 92. 1806.
Ref. Al-Eisawi 159 (2013).

Galium judaicum Boiss., Diagn. Pl. Orient. ser. 1, 10: 69. 1849.
Balqa, on the way to Sbaii coming from Zai, Oak forest (MSB-K 339728); Tal Al-Rumman, *Taifour & Aldamen* (NHJ 2009.0126).

Galium murale (L.) All., Fl. Pedem. i. 8. t. 77. f. 1. 1785.
Syn. *Sherardia muralis* L. (1753).
Dibeen National Park, 2005, *Mahklouf* s.n. (AMM).

Galium nigricans Boiss., Diagn. Pl. Orient. ser. 1, 3: 48. 1843.
Ref. RSCN database; Al-Eisawi 160 (2013).

Galium petrae Oliv. ex Hart, Trans. Roy. Irish Acad. 28: 433. 1885.
Petra, 15 Mar. 1974, *Boulos, Jallad & Al-Eisawi* (AMM).

Galium philistaeum Boiss., Diagn. Pl. Orient. 10: 71. 1849.
Salt, Wadi Shaib, 21 Mar. 1983 (YAR-H).

Galium penicillatum Boiss., Diagn. Pl. Orient. 10: 68. 1849.
Shoubak, 1987 (K).

Galium pisiferum Boiss., Diagn. Pl. Orient. ser. 1, 10: 67. 1849.
Jordan, *Post* 427 (K000772612); Tal Al-Rumman, *Taifour & Aldamen* (NHJ 2009.0127).

Galium rivale (Sibth. & Sm.) Griseb., Spic. Fl. Rumel. 2(4): 156. 1844.
Syn. *Asperula rivalis* Sibth. & Sm. (1806).
Ref. (YAR-H); Al-Eisawi 160 (2013).

Galium rubioides L., Sp. Pl.: 105. 1753.
Syn. *Galium articulatum* Lam. (1792).
Irbid, Al Mansura ((FOR-H).

Galium samuelssonii Ehrend., Notes Roy. Bot. Gard. Edinburgh 22: 387. 1958.
Irbid, Al Mansourah (YAR-H).

Galium setaceum Lam., Encycl. 2(2): 584. 1788.
Syn. *Galium decaisnei* Boiss. (1846).
Wadi Heidan, 26 Apr. 1945, *Davis* (K).

Galium sinaicum Boiss., Fl. Orient. 3: 63. 1875.
Irbid, Al Shounah, *Barsotti* (AMM).

Galium spurium L., Sp. Pl.: 106. 1753.
Ajloun, 5.V.1886 (AUB).

Galium tricornutum Dandy, Watsonia 4: 47. 1957.
Irbid, Howara village, from Irbid city toward Maru (MSB-K 200176); Ajloun, Wadi Orjan, *El-Oqlah et al.* (NHJ 2013.0073).

Galium verrucosum Huds., Philos. Trans. 56: 251. 1767.
Jarash, 1987 (K).

Mericarpaea ciliata (Banks & Sol.) Eig, J. Bot. 75: 190. 1937.
Syn. *Valantia ciliata* Banks & Sol. (1794); *Mericarpaea vaillantioides* Boiss. (1843).
Tal Al-Rumman, *Al- Eisawi, Abu-Sondos, Abu Anza & Shishani* (NHJ 2006.0085).

Oldenlandia capensis L.f., Suppl. Pl. 127. 1782.
Ajloun, Eshtafena (YAR-H).

Plocama calabrica (L.f.) M.Backlund &Thulin, Taxon 56(2): 516. 2007.
Syn. *Asperula calabrica* L. (1782).
Amman, Airport highway. site no 35 (MSB-K 497859).

Plocama calycoptera (Decne.) M.Backlund & Thulin, Taxon 56(2): 323. 2007.
Syn. *Spermacoce calycoptera* Decne. (1834); *Gaillonia calycoptera* (Decne.) Jaub. & Spach (1843); *Pterogaillonia calyoptera* (Decne.) Lincz. (1973); *Jaubertia calycoptera* (Decne.) Täckh. & Boulos (1974).
Wadi Araba (RSCN database).

Rubia danaensis Danin, Willdenowia 27(1–2): 163. 1997.
Endemic. Type: Jordan, Aqaba, Dana Reserve, *Danin* 962509 (holo. HUJ; iso. B, E, K, W); Ma'an, Wadi Musa, Ain-Musa (MSB-K 419246); Juhir, on the way to Wadi Araba, *Taifour, Al Khawaldah & Al Mogherbi* (NHJ 2013.0693).

Rubia peregrina L., Sp. Pl.: 109. 1753.
Jordan, 18 Apr. 1845, *Simpson* (K).

Rubia tenuifolia d'Urv., Mém. Soc. Linn. Paris 1: 273. 1822.
Syn. *Rubia olivieri* A Rich. ex DC. (1830).
Jarash, Souf village, site no 13 (MSB-K 497745); Tafila, Dana Reserve, BOT ERA project (NHJ 2013.0447).

Rubia tinctorum L., Sp. Pl.: 109. 1753.
Tafila, Dana reserve (MSB-K 278308); Ma'an, Petra, *El-Oqlah et al.* (NHJ 2013.0621).

Sherardia arvensis L., Sp. Pl.: 102. 1753.
Irbid, Yarmuk village (YAR-H).

Theligonum cynocrambe L., Sp. Pl. 2: 993. 1753.
Tal Al-Rumman, *Al-Eisawi* (NHJ 2006.0213).

Valantia hispida L., Syst. Nat., ed. 10. 2: 1307. 1759.
Zarqa, on way to Alouk (MSB-K 227867); Tal Al-Rumman, *Al-Eisawi* (NHJ 2006.0125).

Valantia muralis L., Sp. Pl. 2: 1051. 1753.
Irbid, Aqraba (YAR-H).

Ruppiaceae

Ruppia maritima L., Sp. Pl.: 127. 1753.
6–7 km S of Rum rest house, *Boulos* (AMM); Azraq (RSCN database).

Rutaceae

Haplophyllum blanchei Boiss., Fl. Orient. 1: 937. 1867.
Syn. *Haplophyllum rubrum* Vel. (1912); *Haplophyllum rubellulum* Thieb. & Gomb. (1936).
Al Karak, Wadi south of Hasa, *Hunting Aero Survey* 184 (E00437121); Azraq, Wadi Bottom (South west the Castle Amra), *El-Oqlah* (NHJ 2013.0588).

Haplophyllum poorei C.C.Towns., Kew Bull. 20: 97. 1966.
Between Quweira and Rasn Naab (SJordan), *Poore* 172A (E00133263).

Haplophyllum tuberculatum (Forssk.) A.Juss., Mém. Mus. Hist. Nat. 12: 528. 1825.
Syn. *Ruta tuberculata* Forssk. (1775).
Karak, Jebel Ataita, *Frey & Kürschner* VO 5578 (E00437295).

Ruta buxbaumii Poir., Encycl. 6: 336. 1804.
Syn. *Haplophyllum buxbaumii* (Poir.) G.Don (1831).
Irbid, c. 9 km NW Irbid, *Boulos, Jallad, Walid & Lahham* 8332 (E00437165); Tal Al-Rumman, *Taifour & Aldamen* (NHJ 2009.0098).

Ruta chalepensis L., Mant. Pl. 69. 1767.
Syn. *Ruta bracteosa* DC. (1824).; *Ruta chalepensis* L. var. *bracteosa* (DC.) Boiss.
S of Ajloun, *Hunting Aero Survey* 98A (E00437060); Irbid, Korah, *Taifour et al.* (NHJ 2013.0673).

Ruta fruticulosa Labill., Icon. Pl. Syr. 1: 13. 1791.
Syn. *Haplophyllum fruticulosum* G.Don (1831).
Ref. Al-Eisawi 161 (2013) as *Haplophyllum fruticulosum*.

Ruta graveolens L., Sp. Pl.: 383. 1753.
Ajloun Forest Reserve, *Böcker* RB 896 (B100520878).

Salicaceae

Populus alba L., Sp. Pl.: 1034. 1753.
Jordan, Ma'an, Wadi Musa, near Petra, *Boulos & Jallad* 8076 (E00448443).

Populus euphratica Oliv., Voy. Emp. Ottoman. 3: 449. t. 45–46, 449. 1807.
Allen by bridge (banks of Jordan) *Davis* 3788B (E00448504); Wadi Sunfahah, *Ananbeh, Al Khawaldah & Alshwaer* (NHJ 2013.0634).

Salix acmophylla Boiss., Diagn. Pl. Orient. ser. 1, 7: 98. 1846.
Syn. *Salix safsaf* Standl. (1836).
Ma'an, Wadi Musa, near Petra, *Boulos & Jallad* 8077 (E00371126).

Salix acmophylla var. **pseudo-safsaf** (A.Camus & Gomb.) Thiébaut, Fl. Libya 3: 162. 1953.
Syn. *Salix pseuo-safsaf* A.Camus & Gomb. (*1939*).
Wadi Musa, *Frey & Kürschner*.

Salix alba L., Sp. Pl. 2: 1021. 1753.
Wadi Musa, near Ain Mousa, *Nuaimat, Al Khawaldah & Alshwaer* (NHJ 2013.0650).

Salix mucronata Thunb., Prodr. Pl. Cap. 6. 1794.
Syn. *Salix subserrata* Willd. (1806).
Wadi Musa, 22 Apr. 1985 (YAR-H).

Salvadoraceae

Salvadora persica L., Sp. Pl.: 122. 1753.
E side of the Dead Sea, *Hunting Aero Survey* 183A (E00448638); Fifa, *Nuaimat, Al Khawaldah & Alshwaer* (NHJ 2013.0659).

Santalaceae

Osyris alba L., Sp. Pl. 2: 1022. 1753.
Fields near Laqhtar, *Hunting Aero Survey* 159 (E00454792); Dibeen Forest, *RBG-Ire & RBG-Jo* (NHJ 2008.0110).

Thesium bergeri Zucc., Abh. Math.-Phys. Cl. Königl. Bayer. Akad. Wiss. 2: 324. 1837.
Ajloun, Wadi Zuka, *Davis* 3788A (E00454727).

Thesium humile Vahl, Symb. Bot. 3: 43. 1794.
Tal Al-Rumman, *Taifour & Aldamen* (NHJ 2009.0130).

Viscum cruciatum Sieber ex Boiss., Voy. Bot. Espagne 2: 274. 1840.
On the way to Wadi Araba, *Taifour, Al Khawaldah & Al Mogherbi* (NHJ 2013.0692).

Sapindaceae

Dodonaea viscosa (L.) Jacq., Enum. Syst. Pl. 19 1760.
Irbid, outskirts of Irbid towards Um Qeis at Baitras (MSB-K 151647).

Saxifragaceae

Saxifraga hederacea L., Sp. Pl.: 405. 1753.
Zubia reserve, north of Ajloun, 16 Mar. 1996, *Danin* (HUJ); Ajloun (RSCN database).

Scrophulariaceae

Scrophularia deserti Delile, Descr. Egypte, Hist. Nat. 240. 1813.
Petra, 28.vii.1907 (AUB); Jordan valley, Wadi Kofrinqi (YAR-H); 30 km SW Rum Police Station, 13 Mar. 1975, *Boulos et al.* (AMM).

Scrophularia hierochuntina Boiss., Diagn. Pl. Orient. Ser. 1, 12: 35. 1853.
Ajloun, Um e Yanabia 26 Mar. 1982 (YAR-H).

Scrophularia hypericifolia Wydler, Mém. Soc. Phys. Genève 4: 166. 1828.
Lemranne, *Barsotti* (AMM).

Scrophularia libanotica Boiss., Diagn. Pl. Orient. ser. 1, 12: 36. 1853.
Petra (YAR-H).

Scrophularia macrophylla Boiss., Diagn. Pl. Orient. ser. 1, 12: 32. 1853.
Salt, From the centre of Salt town toward Wadi Shuaib, near the first water spring (MSB-K 200338).

Scrophularia rubicaulis Boiss., Diagn. Ser. 1, 12: 34. 1853.
Syn. *Scrophularia michoniana* Coss. et Kral. (1854).
Jerash, 3.v.1911, *Dinsmore & Meyers* (HUJ); Souf, from town towards Civil defence station (NCARE); Jerash, from the centre of Souf village 50–100 m before Civil Defence building, opposite olive extraction factory (MSB-K 200372).

Scrophularia nabataeorum Eig, Palestine J. Bot., Jerusalem Ser. 3: 82. 1944.
Petra, *Dinsmore* 11313 (K001096314).

Scrophularia peyronii Post, Bull. Herb. Boissier 1. 28. 1893.
Jarash, 23 May 1983 (YAR-H).

Scrophularia pinardii Boiss., Diagn. Pl. Orient. 4: 70. 1844.
Irbid, Ain Al Hammam (YAR-H).

Scrophularia xanthoglossa Boiss., Diagn. Pl. Orient. ser. 1, 12: 38. 1853.
Syn. *Scrophularia gileadensis* Post (1888).
Jarash, Souf, 5.V.1886 (AUB); Salt, from Salt city toward Wadi Shuaib at the beginning of Aira (MSB-K 200073); Tal Al-Rumman, *Al-Eisawi* (NHJ 2006.0136).

Scrophularia xylorrhiza Boiss. & Hausskn., Fl. Orient. 4(2): 406. 1879.
Tafila, Site 17, Dana reserve, around the water spring, hanging on a cliff (MSB-K 554477); Tafila, Dana Reserve, plateau near top of Rumana mountain, SW aspect. BOT ERA project (NHJ 2013.0444).

Verbascum agrimoniifolium (K.Koch) Hub.-Mor., Fl. Turkey 6: 487. 1978.
Syn. *Celsia agrimoniifolia* K.Koch (1849).
Upper Jordan Valley (YAR-H).

Verbascum decaisneanum Kuntze, Revis. Gen. Pl. 2: 468. 1891.
Madaba, 12 Aug. 1907 (AUB); Dead Sea area (RSCN database).

Verbascum eremobium Murb., Acta Univ. Lund., 2 29(2): 458. 1933.
40 km west of Azraq, *Dinsmore* 11810 (E00359594); Wadi Araba, BOT ERA Project (NHJ 2013.0518)

Verbascum fruticulosum Post, Bull. Herb. Boissier 1: 405. 1893.
Madaba, from Madaba City toward Maien by 10 km. Roadside on right (MSB-K 212218).

Verbascum gaillardotii Boiss., Diagn. Pl. Orient. 2, 6: 128. 1859.
Ajloun (RSCN database).

Verbascum galilaeum Boiss., Diagn. Pl. Orient. 12: 8. 1853.
Upper Jordan Valley (RSCN database).

Verbascum jordanicum Murb., Acta Univ. Lund., 2, 29(2): 503. 1933.
Irbid, Zobia and Al Salt, Wadi Shuaib (YAR-H).

Verbascum orientale (L.) All., Fl. Pedem. 1: 106. 1785.
Tal Al-Rumman, *Taifour & Aldamen* (NHJ 2010 .0043)

Verbascum petrae P.H.Davis & Hub.-Mor., Notes Roy. Bot. Gard. Edinburgh 21: 49. 1951.
Petra, *Davis* 8754 (E00359577).

Verbascum qulebicum Post, J. Linn. Soc., Bot. 24: 437. 1888.
Jebel El Hadn, *Borzatti* (AMM).

Verbascum schimperianum Boiss., Diagn. Pl. Orient. ser. 1, 12: 11. 1853.
Aqaba (RSCN database).

Verbascum sinaiticum Benth., Prodr. 10: 236. 1846.
Balqa, South Shoneh, on the way to Dead Sea coming form Karameh, (MSB-K 200523).

Verbascum sinuatum L., Sp. Pl.: 178. 1753.
Dibeen National Park, 2005, *Mahklouf* s.n. (AMM).

Verbascum transjordanicum Murb., Acta Univ. Lund., 2 35(1): 5.4 1939.
40 km west of Azraq, *Dinsmore* 21804 (E00327349).

Smilacaceae

Smilax aspera L., Sp. Pl. 2: 1028. 1753.
Ajloun, 5 km from Ajloun Castle down to the valley (MSB-K 200419); Ajloun Reserve, *Taifour et al.* (NHJ 2013.0672).

Solanaceae

Datura inoxia Mill., Gard. Dict., ed. 8. n. 5, 1768.
Irbid (YAR-H).

Datura stramonium L., Sp. Pl.: 179. 1753.
Ma'an, Petra, along the way from the main gate to the siq (MSB-K 186416).

Hyoscyamus albus L., Sp. Pl.: 180. 1753.
Jordan, *Meyers & Dinsmore* 4404 (E00621583); Tal Al-Rumman, *Taifour & Aldamen* (NHJ 2011.0003).

Hyoscyamus aureus L., Sp. Pl.: 180. 1753.
Jordan, *Davis* 8838 (E00621632); Tal Al-Rumman, *Saifan & Fiad* (NHJ 2008.0005).

Hyoscyamus cylindrocalyx Rech.f., Ark. Bot. a.s., 2: 420. 1952.
Azraq Wetland Reserve Herbarium (RSCN); Ref. Guy Mountfort et al. (2008).

Hyoscyamus desertorum (Asch. ex Boiss.) Täckh., Svensk, Bot. Tidskr. 36: 252. 1942.
Madaba, from Dhiban to Wadi Al Mougeb, down to the valley (MSB-K 199717); Rum, on the way from Mnisher to Disi village, *Taifour, Al Khawaldah & Al Mogherbi* (NHJ 2013.0708).

Hyoscyamus insanus Stocks, Hooker's. J. Bot. Kew Garden. Misc. 4: 178. 1852.
Jordan, *Hunting Aero Survey* 89 (E00692885).

Hyoscyamus muticus L., Mant. Pl. 45. 1767.
Um er Ragaba, wadi bottom, 13 Apr. 1955 (FOR-H).

Hyoscyamus pusillus L., Sp. Pl.: 180. 1753.
Jordan, *Jallad, Lahham & Abu Hmaidan* 7760 (E00693142); Dana, *Taifour* (NHJ 2013.0412).

Hyoscyamus reticulatus L., Sp. Pl., ed. 2. 1: 257. 1762.
Jordan, *Boulos & Al-Eisawi* 6083 (E00693310); Tal Al-Rumman, *Taifour & Aldamen* (NHJ 2010.0048).

Lycium barbarum L., Sp. Pl.: 192. 1753.
Azrak, 1963 (K).

Lycium depressum Stocks, Hook. J. Bot. Kew Gard. Misc. 4: 179. 1852.
Jordan, *Boulos & Jallad* 8071 (E00610572).

Lycium europaeum L., Sp. Pl.: 192. 1753.
Negev, between Aqaba and Kusseima, 33 km from Aqaba (HUJ).

Lycium petraeum Feinbrun, Collect. Bot. (Barcelona) 7: 373. 1968.
Dinsmore 10296 (K000759449).

Lycium schweinfurthii Dammer, Bot. Jahrb. Syst. xlviii. 224. 1912.
Jordan, Balqa Governorate, Dead Sea, Zara, (NHJ).

Lycium shawii Roem. & Schultes, Sp. Pl. 404. 1753.
Jordan, *Jones* 6 (E00610549).

Mandragora officinalis Mill., Gard. Dict. ed. 8 s.p. 1768.
Syn. Mandragora autumnalis Bertol. (1820).
Balqa, Al Zahra' suburb, 2 km after Um Jozeh & Rmaimein (MSB-K 200040).

Nicotiana glauca Graham, Edinb. N. Phil. Journ.: 175. 1828.
Jordan, *Trought s.n.* (E00617160); Tal Al-Rumman, *Shishani & Alkoran* (NHJ 2011.0004).

Nicotiana rustica L., Sp. Pl.: 180. 1753.
Ref. Al-Eisawi 167 (2013).

Physalis angulata L., Sp. Pl. 183. 1753.
Ref. RBG Jordan checklist (unpubl.)

Physalis peruviana L., Sp. Pl. ed. 2 1670. 1763.
Karak (Moab), 1923, *Dinsmore* (E).

Solanum coagulans Forssk., Fl. Aegypt.-Arab.: 2: 47. 1775.
Ghour Safi, 6.VIII.1907 (AUB).

Solanum dulcamara L., Sp. Pl.: 185. 1753.
Jerash, 5.1886 (31365-HUJ); RSCN database.

Solanum elaeagnifolium Cav., Icon. [Cavanilles] 3: 22, t. 243, 1795.
Wadi-walleh, *Abulaila* JOR3–1 (K000658914); Tal Al-Rumman, *Taifour & Aldamen* (NHJ 2009.0092).

Solanum incanum L., Sp. Pl.: 188. 1753.
Jordan Valley, limited to the Lower Jordan, *Eig, Feinbrun & Zohary* 285 (E00699222).

Solanum luteum Mill. Gard. Dict. Ed. 8, 3. 1768.
Al-Mafraq, Rehab Al-Dajniyah (IBBU-H); Dibeen Forest Reserve, *Taifour & Othman* (RSCN – Dibeen Forest Reserve Herbarium).

Solanum nigrum L., Sp. Pl.: 186. 1753.
Jordan, *Boulos, Jallad, Lahham & Abu Hmaidan* 7563 (E00617882).

Solanum sinaicum Boiss., Diagn. Pl. Orient. ser. 1, 11: 135. 1849.
Al Mafraq, Al Mansoorah, (IBBU-H); Ma'an, 20 km NE of Quweirah, *Frey & Kürschner*; Aqaba (RSCN database).

Solanum villosum Mill., Gard. Dict., ed. 8. n. 2, 1768.
Jordan, *Dinsmore* 11151(E00617948).

Withania obtusifolia Tackh., Svensk Bot. Tidskr. 26: 370. 1932.
Jebel Almud, 15 Nov. 1997, *Borzatti* (AMM).

Withania somnifera (L.) Dunal, Prodr. [A. P. de Candolle] 13(1): 453. 1852.
Syn. *Physalis somnifera* L. (1753).
Jordan, *Meyer* 4254 (E00610235).

Styracaceae

Styrax officinalis L., Sp. Pl.: 444. 1753.
Ajloun, 5 km from Ajloun Castle toward valley (MSB-K 211897); Dair abusaid Kufur rakib, *El-Oqlah* (NHJ 2013.0056).

Tamaricaceae

Reaumuria alternifolia (Labill.) Britten, J. Bot. 54: 110. 1916.
Syn. *Hypericum alternifolium* Labill. (1791); *Reaumuria cistoides* Adams. (1805–1806).
Jordan, *Frey & Kürschner* 5443 (E00651805); Al Azraq, Shoumeri (YAR-H).

Reaumuria hirtella Jaub. & Spach, Ill. Pl. Orient. iii. 54. 1850.
Amman, 20 km W von Ma'in Zarqa, *Frey & Kürschner* 80–763 (E00454992); Tafila, on the way to Dana from Ghour, *Taifour, Al Khawaldah &Al Mogherbi* (NHJ 2013.0685).

Reaumuria negevensis Zohary & Danin, Israel J. Bot. 19: 309. 1970.
Aqaba, 5 km W of Tafila to Fifi, *Danin* (HUJ).

Tamarix amplexicaulis Ehrenb., Linnaea 2: 275. 1827.
Dead Sea Area (RSCN database).

Tamarix aphylla (L.) H.Karst., Deut. Fl. (Karsten) 641. 1882.
Syn. *Thuja aphylla* L. (1755); *Tamarix articulata* Vahl (1791), nom. illeg.
Azraq, *Frey & Kürschner* VO 5877 (E00382264).

Tamarix arborea (Sieber ex Ehrenb.) Bunge, Tent. Gen. Tamar. 67. 1852.
Syn. *Tamarix negevensis* Zohary (1956); *Tamarix sodomensis* Zohary (1956).
Wadi Araba (RSCN database).

Tamarix chinensis Lour., Fl. Cochinch. 182. 1790.
Syn. *Tamarix pentandra* Pall. (1788).
Ma'an, 15 Apr. 1937, *Meyers & Dinsmore* (E).

Tamarix mannifera Kotschy ex Bunge, Tent. Gen. Tamar.: 52. 1852.
Jordan, *Schimper* 455 (K000318348).

Tamarix nilotica (Ehrenb.) Bunge, Tent. Gen. Tamar.: 54. 1852.
Syn. *Tamarix gallica* L. var. *nilotica* Ehrenb. (1827).
Ghor Al Safi, toward Potash factory (MSB-K 412591); Tal Al-Rumman, *Shishani* (NHJ 2011.0123).

Tamarix palaestina Bertol., Misc. Bot. 14: 16. 1853.
Syn. *Tamarix jordanis* Boiss. (1867).
Wadi Malih, *Dinsmore* 12228 (E00382405); Tal Al-Rumman, *Al-Eisawi* (NHJ 2006.0336)

Tamarix parviflora DC., Prodr. [A. P. de Candolle] 3: 97. 1828.
Ref. RSCN database; Al-Eisawi 169 (2013).

Tamarix passerinoides Delile ex Desv., Descr. Egypte, Hist. Nat. 58. 1813.
Ghor Al Safi, main road towards Potash factory, *Abulaila, Saifan & Tehabshem* (NCARE, K).

Tamarix passerinoides Delile ex Desv. var. **macrocarpa** Ehrenb. ex Bunge, Linnaea 2: 276. 1827.
Syn. *Tamarix macrocarpa* (Ehrenb.) Bunge (1852); *Tamarix aravensis* Zohary *(*1956).
Salt flats of Azraq, *Hunting Aero Survey* 125B (E00382481); Araba valley, Al Karak, Ghour Al Mazrah (YAR-H).

Tamarix tetragyna Ehrenb., Linnaea 2: 257, 258. 1827.
Syn. *Tamarix meyeri* Boiss. (1849); *Tamarix tetragyna* Ehrenb. var. *meyeri* (Boiss.) Boiss. (1867); *Tamarix gennessarensis* Zohary *(1956)*.
Tal Al-Rumman, *Al-Eisawi* (NHJ 2006.0335).

Thymelaeaceae

Daphne mucronata Royle subsp. **linearifolia** (Hart) Halda, Acta Mus. Richnov., Sect. Nat. 5(5): 165. 1998.
Syn. *Daphne linearifolia* Hart (1885).
Petra, *Boulos, Al-Eisawi & Jallad* 6305 (E00454154); Shoubak, *Ananbeh, Al Khawaldah & Alshwaer* (NHJ 2013.0642).

Thymelaea hirsuta (L.) Endl., Gen. Pl. [Endlicher] Suppl. 4 (2): 65. 1848.
Syn. *Passerina hirsuta* L. (1753).
Petra, *Boulos & Al-Eisawi* 6241 (E00454348); Wadi Musa, *Ananbeh, Al Khawaldah & Alshwaer* (NHJ 2013.0647).

Thymelaea passerina (L.) Coss. & Germ., Syn. Anal. Fl. Paris, ed. 2: 360. 1859.
Syn. *Stellera passerina* L. (1753).
Al-Mafraq, Al-Mdawara village (IBBU-H).

Thymelaea pubescens (L.) Meisn., Prodr. [A. P. de Candolle] 14(2): 558. 1857.
Syn. *Daphne pubescens* L. (1753).
Ref. Al-Eisawi 169 (2013).

Typhaceae

Typha anugustifolia L., Sp. Pl.: 971. 1753.
Wadi Siyagh, Apr. 1983, *Holzapfe* (AMM).

Typha domingensis Pers., Syn. Pl. 2: 532. 1807.
Amman, Wadi Essir towards Adaseyah, Bahath & Buhaira villages (MSB-K 211853).

Typha elephantina Roxb., Fl. Ind. 3: 566. 1832.
Ref. RSCN database; Al-Eisawi 170 (2013).

Typha latifolia L., Sp. Pl. 2: 971. 1753.
Al-Mafraq, Al-Swelmiyeh (IBBU-H).

Ulmaceae

Celtis australis L., Sp. Pl.: 1043. 1753.
Ma'an, Petra, *Frey, Hilger & Kürschner* VO5521 (E00405512); Wadi Musa, *Ananbeh &Alshwaer* (NHJ 2013.064).

Celtis laevigata Willd., Berlin. Baumz., ed. 2 81. 1811.
Cultivated.

Ulmus minor Mill., Gard. Dict., ed. 8. n. 6. 1768.
Tafileh, Rashadieh, 2013, *El-Oqlah* 2013. (YAR-H). Hardwood specimen.

Urticaceae

Forsskaolea tenacissima L., Opobalsamum 18. 1764.
15 km S von Aqaba, *Frey & Kürschner* VO 5168 (E00404563); Tal Al-Rumman, *RBG-Ire & RBG-Jo* (NHJ 2008.0074).

Parietaria alsinifolia Delile, Descr. Egypte, Hist. Nat., 2: 281, pl. 50, f. 2. 1813.
Syn. *Freirea alsinaefolia* (Delile) Gaudich. (*1826).*
Petra, *Frey & Kürschner* 80–208 (E00404695);Tafila, near Dana Reserve, BOT ERA project (NHJ 2013.0494).

Parietaria judaica L., Zohary [C. Linnaeus] 32. 1756.
Syn. *Parietaria diffusa* Merlet & W.D.J.Koch (1823).
Irbid, Al Hemma, near River Yarmouk, *Boulos & Hauke* 5833 (E00404708); Tal Al-Rumman, *RBG-Ire & RBG-Jo* (NHJ 2008.0102).

Parietaria lusitanica L., Sp. Pl. 2: 1052. 1753.
Dibeen Forest Reserve (FOR-H); Irbid, Tabkat-Fahel (YAR-H).

Urtica hulensis Fienbr., Palestine J. Bot., Jerusalem Ser. 4: 117. 1947.
Irbid, Ain Aqraba (IBBU-H).

Urtica pilulifera L., Sp. Pl.: 983. 1753.
Medaba, *Meyers, Fred & Dinsmore* 770 (E00404622); Tal Al-Rumman, *Al-Eisawi* (NHJ 2006.0159).

Urtica urens L., Sp. Pl.: 984. 1753.
Tal Al-Rumman, *Al-Eisawi* (NHJ 2006.0357).

Verbenaceae

Phyla nodiflora (L.) Greene, Pittonia 4: 46. 1899.
Al-Hemma, *Tackholm et al.* 8977 (K000318313).

Verbena officinalis L., Sp. Pl.: 20. 1753.
Amman, on road to Na'our from junction on Dea Sea Highway, site 40 (MSB-K 395472).

Verbena supina L., Sp. Pl.: 21. 1753.
Jarash, Tell Ruman, beside King Talal Dam (MSB-K 341480).

Violaceae

Viola alba Besser, Prim. Fl. Galiciae Austriac. i. 171. 1809.
Ref. Al-Eisawi 172 (2013).

Viola modesta Fenzl, Pug. Pl. Nov. Syr. 12. 1842.
Ma'an, 20 km N of Petra, near El-Hesheh, *Frey & Kürschner* (BSB).

Viola occulta Lehm., Ind. Sem. Hort. Hamburg 9. 1829.
Ma'an, 15 km S von Shoubak, *Frey & Kürschner* VO 4986 (E00371510).

Viola odorata L., Sp. Pl. 2: 934. 1753.
Ref. Al-Eisawi 172 (2013).

Viola parvula Tineo, Pl. Rar. Sicil. Pugill. 1: 5. 1817.
Ref. Ma'an, Danin (2004).

Viola pentadactyla Fenzl, Pug. Pl. Nov. Syr. 12. 1842.
Hisheh, *Schmida* (HUJ).

Zygophyllaceae

Balanites aegyptiacus (L.) Delile, Descr. Égypte, Hist. Nat. 221, t. 28, f. 1, 221. 1813.
Syn. *Ximenia aegyptiaca* L. (1753).
Ghour Fifa, *RBG-Ire & RBG-Jo* (NHJ 2008.0026).

Fagonia arabica L., Sp. Pl.: 386. 1753.
Azraq, Wadi Bottom (South west the Castle Amra), *El-Oqlah* (NHJ 2013.0593).

Fagonia bruguieri DC., Prodr. 1: 704. 1824.
Syn. *Fagonia parviflora* Zohary (1849); *Fagonia schimperi* C.Preesl (1845); *Fagonia myriacantha* Boiss. (1849); *Fagonia bruguieri* DC. var. *rechingeri* Hadidi (1972); *Fagonia olivieri* DC. var. *glandulosa* Hadidi (1972).
11 km south of Aqaba, *Boulos & Jallad* 7438 (E00333845).

Fagonia cretica L., Sp. Pl.: 386. 1753.
Mt. Hor east side, *Davis* 8818 (E00333659).

Fagonia glutinosa Delile, Descr. Egypte, Hist. Nat. 230. 1813.
(incl. var. **grandiflora** Boiss.)
Aqaba, Wadi Araba, on the way to Amman (MSB-K 275064); nr. Aqaba, *Boulos & Lahham* 7390 (AMM, BM, E).

Fagonia indica Burm.f., Fl. Indica 102, t. 34, f. 1. 1768.
Syn. *Fagonia olivieri* DC. (1824).
Aqaba, E of Quweira camp, *Hunting Aero Survey* 36A (E00335528).

Fagonia latifolia Delile, Descr. Egypte, Hist. Nat. 230. 1813.
Jebal Gulzan, *Borzatti & Barsotti* (AMM).

Fagonia mollis Delile, Descr. Egypte, Hist. Nat. 220. 1813.
Syn. *Fagonia tristis* Sickenb. (1901).
Wadi Rum, *Davis* 8996 (E00333726); Rum, *Taifour* (NHJ 2013.0395).

Fagonia orientalis C.Presl, Delic. Prag. 23. 1822.
Syn. *Fagonia tenuifolia* Steud. & Hochst. ex Boiss. *(1867); Fagonia bischarorum* Schweinf. *(1899).*
Dead Sea area (YAR-H); Dead Sea Vallley, *Danin* (http://flora.org.il/en/plants/FAGTEN/).

Fagonia scabra Forssk., Fl. Aegypt.-Arab.: 88. 1775.
Syn. *Fagonia sinaica* Boiss. (1843).
Reported from the southern Neqab by Danin (http://flora.org.il/en/plants/FAGSCA/); (YAR-H); likely to be present in Jordan.

Seetzenia lanata (Willd.) Bullock, Kew Bull. 19: 204. 1965.
Syn. *Zygophyllum lanatum* Willd. (1799).
About 8 km S of Aqaba, *Boulos &Lahham* 7413 (E00219096).

Tetradiclis tenella (Ehrenb.) Litv., Trudy Bot. Muz. Imp. Akad. Nauk 3: 122. 1907.
Syn. *Anatropa tenella* Ehrenb., Linnaea (1829).
Zarqa, Azraq Oasis, Wetland Reserve (E00009240).

Tetraena alba (L.) Beier & Thulin, Pl. Syst. Evol. 240(1–4): 35. 2003.
Syn. *Zygophyllum album* L. (1753).
Leonard 7468 (E00219115); nr. Aqaba, *Boulos & Lahham* 5379 (AMM, BM, E).

Tetraena coccinea (L.) Beier & Thulin, Pl. Syst. Evol. 240(1–4): 35. 2003.
Syn. *Zygophyllum coccineum* L. (1753).
Ref. Wadi Araba and Dead Sea, Danin (2004).

Tetraena dumosa (Boiss.) Beier & Thulin, Pl. Syst. Evol. 240(1–4): 35. 2003.
Syn. *Zygophyllum dumosum* Boiss. (1849).
Zerka Main to Jebel Attarus, *Davis* 9407 (E00338694).

Tetraena simplex (L.) Beier & Thulin, Pl. Syst. Evol. 240(1–4): 36. 2003.
Syn. *Zygophyllum simplex* L. (1767).
On road to Mudawwara, *Hunting Aero Survey* 40 (E00339492); nr. Aqaba, *Boulos & Lahham* 5400 (AMM, BM, E).

Tribulus bimucronatus Viv., Pl. Aegypt. Dec. 4: 9. 1830.
Aqaba, Aqaba-Sudi border highway (MSB-K 412627).

Tribulus macropteris Boiss., Diagn. Pl. Orient. 1: 161. 1843.
Aqaba, 5 km after Beer Maz Kor rest house. Site no 3 (MSB-K 496092).

Tribulus pentandrus Forssk., Fl. Aegypt.-Arab.: 81 (sp. dub.). 1775.
Syn. *Tribulus longipetalus* Viv. (1830).
Aqaba, *Leonard* 7482 (E00332425); nr. Aqaba, *Boulos & Lahham* 7405 (AMM, BM, E).

Tribulus terrestris L., Sp. Pl.: 387. 1753.
Syn. *Tribulus terrestris* L. var. *robustus* (Boiss. & Noë) Boiss.; *Tribulus terrestris* L. var. *orientalis* (A.Kern.) Beck.
Ma'an, Wadi Musa, Petra, *Frey, Hilger & Kürschner* VO5611 (E00332483).

Zygophyllum fabago L., Sp. Pl.: 385. 1753.
Zarqa, Azraq, *Dinsmore* 11805 (E00333647).

References and selected bibliography

Abu Yahya, A., Eid, E. & Ananbeh, Y. 2010. Qatar Reserve Herbarium. Flora Baseline Survey. Qatar Protected Area. *Royal Society for the Conservation of Nature.* Amman, Jordan.

Akhani, H., Edwards, G. and Roalson, E. H. 2007. Diversification of the old world Salsoleae s.l. (Chenopodiaceae): Molecular phylogenetic analysis of nuclear and chloroplast data sets and a revised classification. *Int. J. Plant Sci.* 168(6): 931–956.

Akhani, H. & Khoshravesh, R. 2013. The relationship and different C4 Kranz anatomy of Bassia eriantha and Bassia eriophora, two often confused Irano-Turanian and Saharo-Sindian species. Phytotaxa 93(1): 1–24.

Al-Dwiri. 2004. *Survey of the Plant Biodiversity along Zarqa River in Relation to the quality of water*. Jordan University.

Al-Eisawi, D. M. 1982. List of Jordan Vascular Plants. *Mitt. Bot. Munchen*, 81: 79–182.

Al-Eisawi, D. M. 1985. Vegetation of Jordan. In Hadidi (ed.) *Studies in the history and archaeology of Jordan.* Ministry of Archaeology. Amman. Pp. 45–57.

Al-Eisawi, D. M. 1986a. The orchids of Jordan. *Kew Bull.* 41(2): 359–377.

Al-Eisawi, D. M. 1986b. Studies on the flora of Jordan. 12. Monocotyledons new to Jordan, with notes on some interesting species. *Kew Bull.* 41(2): 349–357.

Al-Eisawi, D. M. 1988. Resedaceae in Jordan. *Bot. Jahrb. Syst.* 110 (1): 17–39.

Al-Eisaw, D. M. 1996. *Vegetation of Jordan*. UNESCO, Cairo Office, Regional Office for Science and Technology for the Arab States.

Al-Eisawi, D. M. 1998. *Field Guide to Wild Flowers of Jordan and neighboring countries*. UNDP, GEF/SGP/JSBS. Commercial Press. Jordan Press Foundation.

Al-Eisawi, D. M. 2013. *Flora of Jordan Checklist – Revised Edition 1*. The University of Jordan Press, Amman.

Al-Eisawi, D. M. & Hatough A. 1987. Ecological analysis of the vegetation of Shaumari Reserve in Jordan. *Dirasat* XIV (12): 81–94.

Al-Eisawi, D. M. & Jury, S. 1988. A taxonomic revision of the genus *Tordylium* L. (Apiaceae). *Bot. Jour. Lin. Soc.*, 97: 357–403.

Ananbeh, Y. A. 2014. *Impact of Ajloun Forest Reserve on Plant Biodiversity Conservation and Supporting Local People Livelihood.* Ajloun Forest Reserve Herbarium.

APG III. 2009. An update of the Angiosperm Phylogeny Group classification for the orders and families of flowering plants: APG III. *Botanical Journal of the Linnean Society*, 161: 105–121.

APG IV. 2016. An update of the Angiosperm Phylogeny Group classification for the orders and families of flowering plants: APG IV. *Botanical Journal of the Linnean Society*, 181: 1–20.

Baierle, H. U., El-Oqlah, A. & Kürschner, H. 1988. Some new and interesting records from the flora of Jordan. *Notes RBG Edinb.* 45(3): 457–468.

Boissier, E. 1867–1888. *Flora Orientalis*, volumes 1–5 and supplement. Basle and Geneva.

Borzatti, von Löwenstern, A., Al-Eisawi, D. M. & Garbani, F. 2000. Studies on the Flora of Jordan 14. The species of the Hisma Basin (Wadi Rum desert). *Webbia* 55(2): 195–277.

Boulos, L. 1977. Studies on the flora of Jordan: 5. Candollea 32(1): 99–110.

Boulos, L. & Lahham, J. 1977. Studies on the flora of Jordan: 3. Candollea 32(1).

Boulos, L. & Lahham, J. 1977. Studies on the flora of Jordan: 4. Candollea 32(1): 81–98.

Boulos, L., Jallad, W. & Lahham, J. 1975. Studies on the flora of Jordan: 2. Bot. Not.

Boulos, L., Lahham, J. & Jallad, W. 1977. 1977. Studies on the flora of Jordan: 7. Candollea 32(2): 255–268.

Cope, T. A. 1985. A Key to the Grasses of the Arabian Peninsula. *Arab Gulf Journal of Scientific Research Special Publication No. 1*: 1–82.

Cope, T. A. 2006. *Flora of the Arabian Peninsula: Vol. 5. Grasses.* RBG Kew.

Cribb, P. J. 1979. The orchids of Arabia. *Kew Bulletin* 33: 651–678.

Cribb, P. J. 1987. New records of Orchidaceae for Arabia. *Kew Bulletin* 42: 401–404.

Danin, A. 1983. *Desert Vegetation of Israel and Sinai*. Cana Publishing House, Jerusalem.

Danin, A. 1990. Two new species of Origanum (Labiatae) from Jordan. *Willdenowia* 19: 181–184.
Danin, A. 1991. Kickxia petraena (Scrophulariaceae), a new species from Jordan. *Willdenowia* 21: 181–184.
Danin, A. 1996. *Plants of Desert Dunes.* Springer-Verlag, Berlin.
Danin, A. 2005. *Distribution Atlas of Plants in the Flora Palaestina Area.* Published by The Israel Academy of Sciences and Humanities.
Danin, A. & Kunne, I. 1996. *Origanum jordaicum* (Labiatae), a new species from Jordan, and notes on the other species of O. sect. Campanulaticalyx. *Willdenowia* 25: 601–611.
Danin. A. 1997. Contributions to the flora of Jordan: new and interesting plants from Dana Nature Reserve, SW Jordan. *Willdenowia* 27: 147–153..
Danin, A. & Hedge, I. 1998. Contributions to the flora of Jordan 2. A new species of *Satureja* (Labiatae) and some new records. *Willdenowia* 28: 135–142.
Danin, A. 1999. Contributions to the flora of Jordan 3: A new species of *Artemisia* (Compositae, Anthemideae) from S Jordan. *Willedenowia* 29: 181–184.
Danin, A. 2013. *Flora of Israel.* [http://flora.org.il/en/about-analytical-flora-of-isarel/].
El-Oqlah, A. A. & Lahham, L. N. 1985. A checklist of vascular plants of Ajloun mountain. *Candollea* 40(2): 377.
Falahat. 1999. *Petra: Access list on the plants of Petra.*
Feinbrun, N. & Zohary, M. 1955. A geobotanical survey of Transjordan. Bull. Res. Counc. Isr. 5D, 1: 5–28.
Feinbrun-Dothan, N. 1978. *Flora Palaestina*, Vol. 3. Israel Academy of Science and Humanities. Jerusalem.
Feinbrun-Dothan, N. 1986. *Flora Palaestina*, Vol. 4. Israel Academy of Science and Humanities. Jerusalem.
Forsskål, P. 1775. *Flora Aegyptiaco-Arabica.* Sive Descriptiones Plantarum, quas per Aegyptum Inferiorem et Arabiam Felicem detexit, Illustravit Petrus Forsskål. Post mortem auctoris edidit Carsten Niebuhr. Accedit tabula Arabiae Felicis Geographixo Botanica, Copenhagen.
Freitag, H. 1991. The distribution of some prominent Chenopodiaceae in SW Asia and their phytogeographical significance. *Flora et Vegetatio Mundi* 9: 281–292.
Gharaibeh, M. 2005. *An Ecosystematic Study of a Semi-Arid Ecotone in the Northern Part of Jordan.* Yarmouk University, Department of Earth Sciences and Environment. Irbid, Jordan.
Ghazanfar, S. A. & Fisher, M, (eds) 1998. *Vegetation of the Arabian Peninsula.* 362 pp. Dordrecht: Kluwer Academic Press.
Ghazanfar, S. A. 2006. Saline and alkaline vegetation of NE Africa and the Arabian Peninsula: an overview. In: M. Orzturk, Y. Waisel, M.A. Khan & G. Gork (eds), *Biosaline agriculture and salinity tolerance in plants.* Birkhaeuser Publishing Ltd. Pp. 101–108.
Guy Mountfort, et al. 2008. *Azraq Wetland Reserve Herbarium*, Royal Society for the Conservation of Nature. Amman, Jordan.
Hepper, F. N. & Friis, I. 1994. *The Plants of Pehr Forsskål's 'Flora Aegyptiaco-Arabica'.* Royal Botanic Gardens, Kew.
Jonsell, B. 1986. A monograph of Farsetia (Cruciferae). *Acta Universitatis Upsalensis* 25:1–107.
Kasapligil, B. 1956. Report to the Government of the Hashmite Kingdom of Jordan on ecological survey of the vegetation in relation to forestry and grazing. *FAO Report* 56/9/7164, Rome, Italy.
Léonard, J. 1989. *Contribution à l'étude de la flore et de la vegetation des deserts d'Iran (Dasht-e-Kavir, Dasht-e-Lut, Jaz Kurian).* Fascicule 9. Considérations phytogéographiques sur les phytochories irano –touranienne, saharo-sindienne et de la Somalie-pays Masai. Jardin botanique national de Belgique.
Maani, Jarir 2008. *Field Guide to Jordan.* National Press. Jordan.
Mahklouf, El-Hadi 2004. *Plant Biodiversity and Flora of Dibeen National Park.* University of Jordan.
Mandaville, J. P. 1984. Studies in the flora of Arabia XI: Some historical and geographical aspects of a principal floristic frontier. *Notes from the Royal Botanic Garden Edinburgh* 42: 1–15.
Miller, A. G. & Cope, T. A. 1996. *Flora of the Arabian Peninsula and Socotra*, Volume 1. Edinburgh University Press, Edinburgh.
Qishawi, M. 1999. *Wadi Rum Nature Reserve Flora Survey.* Royal Society for the Conservation of Nature. Amman, Jordan.
Qishawi, M., El-Moghrabi, L. & Fagir, O. 2013. *Shaumari Reserve Herbarium*
RSCN Database. Royal Society for the Conservation of Nature. Amman. Jordan.
Schwartz, O. 1939. Flora des Tropischen Arabien. *Mitteilungen aus dem Institut für Allgemeine Botanik Hamburg* 10: 1–393.

Uotila, P. 2011. Chenopodiaceae (pro parte majore). Euro+Med Plantbase - the information resource for Euro-Mediterranean plant diversity. https://ww2.bgbm.org/EuroPlusMed.

Wadi Ibn Hammad Herbarium 2011. Royal Society for the Conservation of Nature. Amman, Jordan.

White, F. 1983. *The Vegetation of Africa. A Descriptive Memoir to Accompany the UNESCO, AETFAT, UNSO Vegetation Map of Africa*. UNESCO, Paris.

White, F. & Léonard, J. 1991. Phytogeographical links between Africa and Southwest Asia. *Flora et Vegetatio Mundi* 9: 229–246.

Wolfgang, F. and Harald, K. 1980–1993. *Collections from Jordan*. Berlin, Germany.

Zohary, M. 1966. *Flora Palaestina*, Vol. 1. Israel Academy of Science and Humanities. Jerusalem.

Zohary, M. 1972. *Flora Palaestina*, Vol. 2. Israel Academy of Science and Humanities. Jerusalem.

Zohary, M. 1973. *Geobotanical Foundations of the Middle East*. Fischer Verlag, Stuttgart.

Index of genera and families